Lecture Notes in Physics

The Editorial Policy for Proceedings

The series Lecture Notes in Physics reports new developments in physical research and teaching – quickly, informally, and at a high level. The proceedings to be considered for publication in this series should be limited to only a few areas of research, and these should be closely related to each other. The contributions should be of a high standard and should avoid lengthy redraftings of papers already published or about to be published elsewhere. As a whole, the proceedings should aim for a balanced presentation of the theme of the conference including a description of the techniques used and enough motivation for a broad readership. It should not be assumed that the published proceedings must reflect the conference in its entirety. (A listing or abstracts of papers presented at the meeting but not included in the proceedings could be added as an appendix.)

When applying for publication in the series Lecture Notes in Physics the volume's editor(s) should submit sufficient material to enable the series editors and their referees to make a fairly accurate evaluation (e.g. a complete list of speakers and titles of papers to be presented and abstracts). If, based on this information, the proceedings are (tentatively) accepted, the volume's editor(s), whose name(s) will appear on the title pages, should select the papers suitable for publication and have them refereed (as for a journal) when appropriate. As a rule discussions will not be accepted. The series editors and Springer-Verlag will normally not interfere with the detailed editing except in fairly obvious cases or on technical matters.

Final acceptance is expressed by the series editor in charge, in consultation with Springer-Verlag only after receiving the complete manuscript. It might help to send a copy of the authors' manuscripts in advance to the editor in charge to discuss possible revisions with him. As a general rule, the series editor will confirm his tentative acceptance if the final manuscript corresponds to the original concept discussed, if the quality of the contribution meets the requirements of the series, and if the final size of the manuscript does not greatly exceed the number of pages originally agreed upon.

The manuscript should be forwarded to Springer-Verlag shortly after the meeting. In cases of extreme delay (more than six months after the conference) the series editors will check once more the timeliness of the papers. Therefore, the volume's editor(s) should establish strict deadlines, or collect the articles during the conference and have them revised on the spot. If a delay is unavoidable, one should encourage the authors to update their contributions if appropriate. The editors of proceedings are strongly advised to inform contributors about these points at an early stage.

The final manuscript should contain a table of contents and an informative introduction accessible also to readers not particularly familiar with the topic of the conference. The contributions should be in English. The volume's editor(s) should check the contributions for the correct use of language. At Springer-Verlag only the prefaces will be checked by a copy-editor for language and style. Grave linguistic or technical shortcomings may lead to the rejection of contributions by the series editors.

A conference report should not exceed a total of 500 pages. Keeping the size within this bound should be achieved by a stricter selection of articles and not by imposing an upper limit to the length of the individual papers.

Editors receive jointly 30 complimentary copies of their book. They are entitled to purchase further copies of their book at a reduced rate. As a rule no reprints of individual contributions can be supplied. No royalty is paid on Lecture Notes in Physics volumes. Commitment to publish is made by letter of interest rather than by signing a formal contract. Springer-Verlag secures the copyright for each volume.

The Production Process

The books are hardbound, and quality paper appropriate to the needs of the authors is used. Publication time is about ten weeks. More than twenty years of experience guarantee authors the best possible service. To reach the goal of rapid publication at a low price the technique of photographic reproduction from a camera-ready manuscript was chosen. This process shifts the main responsibility for the technical quality considerably from the publisher to the authors. We therefore urge all authors and editors of proceedings to observe very carefully the essentials for the preparation of camera-ready manuscripts, which we will supply on request. This applies especially to the quality of figures and halftones submitted for publication. In addition, it might be useful to look at some of the volumes already published.

As a special service, we offer free of charge LATEX and TEX macro packages to format the text according to Springer-Verlag's quality requirements. We strongly recommend that you make use of this offer, since the result will be a book of considerably improved technical quality.

To avoid mistakes and time-consuming correspondence during the production period the conference editors should request special instructions from the publisher well before the beginning of the conference. Manuscripts not meeting the technical standard of the series will have to be returned for improvement.

For further information please contact Springer-Verlag, Physics Editorial Department V, Tiergartenstrasse 17, W-6900 Heidelberg, FRG

M. J. R. Hoch   R. H. Lemmer (Eds.)

Low Temperature Physics

Proceedings of the Summer School
Held at Blydepoort, Eastern Transvaal
South Africa, 15-25 January 1991

Springer-Verlag Berlin Heidelberg GmbH

Editors

Michael J. R. Hoch
Richard H. Lemmer
Physics Department, University of the Witwatersrand
Private Bag 3, Johannesburg, South Africa

ISBN 978-3-662-13809-0 ISBN 978-3-540-46512-6 (eBook)
DOI 10.1007/978-3-540-46512-6

Originally published by Springer-Verlag Berlin Heidelberg New York in 1991
Softcover reprint of the hardcover 1st edition 1991

Typesetting: Camera ready by author

58/3140-543210 - Printed on acid-free paper

PREFACE

The discovery of high T_c superconductivity in 1986 has resulted in a significant resurgence in the theoretical and experimental study of many–body effects in solids at low temperatures. This renewed activity has led to fresh insights into the behaviour of correlated fermion systems in a variety of different physical environments. It therefore seemed appropriate, in organizing the Seventh South African Theoretical Summer School, to gather together a panel of experts to discuss the latest developments in the behaviour of matter from very low temperatures (liquid ^{3}He and heavy fermion systems) up to temperatures in the regime of high–T_c superconductivity.

The present school also departed from the usual format in that we included a series of lectures by two leading low–temperature experimental physicists in order to emphasize the interlocking nature of current advances in experiment and theory in the low temperature field.

We were fortunate to have been able to assemble an excellent panel of lecturers: D Einzel (Walther–Meissner Institut, Garching), D S Fisher (Harvard), P Fulde (Max–Planck–Institut, Stuttgart), J Hook (Manchester) and R C Richardson (Cornell). Special seminars were presented by P de V du Plessis (Rand Afrikaans University), M J R Hoch and D S McLachlan (University of the Witwatersrand) and F Solms (University of Pretoria). The organizers would like to thank the contributors for the considerable time and effort they put into the presentation of lectures and seminars, and the subsequent preparation of those manuscripts which appear in this volume.

The South African Theoretical Summer School Series, which is administered by the Organization for Theoretical Physics, is indebted to the Foundation for Research Development (FRD) for generous financial support. Additional support for the Summer School was obtained from the University of the Witwatersrand and from individual FRD grant holders. The organizing committee would like to thank all persons involved for their contributions to a very rewarding Summer School. In particular we would like to acknowledge Mrs Renée Hoch for painstaking assistance in many editorial activities.

We are grateful to the editors of Springer–Verlag who have agreed to publish these lectures in their "Lecture Notes in Physics" series.

Johannesburg
South Africa
June 1991

M J R Hoch
R H Lemmer

CONTENTS

J Hook

D Einzel

ORGANIZING COMMITTEE

F J W Hahne	University of Stellenbosch
M J R Hoch	University of the Witwatersrand (Co–Chairman)
R H Lemmer	University of the Witwatersrand (Chairman)
E Pheiffer	Foundation for Research Development (Secretariat)
P du Toit van der Merwe	Atomic Energy Corporation
W S Verwoerd	University of South Africa

PARTICIPANTS

Albers, Mr A	University of the Witwatersrand
Albers, Mrs M C P	University of the Witwatersrand
Badziag, Dr P	University of South Africa
Baran, Dr A	Rand Afrikaans University
Blieden, Mr G S	University of the Witwatersrand
Carter, Dr R M	University of Pretoria
Comins, Professor J D	University of the Witwatersrand
Davidson, Mr N J	University of Pretoria
de Beer, Mr W	University of the Witwatersrand
de Haas, Professor W	University of the North, Qwa–Qwa Branch
de Lange, Professor O L	University of Natal, Pietermaritzburg Branch
Dixon, Mr P J	University of the Witwatersrand
Doyle, Mr R A	University of the Witwatersrand
du Plessis, Professor P de V	Rand Afrikaans University
Einzel, Dr D	Walther–Meissner–Institut für Tieftemperaturforschung, Garching, Germany
Every, Professor A G E	University of the Witwatersrand
Fisher, Dr D S	Harvard University, Cambridge, Ma, USA
Fulde, Professor P	Max–Planck–Institüt für Festkörperforschung, Stuttgart, Germany
Geyer, Professor H B	University of Stellenbosch
Grammatica, Miss N	University of the Witwatersrand
Gridin, Dr V V	University of the Witwatersrand
Hahne, Professor F J W	University of Stellenbosch
Hearne, Mr G R	University of the Witwatersrand
Heiss, Professor W D	University of the Witwatersrand
Hoch, Professor M J R	University of the Witwatersrand
Hook, Dr J	University, Manchester, U K

Joubert, Dr D P	University of the Witwatersrand
Kasl, Mr C	University of the Witwatersrand
Keartland, Mr J M	University of the Witwatersrand
Leggett, Professor A J	University of Illinois at Urbana–Champaign, Urbana, Ill. USA
Lemmer, Professor R H	University of the Witwatersrand
Manyala, Mr N I	University of the Witwatersrand
McLachlan, Professor D S	University of the Witwatersrand
Miller, Professor H G	University of Pretoria
Mjwara, Mr P M	University of the Witwatersrand
Mjojo, Professor C C	University of Malawi
Mphahlele, Mr J M	University of the Witwatersrand
Müller–Nedebock, Mr K K	University of Stellenbosch
Nabarro, Professor F R N	University of the Witwatersrand
Nettleton, Professor R E	University of the Witwatersrand
Pollak, Dr H	University of the Witwatersrand
Richardson, Professor R C	Cornell University, Ithaca, NY, USA
Ritchie, Mr R A	University of Pretoria
Scholtz, Dr F G	University of Stellenbosch
Sideras Haddad, Mr E	University of the Witwatersrand
Solms, Mr F	University of Pretoria
Spits, Mr R A	University of the Witwatersrand
Stoddart, Mr S T	University of the Witwatersrand
Strydom, Mr A M	Atomic Energy Corporation
Theron, Mr A N	University of Stellenbosch
Tshelane, Mr G P	University of the Witwatersrand
Tshepe, Mr T F	University of the Witwatersrand
van der Merwe, Dr P du T	Atomic Energy Corporation
Verwoerd, Professor W S	University of South Africa
von Oertzen, Dr D W	University of Pretoria
Yen, Dr G D	University of Pretoria

TOPICS IN SUPERFLUIDITY AND SUPERCONDUCTIVITY

Anthony J. Leggett
Department of Physics
University of Illinois at Urbana-Champaign
1110 West Green Street, Urbana, IL 61801 USA

Preface

These lectures were given to an audience most of whom were not specialists in the field of superconductivity or superfluidity, and some of whom had indeed had very little previous exposure to these topics. I have therefore tried to give a general overview of the field, with particular emphasis on those considerations which I believe may be specially relevant to the currently most exciting problem in the area, that of high-temperature superconductivity (briefly discussed in the last lecture). Most of these considerations have actually been known for many years, and thus the bulk of these lectures is textbook material. However, it is easy for the beginning student in this area to get lost in a forest of Green's functions and Feynman diagrams, and I have therefore tried to present the essentials in as simple a language as possible. A particular danger is that students who learn their superconductivity and superfluidity by the elegant field-theoretic techniques currently fashionable may end up unaware of the classic and actually very simple results on the general phenomenology of, and conditions for, these phenomena due to the Londons, Landau, Bloch, Yang and others, and I have therefore spent a substantial portion of these lectures (1–4) on these topics. Most of the rest (Lectures 5–8) is on BCS theory as applied to superconductors and superfluid ^{3}He; again, the results are standard textbook material, but I have tried where possible to give alternative and, I hope, simpler derivations to the standard ones. Finally, Lecture 9 tries to pull together various generalizations of BCS theory which may be found in the literature, but rarely collected in one place, with an eye on their possible relevance to the high-T_c problem.

Because of the nature of these lectures, I have not attempted to give detailed references for each point discussed; rather, I list at the end some suggested general supplementary reading on the subject-matter of each lecture, with references on specific points only when they are not covered in standard texts.

This work has been supported by the National Science Foundation under grant no. DMR88-22688.

Lecture 1: Phenomenology of Superfluidity

"Superfluidity" and "superconductivity" each describe a whole complex of phenomena which seem, at least up to now, to occur together. The general belief is that superconductivity is in effect simply superfluidity in an electrically charged system, so I will start in this lecture with a discussion of the phenomenology of superfluidity in a neutral system such as ^4He and subsequently discuss, in Lecture 2, the special features of a charged system.

The property of "superfluid" ^4He which gives it its name is of course the ability to flow through narrow channels, where any ordinary liquid would be clamped, without apparent friction. However, we shall see that, because this phenomenon involves "open" boundary conditions, it is actually rather subtle conceptually, and it is much more convenient to work in a closed geometry whenever possible. In fact, for convenience let us consider a narrow annulus of thickness ΔR very small compared to its mean radius R (but still long compared to microscopic lengths). In what follows we shall ignore effects which tend to zero with ΔR (such as the formation of a meniscus in an ordinary liquid rotating in this geometry). Then the two properties which we take as fundamental to "superfluidity" are

(1) Nonclassical rotational inertia (NCRI). We recall that above the so-called lambda temperature T_λ, liquid ^4He behaves qualitatively like any other liquid. Suppose then that we start with $T > T_\lambda$ and rotate the container very slowly (just how slowly we need to do it for the purposes of this experiment will be seen below). The helium will then come into equilibrium with the rotating walls, i.e. it will acquire a mean velocity $\underset{\sim}{\omega} \times \underset{\sim}{R}$ where $\underset{\sim}{\omega}$ is the vector angular velocity (directed along the axis of the annulus), hence an angular momentum $MR^2\underset{\sim}{\omega} \equiv I_{cl}\underset{\sim}{\omega}$. Now suppose that, while still rotating the container, we cool below T_λ and on down to zero (or in practice to $T << T_\lambda$). Then we find that the angular momentum of the liquid (observable, e.g., by gyroscopic methods) drops below $I_{cl}\underset{\sim}{\omega}$ as T falls below T_λ and tends to zero at very low temperatures.

We can define a normal density ρ_n (or normal fraction ρ_n/ρ, with ρ the total density) by writing the observed angular momentum $\underset{\sim}{L}(T)$ as $I(T)\underset{\sim}{\omega}$, and putting

$$\rho_n(T)/\rho \equiv I(T)/I_{cl} \tag{1.1}$$

i.e. the normal fraction is the apparent density of the liquid if we were to assume that it was rotating with the walls. We then have

$$\rho_n(T) \to 0, \quad T \to 0: \quad \rho_n(T) \to \rho, \quad T \to T_\lambda. \tag{1.2}$$

The superfluid density $\rho_s(T)$ is simply defined as $\rho - \rho_n(T)$. Phenomenologically, then, the liquid behaves as if a fraction ρ_s/ρ is at rest in the lab. frame (though see below), while the remaining ρ_n/ρ rotates with the container like a normal liquid. The quantity $\rho_n(T)$ is a uniquely defined function of temperature and the behavior is entirely reversible on re-heating. It is clear that the phenomenon of NCRI cannot be explained in terms of very long relaxation times, since the liquid actively changes its state on cooling; it is a true thermodynamic equilibrium phenomenon.

(2) Persistent currents: Suppose we now do a different experiment. Again we start above T_λ with the walls of the container rotating, but this time rather "fast" (cf. below). We cool below T_λ with the walls still rotating (at angular velocity $\underset{\sim}{\omega}$), and then stop the rotation. What we now find is that the liquid continues, even long after rotation has stopped, to possess an angular momentum $\underset{\sim}{L}$ which turns out, at temperature T, to be given by (approximately, see below)

$$\underset{\sim}{L}(T) = (\rho_s(T)/\rho)\, I_{cl}\underset{\sim}{\omega} \tag{1.3}$$

i.e. this time it is as if the "normal component" ("normal fraction") has come to rest and the superfluid component is still rotating. A striking feature of this experiment is that once this rotation is established, it is possible to increase or decrease L at will by varying the temperature and thereby $\rho_s(T)$, so long as we always stay below T_λ. Warming up above T_λ (with the walls at rest), however, immediately destroys the effect. We will see below that this phenomenon cannot represent the true thermodynamic equilibrium behavior but must indicate a very long-lived metastable state.

A phenomenological understanding of both the above effects (NCRI and persistent currents) may be obtained as follows: We assume that there exists in the liquid below T_λ a quantity with the dimensions of velocity, which we write $\underset{\sim}{v}_s$ and call the "superfluid velocity": if this velocity is finite but otherwise thermodynamic equilibrium is attained, then the mass current associated with this velocity is $\rho_s \underset{\sim}{v}_s$. In addition, if a normal liquid under the conditions in question would move with velocity $\underset{\sim}{v}_n$ (the so-called "normal velocity") then there is an additional contribution $\rho_n \underset{\sim}{v}_n$, i.e. in general the mass current $\underset{\sim}{j}$ is given by

$$\underset{\sim}{j} = \rho_s \underset{\sim}{v}_s + \rho_n \underset{\sim}{v}_n. \tag{1.4}$$

The fundamental ansatz, now, is that the integral of the superfluid velocity around a closed loop can take only certain quantized values. Assuming (as seems reasonable and in fact turns out to be true) that these values are equally spaced and include zero we have

$$\oint \underset{\sim}{v}_s \cdot d\underset{\sim}{l} = n\kappa_o \tag{1.5}$$

where κ_o is the "quantum" of circulation (which at the present stage is unknown).

We now immediately see the reason for the phenomenon of NCRI: In general, the system would "like" to make its angular momentum as close to the classical value $I_{cl}\underset{\sim}{\omega}$ as possible (we will show this formally at the end of this lecture). However, if ω is so small that, if we substitute $\underset{\sim}{v} = \underset{\sim}{\omega} \times \underset{\sim}{R}$, the value of $\oint \underset{\sim}{v} \cdot d\underset{\sim}{l}$ is less than $\frac{1}{2}\kappa_o$, i.e. if $\omega < \kappa_o/4\pi R^2$, then the best it can do is to take $n = 0$ in eq. (1.5), i.e. to take $\underset{\sim}{v}_s = 0$ everywhere. Arguing in a similar way, we should expect that for $\kappa_o/4\pi R^2 < \omega < 3\kappa_o/4\pi R^2$ the "best" value of n is 1, and so on. We will see below that in real ^{4}He the value of κ_o is h/m (h = Planck's constant, m = mass of ^{4}He atom), so the "critical" value ω_c of ω below which the liquid stays at rest in the lab is $\hbar/2mR^2$, which for R ~ 1 cm is about 10^{-4} rad/sec – a very slow but observable rotation speed.

To explain the phenomenon of persistent currents a further assumption is needed, namely that states corresponding to different values of n in eqn. (1.5) are separated by free energy barriers ΔF large enough that $\exp - \Delta F/kT$ is a very small number (say $\lesssim 10^{-40}$). Since the relevant values of T are $\lesssim$ 2K, this corresponds to $\Delta F \gtrsim$ room temperature. If that is so, then we expect that as the liquid is cooled, with the walls rotating at an angular velocity $\omega >> \omega_c$, the system adopts a state such that n in eqn. (1.5) is as close as possible to $\omega/2\omega_c$: this means that the superfluid velocity will essentially be indistinguishable, without very accurate measurement, from that of the walls. When the walls are brought to rest, the large value of ΔF which separates states corresponding to (for example) n and n–1 guarantees the metastability of the superfluid part of the current on all practical timescales.

At this stage let us return briefly to the originally defining property of superfluidity, namely flow through small pores without apparent friction. On the basis of the above discussion, we may anticipate that there is actually an important conceptual distinction between superflow at "very low" velocities $\underset{\sim}{v}_s$ such that $\int \underset{\sim}{v}_s \cdot d\underset{\sim}{l}$ along the length of the tube in question is less than $\frac{1}{2}\kappa_0$, and at higher velocities where it is greater. The former is in some sense a variant of NCRI, the latter of the phenomenon of persistent supercurrents in an annulus. This distinction, which seems to have been made rather rarely in the literature of the subject, seems to me an important one, and I will return to it later.

Let us now try to see what the phenomenon of NCRI tells us formally about the properties of the system. As usual, we specialize to an annulus with $\Delta R <<$ R, and for the moment confine ourselves to zero temperature. We first write down the Hamiltonian of the system, as observed from the lab. frame of reference, when the walls are at rest in this frame:*

$$\hat{H}_{(lab)} = \sum_i p_i^2/2m + \frac{1}{2}\sum_{ij} V(|\underset{\sim}{r}_i - \underset{\sim}{r}_j) + \sum_j U_i(\underset{\sim}{r}) \tag{1.6}$$

where the one-particle potential energy U(r) expresses the potential due to the container walls (which we do <u>not</u> assume are perfectly smooth**). Note that in writing eqn. (1.6) we have assumed that the two-particle interaction $V(\underset{\sim}{r}_i - \underset{\sim}{r}_j)$ is isotropic, i.e. it is a function only of $|\underset{\sim}{r}_i - \underset{\sim}{r}_j|$; if we were dealing with superfluid ^{3}He rather than ^{4}He, so that dipole forces are important (cf. Lecture 8) some care would be necessary at this point. The ground state of the system

---

* For the moment we assume that the laboratory frame of reference is an inertial frame: cf. however below.

** Even if the walls are <u>on average</u> smooth, we still have to take into account the quantum and thermal fluctuations of the constituent atoms.

corresponds, as usual, to the lowest eigenvalue of the Hamiltonian (1.6) (which is, trivially, time-independent).

Now we ask: What is the correct description of the system when the walls are rotating with angular velocity $\underset{\sim}{\omega}$? If we now write down the Hamiltonian as observed from the lab. frame of reference, the first two terms are identical but the one-particle potential U will in general depend on time (because the rugosities on the walls are moving). This time-dependent potential can do work on the system, and therefore we cannot do equilibrium thermodynamics in the lab. frame. The correct procedure is to minimize the Hamiltonian as observed from the frame of reference rotating with the walls: marking the dynamical variables as observed from this frame by primes, we have

$$H_{(rot)} = \sum_i p_i'^2/2m + \frac{1}{2}\sum_{ij} V(|\underset{\sim}{r}'_i - \underset{\sim}{r}'_j|) + \sum_i U(\underset{\sim}{r}'_i) \tag{1.7}$$

which is time-independent. Now in classical mechanics the rotating frame is completely equivalent (up to terms of order $\Delta R/R$) to the laboratory one and there is no difficulty in principle about doing a calculation in the rotating frame so as to minimize $H_{(rot)}$. However, in quantum mechanics there is a crucial difference, in that the simple boundary condition we are used to specifying on the wave function is valid only in the lab. frame (or more precisely in an inertial frame: see below). Consequently, it is usually most convenient to carry out calculations in the lab. frame, which requires us to express $H_{(rot)}$ in terms of the variables $\underset{\sim}{r}_i$, $\underset{\sim}{p}_i$ seen from this frame. At this point we make the crucial observation that if the walls are reasonably smooth, the one-particle potential $U(\underset{\sim}{r})$, or more precisely that part of it which is time-dependent, is a small perturbation whose only function is to ensure that the system reaches thermal equilibrium in the frame of the walls. We therefore really need only minimize explicitly the first two terms in (7), and it is therefore these which we should express in lab. frame variables. The resulting <u>time-independent</u> Hamiltonian H' is therefore, since $\underset{\sim}{p}'_i = \underset{\sim}{p}_i - m\underset{\sim}{\omega}\times\underset{\sim}{r}_i$ and $|r'_i - r'_j| = |\underset{\sim}{r}_i - \underset{\sim}{r}_j|$,

$$H' = \sum_i \frac{(\underset{\sim}{p}_i - m\underset{\sim}{\omega}\times\underset{\sim}{r}_i)^2}{2m} + \frac{1}{2}\sum_{ij} V(|\underset{\sim}{r}_i - \underset{\sim}{r}_j|). \tag{1.8}$$

We can rewrite this, using the invariance of the triple scalar product and the condition $\Delta R \ll R$, in the form

$$H' = H_0 - \underset{\sim}{\omega}\bullet\underset{\sim}{L} + \frac{1}{2} I_{cl}\omega^2 \qquad (I_{cl} \equiv MR^2) \tag{1.9}$$

where H_0 is the laboratory-frame Hamiltonian (1.6) with the perturbation U (or more precisely the part of it which rotates with the walls) dropped.* Thus, to obtain the ground state of the rotating system by a lab. frame calculation we need to minimize H'. Similarly, at finite temperature we need to minimize, not the simple free energy $F = \langle H_0\rangle - TS$ (S = entropy) which would be appropriate for the walls stationary, but rather the modified free energy

$$F' \equiv \langle H'\rangle - TS. \tag{1.10}$$

Although the last term in (1.9) is a constant (for our condition $\Delta R << R$), it is convenient to keep it for certain formal purposes.

An ordinary liquid when rotated is of course stationary in the frame of the rotating walls, which is equivalent to saying that the angular momentum corresponding to the minimum of H' (or at finite T, F') is equal to $I_{cl}\,\underset{\sim}{\omega} \equiv MR^2\underset{\sim}{\omega}$. Thus, if we regard the term $-\underset{\sim}{\omega}\bullet\underset{\sim}{L}$ in H' as a perturbation on the original Hamiltonian H_0, and the (lab. frame) angular momentum $\langle\underset{\sim}{L}\rangle$ as a "response" induced by this perturbation, we can write formally for the corresponding "susceptibility"

$$\chi_{LL} \equiv \frac{\partial\langle\underset{\sim}{L}\rangle}{\partial\underset{\sim}{\omega}} = I_{cl} \qquad \text{(normal system)} \tag{1.11}$$

whereas for a superfluid system we have,

$$\chi_{LL} \equiv I(T) \equiv (\rho_n(T)/(\rho)\, I_{cl}. \tag{1.12}$$

It is instructive to put this in a slightly different way. Rather than writing the perturbation as $-\underset{\sim}{\omega}\bullet\underset{\sim}{L}$, we write $\underset{\sim}{L} = RJ_\perp\hat{\underset{\sim}{\omega}}$, $\underset{\sim}{v}_\perp \equiv \omega R\hat{\underset{\sim}{\omega}}$, where $\underset{\sim}{J}_\perp$ is the total mass current circulating around the loop and $\hat{\underset{\sim}{\omega}}$ is a vector along $\underset{\sim}{\omega}$. Then the perturbation to H_0 can be written $-v_\perp J_\perp$, and the normal-liquid result $\langle\underset{\sim}{L}\rangle = I_{cl}\underset{\sim}{\omega}$ becomes

---

* The part of U which does not rotate with the walls is, of course, large (since it confines the He in the annulus!) and should strictly speaking be kept in both H' and H (eqn. (1.9) is of course still valid if this is done). However, it is often adequate to replace it by appropriate radial boundary conditions.

$$\langle \underset{\sim}{J}_\perp \rangle = Mv_\perp \equiv Nmv_\perp \tag{1.13}$$

or

$$\chi_{J_\perp J_\perp} = Nm \qquad \text{(normal system)} \tag{1.14}$$

whereas for a superfluid system we have

$$\chi_{J_\perp J_\perp} = \left(\frac{\rho_n}{\rho}\right) Nm. \tag{1.15}$$

Note that $J_\perp$ here has the significance of a current circulating around the loop; this is divergence-free but has finite curl, and may therefore be regarded, when viewed locally, as a "transverse" current. On the other hand, consider the "longitudinal" part of the current (which is defined as having finite divergence but zero curl). Using the continuity equation div $\underset{\sim}{J} + \partial\rho/\partial t = 0$ and the well-known "f-sum rule" (Thomas-Reich-Kuhn sum rule) it is straightforward to show that for <u>any</u> system which conserves particle number

$$\chi_{J_{||}J_{||}} = Nm. \tag{1.16}$$

Hence, for a superfluid system we have

$$\chi_{J_{||}J_{||}} \neq \chi_{J_\perp J_\perp} \tag{1.17}$$

i.e. the system "knows the difference" between a longitudinal and a transverse current even in the limit of a homogeneous perturbation.

We finally note that since the expectation value of the current <u>in the rotating frame</u>, $\langle j' \rangle$, is given from (1.8) by

$$\langle \underset{\sim}{j}' \rangle = \partial\langle H' \rangle / \partial\underset{\sim}{\omega} \tag{1.18}$$

it follows that for a normal system, for which $\langle \underset{\sim}{j}' \rangle$ is zero, we have

$$\partial^2\langle H' \rangle / \partial\underset{\sim}{\omega}^2 = 0 \qquad \text{(normal system)} \tag{1.19}$$

whereas for a superfluid system

$$\partial^2\langle H'\rangle / \partial \underset{\sim}{\omega}^2 = I_{cl}\,(\rho_s(T)/\rho). \tag{1.20}$$

All the above formulations of the property of NCRI are essentially equivalent.

We now sketch a proof, which is essentially the celebrated Bohr-van Leeuwen theorem of atomic physics, that no system of particles which obeys classical <u>mechanics</u> can show NCRI. (Recall that NCRI is <u>defined</u> only when the system is in thermal equilibrium.) The proof simply consists in writing the classical expression for the thermal average value of $\langle \underset{\sim}{j}'\rangle$, namely

$$\langle \underset{\sim}{j}'\rangle = Z^{-1} \iint \{\sum_i (\underset{\sim}{p}_i - m\underset{\sim}{\omega}\times\underset{\sim}{r}_i)\ \exp-H'(r_i,p_i)\} dr_i dp_i \tag{1.21}$$

with H' given by (1.8), and transforming from the lab. frame variables r_i,p_i to the rotating-frame ones $\underset{\sim}{r}'_i, \underset{\sim}{p}'_i$. Since in classical mechanics the two sets of variables are completely equivalent, then changing the <u>names</u> of the variables back to r_i,p_i we have

$$\langle j'\rangle = Z^{-1} \iint \{\sum_i (\underset{\sim}{p}_i\ \exp-H_o(r_i,p_i)/kT\ dr_i dp_i = 0 \tag{1.22}$$

since the expression on the RHS is precisely the lab. frame current when the walls are at rest, which (for ^4He) is zero. Thus the system cannot show NCRI, QED. In quantum mechanics the argument breaks down because the problem is defined by the Hamiltonian plus the commutation relations, and we have

$$[p_i', p_j'] = 2i\hbar\omega \neq [p_i,p_j] \quad (= 0). \tag{1.23}$$

An equivalent statement, as we shall see below, is that the boundary conditions on the Schrödinger equation are different in the two frames. This strongly suggests, though it does not prove, that this boundary condition plays a crucial role in superfluid behavior.

Now we turn more briefly to the question of supercurrent stability. All that we want to prove at present in this connection is that when the walls are at rest a sufficiently large angular momentum cannot correspond to the ground state but must be metastable. This may seem obvious, and indeed it is easily seen from symmetry considerations that the ground state, or normal equilibrium state, of any <u>classical</u> system must correspond to zero current and hence zero angular momentum. However, this is less obvious, in quantum mechanics, and indeed is almost certainly false, for example, for superfluid ^3He–A in certain geometries. However, it is possible to prove that even in QM the expectation value of the angular momentum in the ground state (or thermal equilibrium state) cannot be greater than an upper limit which in general is a function of the geometry and for our particular geometry is $N\hbar/2$. Since the trick used in the proof recurs in various guises below, we sketch it briefly. The ground state of the system minimizes the expectation value of the Hamiltonian H_o, i.e. the quantity

$$\langle H_o\rangle_{\Psi_o} = \langle \sum_i p_i^2/2m\rangle + \frac{1}{2}\langle \sum_{ij} V(\underset{\sim}{r}_i - \underset{\sim}{r}_j)\rangle \equiv E_o \tag{1.24}$$

subject to the boundary condition (of which more below) that one turn around the annulus leaves the wave function unchanged, i.e. if ϕ_i represents the angle around the annulus measured from some arbitrary zero, then

$$\psi(\phi_i + 2\pi) = \psi(\phi_i)\,, \quad \forall_i. \tag{1.25}$$

The proof relies on a reductio ad absurdum: Suppose that the ground state Ψ_o carries an angular momentum $\langle L\rangle > N\hbar/2$. Then we construct a new trial ground state Ψ_o' as

$$\Psi_o' \equiv \exp(i \sum_i \phi_i)\, \Psi_o. \tag{1.26}$$

This new wave function satisfies (1.25) and clearly has the property that $|\Psi_o'\{r_i\}|^2 = |\Psi_o\{r_i\}|^2$ for all $\{r_i\}$, and hence when we insert it in (1.24) the potential energy is unaffected. However, using the explicit quantum-mechanical expression $-i\hbar\underset{\sim}{\nabla}_i$ for $\underset{\sim}{p}_i$, we see that the kinetic-energy term changes relative to $\langle H_o\rangle_{\Psi_o}$ by an amount

$$\Delta E \equiv \langle H_o\rangle_{\Psi_o'} - \langle H_o\rangle_{\Psi_o} = -\frac{\hbar\langle L\rangle}{mR^2} + \frac{N\hbar^2}{2mR^2}$$

which is negative for $\langle L\rangle > N\hbar/2$. Consequently, no state satisfying this condition can be the ground state. A similar argument shows that in a <u>rotating</u> system the equilibrium state must satisfy $|\underset{\sim}{L} - I_{cl}\underset{\sim}{\omega}| \lesssim N\hbar/2$. (This mini-theorem is easily generalized to any multiply-connected topology.)

Lecture 2: Phenomenology of Superconductivity

Next we turn to the case of a charged system. In this case the phenomenon of superconductivity, as originally named, is most naturally defined as the ability of a piece of superconducting metal to carry a finite current under zero voltage drop. Just as in the case of superfluidity, however, it is more convenient to work in closed geometry and define two phenomena which are the analogs of NCRI and persistent-current flow, respectively. For clarity of exposition let us first consider the case where we can neglect the Ampère self-interaction of the currents in the superconductor, that is, treat the magnetic field at any point as simply the externally imposed one. It turns out (cf. below) that this is legitimate if we consider the static or steady-state properties of a hollow cylinder (annulus) whose thickness is small compared not only to its radius but to the "London penetration depth" introduced below (in practice, a length of a few hundred Å) and provided also that all currents are small compared to the critical current of the system. Let us, then, consider such a cylinder. In zero external field the Hamiltonian of the electrons* is identical in form to that of a superfluid when the container is at rest, eqn. (1.6):

$$H = \sum_i \frac{\underset{\sim}{p}_i^2}{2m} + \frac{1}{2}\sum_{ij} V(\underset{\sim}{r}_i - \underset{\sim}{r}_j) + \sum_i U(\underset{\sim}{r}_i), \tag{2.1}$$

In the present case we do not need to assume that V is a function only of $|\underset{\sim}{r}_i - \underset{\sim}{r}_j|$.

Now suppose that we apply a magnetic field, $\underset{\sim}{B}$, constant in space and time, parallel to the z-axis (axis of symmetry of the cylinder). If we neglect the Zeeman interaction term, which plays no particular role in the ensuing considerations, then with a particular choice of gauge the resulting Hamiltonian as observed from the lab. frame may be written

$$H = \sum_i \frac{(\underset{\sim}{p}_i - e\underset{\sim}{A}(r_i))^2}{2m} + \frac{1}{2}\sum_{ij} V(\underset{\sim}{r}_i - \underset{\sim}{r}_j) + \sum_i U(\underset{\sim}{r}_i), \tag{2.2}$$

$$\underset{\sim}{A}(\underset{\sim}{r}_i) \equiv -\frac{1}{2}(\underset{\sim}{r}_i \times \underset{\sim}{B}). \tag{2.3}$$

---

* For simplicity we shall neglect for present purposes the motion of the ions. However, taking it into account would not affect any of our conclusions.

where $\underset{\sim}{p}_i$ is the canonical momentum. This is formally identical to the Hamiltonian H' of the superfluid as observed in the rotating frame, eqn. (1.8), with the correspondence

$$\underset{\sim}{B} \rightleftarrows -2m\,\underset{\sim}{\omega}/e. \tag{2.4}$$

Now consider the (kinematic) angular momentum of the electrons in the lab. frame. Since the kinematic velocity of an electron is $(p_i - e\underset{\sim}{A}(\underset{\sim}{r}_i)/m$, the angular momentum is given by the expression

$$\langle \underset{\sim}{L} \rangle = \sum_i \underset{\sim}{r}_i \times (\underset{\sim}{p}_i - e\underset{\sim}{A}(\underset{\sim}{r}_i) \tag{2.5}$$

which, with the term (2.3) for $\underset{\sim}{A}$, is formally identical to the expression for the angular momentum of the neutral superfluid as observed from the rotating frame, (eqn. (1.8)) with the correspondence (2.4). Consequently, the phenomenon in a charged system which is the exact analog of NCRI in a neutral one is that a static magnetic field $\underset{\sim}{B}$ induces a finite angular momentum in the lab frame. Using the above correspondence, we may define the superfluid density $\rho_s(T)$ by the relation

$$\frac{\partial \langle \underset{\sim}{L} \rangle}{\partial \underset{\sim}{B}} = \frac{e}{2m} \bullet \frac{\rho_s(T)}{\rho} \bullet I_{cl} \tag{2.6}$$

so that a normal system has $\rho_s = 0$ (compare and contrast eqns. (11)–(12)).* Just as in the neutral case, we can re-express these results in terms of the static transverse current susceptibility $\chi_{J_\perp J_\perp}$. However, there is a slight catch here: the expression for the lab. frame current explicitly contains the vector potential $\underset{\sim}{A}(\underset{\sim}{r}_i)$, and therefore if we define $\chi_{J_\perp J_\perp}$ in the conventional way in terms of the matrix elements and energy levels of the unperturbed ($\underset{\sim}{A} = 0$) system, i.e.

$$\chi_{J_\perp J_\perp} = \sum_n \frac{2\,|\langle n\,|\,\underset{\sim}{J}_\perp\,|\,0\rangle|^2}{E_n - E_o} \tag{2.7}$$

then, just as in the neutral system, (eqn. (1.15)) we have

---

* Note that the equation for a neutral system which is formally analogous to (2.6) is $\partial\langle \underset{\sim}{L}' \rangle/\partial\,\underset{\sim}{\omega} = (\rho_s(T)/\rho)I_{cl}$ where L' is the angular momentum observed from the rotating frame.

$$\chi_{J_\perp J_\perp} = (\rho_n(T)/\rho)\, Nm \tag{2.8}$$

whereas for the "longitudinal" susceptibility we again have (cf. (1.16))

$$\chi_{J_\perp J_\perp} = Nm \tag{2.9}$$

so that in this case also the system can "tell the difference" between longitudinal and transverse currents in the homogeneous limit.

Finally we note that the analog of eqns. (1.19)–(1.20) for the charged system would be

$$\frac{\partial^2 F}{\partial \underset{\sim}{B}^2} = 0 \quad \text{(normal system)},$$
$$\frac{\partial^2 F}{\partial \underset{\sim}{B}^2} = \left(\frac{e}{2m}\right)^2 I_{cl}\, \rho_s(T)/\rho \quad \text{(superconducting system)}. \tag{2.10}$$

Actually, neither eqn. (1.19)–(1.20) nor (2.10) are exactly correct, for a ring of finite thickness, even in the absence of Ampère self-coupling, because of small effects associated with centrifugal compression in the neutral case and Landau diamagnetism in the charged case. For this reason it is convenient, in considering the phenomenon of NCRI in a finite ring, to apply the magnetic field under "Aharonov-Bohm conditions", i.e. to prescribe that $\underset{\sim}{B}$ be finite within the inner radius of the cylinder but zero everywhere in the cylinder itself. This corresponds to a magnetic vector potential $\underset{\sim}{A}(\underset{\sim}{r})$ which is given everywhere in the body of the cylider (in an appropriate gauge) by

$$A_r = A_z = 0, \qquad A_\phi = \Phi/2\pi r$$

where $\Phi \equiv \int \underset{\sim}{B} \bullet d\underset{\sim}{S}$ is the total magnetic flux through the inside of the cylinder. The corresponding, rather artificial, thought-experiment for a neutral superfluid would consist in rotating the end walls of the cylindrical annulus in such a way that their velocity $\underset{\sim}{v}(\underset{\sim}{r})$ is given by $v_r = v_z = 0$, $v_\phi = \text{const.}/2\pi r$. It is perhaps simpler to endow the "neutral" superfluid with a small fictitious charge and apply a magnetic flux Φ as above. One may then define a normal system unambiguously by the statement that under these conditions

$$\frac{\partial^2 F}{\partial \Phi^2} = 0 \qquad \text{(normal system)} \tag{2.11}$$

which for any system which carries no current in the absence of the flux* is equivalent to the statement that the free energy F is independent of flux Φ under Aharonov-Bohm conditions. For a superfluid (superconducting) system, by contrast, the free energy F is a nontrivial function of Φ even under such conditions, and in fact we can define the superfluid density $\rho_s(T)$ by the statement

$$\left.\frac{\partial^2 F}{\partial \Phi^2}\right|_{\Phi=0} = \frac{Ne^2}{m} \bullet \frac{1}{(2\pi R)^2} \rho_s(T)/\rho. \tag{2.12}$$

The circulating electric current I_ϕ is $-\partial F/\partial\Phi$, and therefore the (transverse) electric current density $\underset{\sim}{J}_{el}\underset{\sim}{r}$ is related to the vector potential $\underset{\sim}{A}(\underset{\sim}{r})$ by

$$\underset{\sim}{J}_{cl}(\underset{\sim}{r}) = \frac{-n\rho_s(T)}{\rho}\frac{e^2}{m}\underset{\sim}{A}(\underset{\sim}{r}) \quad (\underset{\sim}{A} \to 0) \tag{2.13}$$

an expression which is often taken as the fundamental definition of the superfluid density.** Eqn. (2.13) should be regarded as more fundamental than (2.10), since in contrast to the latter it applies even when the magnetic field is non-uniform, indeed even under AB conditions. It is characteristic of a superfluid system that it is sensitive not only to the electric and magnetic fields (which are the only quantities which are physically meaningful in classical electromagnetic theory) but also to the electromagnetic potentials.

We turn now more briefly to the phenomenon of metastable current flow in a charged system. To the extent that we neglect Ampère self-coupling,† this phenomenon is the exact analog of the persistent currents in a neutral system and can be discussed in exactly the same terms. Further, it is clear that just as in the neutral case we may obtain a phenomenological description by assuming that there exists in the system, in the absence of a magnetic vector potential, a velocity $\underset{\sim}{v}_s$ such that in analogy to eqn. (1.5) we have

$$\oint \underset{\sim}{v}_s \bullet d\underset{\sim}{l} = n\kappa_0' \tag{2.14}$$

---

* It should be emphasized that the class of systems which violate this condition is probably not empty: cf. the remark above about ^{3}He–A.

** Note that there is an implicit assumption here that J and A are not appreciably varying in the radial direction. This is true (cf. below) provided the thickness of the cylinder is indeed small compared to the London penetration depth and it is homogeneous.

† In a ring of the geometry described, self-coupling effects simply add to the total free energy the term $1/2\ LI^2$, which is usually of negligible importance for the considerations below.

(where the constant κ_0' is of course not expected to be the same as that (κ_0) occurring for superfluid He), and moreover that a high free energy barrier separates states of different n. The explanation of the metastability of supercurrent flow then follows the lines set out in the neutral case.

What about the phenomenon which we have identified as the analogy of NCRI? This will occur naturally if we identify the $\underset{\sim}{v}_s$ which satisfies the condition (2.14) with the <u>canonical</u> "velocity", that is the quantity $\underset{\sim}{p}_c/M^*$ where $\underset{\sim}{p}_c$ is the canonical momentum and M* the mass of the relevant unit (which need not necessarily be a single electron). Then the <u>kinematic</u> velocity $\underset{\sim}{v}_k$ and electric current $\underset{\sim}{j}$ are related to $\underset{\sim}{v}_s$ by

$$\underset{\sim}{v}_k = \underset{\sim}{v}_s - e^*\underset{\sim}{A}/M^* \quad , \quad \underset{\sim}{j} = e\rho_s \underset{\sim}{v}_k . \tag{2.15}$$

where e* is the charge of the relevant unit, and we see that in a sufficiently small but finite potential $\underset{\sim}{A}$, where the optimum value of n is zero, the kinematic velocity and hence the electrical current is proportional to $\underset{\sim}{A}$, in accordance with the definition of a superfluid system given above. Moreover, just as in the case of a neutral system, we can classify the original phenomenon of "superconductivity" (current flow under zero potential drop) into two conceptually distinct regimes, depending on whether $\int \underset{\sim}{v}_s \cdot d\underset{\sim}{\ell}$ across the superconducting circuit element is small or large compared to κ_0'. It is noteworthy that for a typical sample dimension (length say 1 cm) it will turn out that the crossover between these two regimes occurs at a current density of the order of 10^5 A/cm^2, which lies in an experimentally interesting region. Strictly speaking the phenomenon of flow under zero potential drop needs to be discussed separately below and above this threshold.

We now relate the definition of "superconductivity" outlined above to the two equilibrium phenomena most commonly observed in superconductors, namely the Meissner effect and flux quantization. This requires that we now take into account the Ampère self-coupling of the currents, i.e. the fact that the local magnetic vector potential $\underset{\sim}{A}(\underset{\sim}{r})$ has a contribution from the currents flowing in the metal itself; up to now we have ignored this feature.

We consider first a bulk, simply connected piece of superconducting metal. The simplest way to incorporate the self-coupling is to write down the Maxwell equation

$$\underset{\sim}{\nabla} \times \underset{\sim}{H} = \underset{\sim}{j} + \partial \underset{\sim}{D}/\partial t. \tag{2.16}$$

Since we consider an equilibrium situation and since* $\underset{\sim}{H} = \mu_o \underset{\sim}{B} = \mu_o(\underset{\sim}{\nabla}\times \underset{\sim}{A})$, this can be written

$$\underset{\sim}{\nabla}\times(\underset{\sim}{\nabla}\times \underset{\sim}{A}) = \mu_o \underset{\sim}{j}\,. \tag{2.17}$$

Combining this with eqn. (2.13), namely

$$\underset{\sim}{j} = - n\left(\frac{\rho_s}{\rho}\right)\frac{e^2}{m}\underset{\sim}{A} \tag{2.18a}$$

or

$$\underset{\sim}{\nabla}\times\underset{\sim}{j} = - n\left(\frac{\rho_s}{\rho}\right)\frac{e^2}{m}\underset{\sim}{B} \tag{2.18b}$$

and choosing a gauge in which $\underset{\sim}{\nabla}\bullet\underset{\sim}{A} = 0$, we obtain

$$\nabla^2\underset{\sim}{j} = \lambda_L^{-2}\underset{\sim}{j}\,, \tag{2.19a}$$

$$\nabla^2\underset{\sim}{A} = \lambda_L^{-2}\underset{\sim}{A} \tag{2.19b}$$

where λ_L, the so-called London penetration depth, is given by the expression

$$\lambda_L^{(T)} \equiv \left[\frac{ne^2\mu_o}{m}\left(\frac{\rho_s(T)}{\rho}\right)\right]^{-1/2} \equiv \delta_o(\rho_s/\rho)^{-1/2} \tag{2.20}$$

where $\delta_o \equiv (ne^2\mu_o/m)^{-1/2}$ is the classical high-frequency skin depth. In a typical metal, $\lambda_L(0)$ ($\equiv \delta_o$) is of the order of a few hundred angstroms: quite generally, $\lambda_L(T)$ diverges as $(T_c-T)^{-1/2}$ as $T \rightarrow T_c$.

If we apply eqn. (2.19) to a flat semi-infinite slab of superconductor with the magnetic field outside it parallel to the surface, and require that the field not increase indefinitely in the bulk metal, we find that both the current $\underset{\sim}{j}$ and the field $\underset{\sim}{H}$ die off exponentially as we go into the metal, with an attenuation length λ_L. Thus, for points much more distant than λ_L from the surface, no current flows and no magnetic field penetrates. The situation is qualitatively similar in any reasonable simply connected geometry: the magnetic field is excluded from the bulk of the superconductor, over a length of order λ_L. This is the so-called Meissner effect. A simple application is to the case of a

---

* The magnetic permeability can be taken equal to unity to a very good approximation.

superconducting wire of diameter $R >> \lambda_L$ carrying a d.c. current: the current flows entirely within a "skin depth" λ_L of the surface of the wire (just as would a high-frequency current in a normal conductor, with λ_L replaced by δ_o). For practical reasons it is usually much easier to test a putative superconductor for the Meissner effect than directly for the property expressed by eqn. (2.13). However, in view of some recent discussions of so-called "anyon superconductivity" in the literature, it should be emphasized that while eqn. (2.13) (plus, of course, Maxwell's equations) certainly implies the Meissner effect, I know of no rigorous proof that the converse is true;* and it seems entirely conceivable that there might exist systems which display the Meissner effect but not one or more of the other characteristic features of ordinary superconductors.

Let us now consider a superconductor of multiply connected topology such as a hollow cylinder. It now becomes relevant that (2.18) is not, in fact, the most general form of the relation between the current $\underset{\sim}{j}(\underset{\sim}{r})$ and electromagnetic vector potential $\underset{\sim}{A}(\underset{\sim}{r})$ (though it always applies, of course, in the limit $\underset{\sim}{A} \to 0$). In fact, from eqns. (2.14) and (2.15) we see that (2.13) is actually the special case corresponding to $\underset{\sim}{v}_s = 0$, i.e. $n = 0$ in eqn. (2.14). If we consider a general value of n, the generalization of (2.13) is:

$$\underset{\sim}{j}(r) = -\frac{ne^2}{m}\frac{\rho_s}{p}(\underset{\sim}{A}(r) - \underset{\sim}{A}_o(\underset{\sim}{r})) \tag{2.21}$$

where $A_o(\underset{\sim}{r})$ satisfies the conditions

$$\underset{\sim}{\nabla} \times A_o(\underset{\sim}{r}) = 0 \tag{222a}$$

$$\oint A_o(\underset{\sim}{r}) \bullet d\underset{\sim}{l} = n\frac{M^*}{e^*}\kappa_o' \tag{2.22b}$$

where $\oint$ represents an integral along any closed circuit lying entirely within the superconducting metal. As a result, while in the bulk of the metal eqns. (2.118a) and (2.19) remain valid, so that the electric current $\underset{\sim}{j}$ and magnetic field $\underset{\sim}{B}$ fall off exponentially as before with a characteristic length λ_L, (2.19b) fails and the magnetic vector potential may fall off not to zero but to a value $A_o(\underset{\sim}{r})$ which satisfies eqns. (2.22). Since the left-hand side of this equation is, by Stoke's theorem, simply the flux trapped through the relevant circuit, the flux

* In the sense that (2.13) is true in the weak-field limit for arbitrary directions, topologies etc.

trapped through the hole(s) in a hollow superconductor is quantized in units of $(M^*/e^*)\kappa_0'$. We will see that for a standard BCS superconductor we have $M^* = 2m$, $e^* = 2e$, $\kappa_0' = h/2m$, so the "quantum" of flux is $h/2e \equiv \phi_0$. If we apply an arbitrary external flux ϕ through the hole in such a superconductor and cool it through T_c, what happens is that screening currents are set up in the inner as well as the outer surface so that the total trapped flux (external plus self-generated) is $n\phi_0$, where n is the integer closest to ϕ/ϕ_0. Note that for a simply connected superconductor which remains everywhere superconducting we must have $n = 0$ in (2.22), and thereby recover the results described above (i.e. the special case $\underset{\sim}{A}_0 = 0$): if n were different from zero, this would imply that $\underset{\sim}{B}$ is finite at some points deep in the bulk, which we have seen is excluded by the eqns. (2.18b) and (2.17). Even in a simply connected piece of superconducting metal, however, it may be possible to form normal regions which in effect act as "holes" and thereby obtain a finite penetration of magnetic flux; this is, of course, what happens in a "type-II" superconductor, the "holes" being the cores of the magnetic vortices formed in such systems.

To conclude this lecture, let us briefly discuss a few amusing questions related to the concepts introduced here. First, we have so far implicitly assumed that the "laboratory" frame of reference is an inertial frame, and, as we shall see in the next lecture, this assumption is essential in the proof that eqn. (2.14) is valid in the lab. frame. However, strictly speaking the lab. frame is not a local inertial frame in the sense of general relativity, since it rotates once a day with respect to the fixed stars. It is in the frame of the latter that we should, strictly speaking, apply eqn. (2.14), and the corollary is that if we indeed place a bucket of superfluid helium at the North Pole (the simplest thought-experiment) the helium as observed by the (earth-based) experimenter should rotate with a period of 1 day! This rotational frequency, small as it is, is not all that small by comparison with the characteristic angular velocity ω_c defined in Lecture 1, and attempts have been made to observe it.

A second question concerns the combination of rotational and magnetic effects in a charged system (superconductor). To a first approximation, at least, these effects should be additive, so that the effective Hamiltonian of the electrons in a superconductor rotating with angular velocity $\underset{\sim}{\omega}$, as seen from the frame rotating with the lattice, should be

$$H' = \sum_i \frac{1}{2m} (\underset{\sim}{p}_i - e\underset{\sim}{A}(\underset{\sim}{r}_i) - m\underset{\sim}{\omega} \times \underset{\sim}{r}_i)^2 + V + U \qquad (2.23)$$

(where V and U are respectively the two- and one-particle potentials introduced above). Since for a constant magnetic field $\underset{\sim}{B}$ we have $\underset{\sim}{A}(\underset{\sim}{r}) = -1/2\, \underset{\sim}{B} \times \underset{\sim}{r}$, it follows that when the condition

$$\underset{\sim}{B} = e(2m/e)\underset{\sim}{\omega} \tag{2.24}$$

is fulfilled the Hamiltonian is just that of the superconductor at rest in zero field, and we expect that the canonical momentum is zero. If then we consider a situation in which the superconductor is rotated in zero external magnetic field, we might guess that the system would spontaneously generate a magnetic field satisfying (2.24), and that at least in the bulk this would permit the electrons to rotate with the same velocity $\underset{\sim}{\omega}$ as the lattice ions. A more detailed consideration shows this indeed to be the case: the resulting magnetic moment of a rotating superconductor is known as the London moment. This effect is experimentally confirmed: it is an interesting and currently not totally resolved question whether the "mass" m which appears in the relation (2.24) is the free electron mass or should be corrected for relativistic and other effects.

Finally let us enquire whether there is any analog, in an electrically neutral system, of the Meissner effect. In fact we should expect such an effect to exist: just as two neutral atoms at rest interact via the Newtonian gravitational interaction $-Gm^2/r$ (rather than the Coulomb potential $e^2/4\pi\varepsilon_0 r$), so in classical general relativity the Ampère interaction of currents has a gravitational analog (essentially, the so-called Lense-Thirring effect). If we neglect the differences of detail between the two interactions (due to the fact that gravitation is a tensor interaction, etc.) we can at least estimate the conditions under which such an effect should be important. The "gravitational London penetration depth λ_G" is obtained by replacing $e^2/4\pi\varepsilon_0$ in the formula for λ_L by Gm^2, i.e. for $T = 0$

$$\lambda_G = (4\pi\rho G/c^2)^{-1/2}$$

where ρ is the mass density.* For the densities corresponding to ordinary terrestrial matter this gives a λ_G of the order of 10^7 km, much greater than the radius R of any object that is likely to be superfluid, and thus the effect should

---

* We assume here for simplicity that we are dealing with a single type of particle.

be irrelevant under these conditions. On the other hand, the criterion $\lambda_G \sim R$ is easily seen to be, up to a factor ~ 1, precisely the "black-hole" condition $MG/R \sim c^2$. Thus, it is possible that the gravitational analog of the Meissner effect might be significant for a neutron star (which is believed to be at least partially superfluid) during the process of collapse to a black hole.

Lecture 3: The Topological Basis for Superfluidity and Superconductivity: the "Landau criterion"

In the last lecture we saw that a fundamental property which is common to systems which show "superfluidity" (in the sense of NCRI) and "superconductivity" (in the sense of the Meissner effect)† is that if we consider a thin annular cylinder of the material (thickness << R), attach to each particle a (possibly fictitious) "electric charge" e and neglect self-inductance effects, and apply a magnetic flux Φ through the hole under "Aharonov-Bohm" conditions, then the free energy is a nontrivial function of Φ. It is convenient to focus on this property (which we shall call "AB diamagnetism") for the ensuing discussion.

For the purposes of this discussion we shall assume, as we implicitly did above, that we do not need to take explicit account of the dynamics of the "environment" in which the system moves (e.g., the lattice ions in the case of the electrons in a superconductor, the atoms of the vessel walls in the case of liquid He, etc.), but can represent their effect adequately by assigning a temperature to the system. (It would be possible to repeat the arguments given below taking the environment dynamics explicitly into account, but it is unnecessary for our purposes.) Under this assumption the relevant Hamiltonian is

$$\hat{H} = \frac{1}{2m}\sum_i (\underset{\sim}{p}_i - e\underset{\sim}{A}(\underset{\sim}{r}_i))^2 + \frac{1}{2}\sum_{ij} V((\underset{\sim}{r}_i - (\underset{\sim}{r}_j) + \sum_i U(r_i) \tag{3.1}$$

where $\underset{\sim}{A}(\underset{\sim}{r})$ is the externally imposed vector potential, which satisfies the conditions

$$\underset{\sim}{\nabla} \times \underset{\sim}{A} = 0 \text{ everywhere in sample,} \quad \oint^{c} \underset{\sim}{A} \bullet d\underset{\sim}{l} = \Phi \tag{3.2}$$

† Though cf. the caveat on page [17].

where c is any contour lying entirely within the sample and passing once around the cylinder axis in (say) a clockwise direction. In view of the conditions (3.2), we may choose a gauge such that $\underset{\sim}{A}$ has the simple form

$$\underset{\sim}{A}(\underset{\sim}{r}) = (\Phi/2\pi\, r)\,\hat{\underset{\sim}{\phi}} \tag{3.3}$$

$\hat{\underset{\sim}{\phi}}$ being a unit vector in the tangential direction. We wish to solve the time-independent Schrödinger equation

$$\hat{H}\,(\Phi)\psi = E(\Phi)\psi \tag{3.4}$$

for the eigenvalues and eigenfunctions of $\hat{H}$ (eqn. (3.1)) and to use them to calculate the free energy $F \equiv -\beta^{-1} \ln \mathrm{Tr} \exp -\beta\, E_n(\Phi)$ as a function of Φ and $T(\equiv k_0^{-1}\beta^{-1})$. An essential point is that the wave functions $\psi_n(\Phi)$ which satisfy the Schrödinger equation must satisfy the "single-valuedness boundary condition": if we express the position vector $\underset{\sim}{r}$ in terms of the standard cylindrical polar coordinates r, z, θ, and denote the collection of r- and z-coordinates by $\{\xi\}$, then this condition is written

$$\Psi(\theta_1\theta_2\ldots\theta_j+2\pi\ldots\theta_N:\{\xi\}) = \Psi(\theta_1\theta_2\ldots\theta_j\ldots\theta_N:\{\xi\}). \tag{3.5}$$

(In words: if a single particle is taken once around the cylinder, with all the others fixed, the final wave function must coincide with the initial one.)

We can immediately write down two important theorems concerning the dependence of the free energy F on flux Φ (for any temperature T): First, since the Hamiltonian $\hat{H}$ (eqn. (3.1)) with $\underset{\sim}{A} = 0$ is invariant under time reversal, F must be an even function of Φ:

$$F(\Phi) = F(-\Phi). \tag{3.6}$$

Secondly, F must be a periodic function of Φ with period h/e. To establish this, we define a new wave function* Ψ' by

$$\Psi'\{\underset{\sim}{r}_i\} \equiv \Psi\{\underset{\sim}{r}_i\} \exp ie \sum_i \int_0^{r_i} \underset{\sim}{A}(r_i')\bullet d\underset{\sim}{r}_i'/\hbar. \tag{3.7}$$

---

* The operation described by eqn. (3.7) is sometimes called a "singular gauge transformation".

Substituting Ψ' into the time-independent Schrödinger equation, we see that Ψ' solves the "zero-flux" Schrödinger equation

$$\hat{H}_0\Psi' = E\Psi' \qquad (\hat{H}_0 \equiv \hat{H}(\Phi = 0)) \tag{3.8}$$

but with the SVBC modified to read

$$\Psi'(\theta_1\theta_2\ldots\theta_j+2\pi\ldots\theta_N) = \exp\,(ie\,\Phi/\hbar)\Psi'(\theta_1\theta_2\ldots\theta_j\ldots\theta_N). \tag{3.9}$$

Now, it is clear that under the replacement $\Phi \to \Phi + nh/e$ neither the Schrödinger equation (3.8) nor the SVBC, eqn. (3.9), is changed. Consequently we reach the conclusion

$$F(\Phi + nh/e) = F(\Phi) \tag{3.10}$$

as stated. Eqn. (3.10) is perfectly general,† and eqn. (3.6) is general provided the Hamiltonian is invariant under time reversal. One obvious way of satisfying these two equations is to have $F(\Phi)$ a nontrivial function of Φ with a minimum at the origin and a minimum periodicity of h/e (see fig. 1a). However, it should be emphasized that eqns. (3.6) and (3.10) are equally compatible with (at least) the following situations:

(1) $F(\Phi)$ = constant ("normal" system (fig. 1b)
(2) $F(\Phi)$ has minimum periodicity h/ne (n = any integer) (fig. 1c)
(3) $F(\Phi)$ has a maximum at $\Phi = 0$ (fig. 1d)

(or, of course, a combination of (2) and (3)). We will see that each of these conditions can describe a situation of potential experimental interest.

Let us now consider how the system behaves in the presence of a time-varying magnetic flux Φ. For simplicity we confine ourselves to the case of zero temperature, and assume that all relevant values of Φ are small compared to the flux "quantum" h/e (or, in case (2), to h/ne). We exclude for the moment case (3) above. Then either the equilibrium free energy $F(\Phi)$ is independent of Φ, or for small Φ we have $F(\Phi) > F(0)$. In any but pathological

† We have implicitly assumed that all particles carry the same charge e. The theorem is obviously also true if all charges are integral multiples of some basic unit e.

cases $F(\Phi)$ and its first derivative is continuous at the origin, so we may assume in the latter case that $F(\Phi) = F(0) + \frac{1}{2}\alpha\,\Phi^2$, $\alpha > 0$. Now, it is easy to show (using the definition of the current density $\underset{\sim}{j}(r)$ in terms of $\underset{\sim}{A}(\underset{\sim}{r})$) that the current I in the cylinder is given by the expression

$$I = -\,\partial F/\partial\Phi. \tag{3.11}$$

Consequently if α is finite a finite flux Φ, if sufficiently slowly applied, gives rise to a finite current I in the opposite sense, i.e. the system behaves like a diamagnet. This is just what we normally call superconductivity (in the sense related to the Meissner effect), and indeed it is clear that the quantity α is related to the superfluid density by (cf. eqn. (2.12))

$$\alpha = \frac{Ne^2}{m} \bullet \frac{1}{(2\pi R)^2}\,\rho_s(T)/\rho. \tag{3.12}$$

Consider now the case that $F(\Phi) = \text{const}$. At first sight, at least, this can happen in two distinct ways. First, it may be possible to start from the $\Phi = 0$ ground state and deform it (i.e. Ψ', not Ψ) adiabatically with increasing Φ so as to accommodate the changing SVBC (eqn. (3.9)) without any energy cost. Call this situation (1a) (fig. 1b). On the other hand, it may be that the original wave function cannot be deformed without an energy cost, but when the flux reaches a very small value on the scale of h/e (say $\sim h/N^{\gamma}e$, where N is the total number of particles in the system and $\gamma > 0$), then an energy level corresponding to a wave function Ψ' orthogonal to the ground state drops below the "ground state" energy $E_0(\Phi)$ (case (1b, fig. (2)). Thus to all intents and purposes $E(\Phi)$ is independent of Φ. At finite temperatures a third mechanism, somewhat related to the above, may operate: while the individual energy levels E_n may be nontrivial functions of Φ, it may be possible to rearrange the Gibbs thermal occupation factors ($\propto \exp{-\beta E_n(\Phi)}$) so that the free energy itself is only negligibly dependent on Φ (case (1c)). It is fairly clear that the situation with regard to energy dissipation is quite different in case (a) on the one hand, and (b) or (c) on the other: In the first case, any energy dissipation must be at best exponentially small in the voltage $V = -\dot{\Phi}$ applied around the ring, and the system is an <u>insulator</u>. In case (b) or (c), on the other hand, there exists the possibility of "overshoot": if the voltage is finite, i.e. $\Phi(t)$ is varied at a finite rate, then the system may stay (at $T = 0$) in the adiabatically evolving state $\Psi_0'(\Phi)$ even though a value of Φ has been reached for which this is no longer the ground state. Subsequently, through some small interaction not explicitly included in the Hamiltonian, the system will make a transition back to a more stable state, emitting in the process a phonon or something similar. Thus in this case the time-varying flux (voltage) will produce a finite rate of energy dissipation, which may or may not be proportional to V^2 ("ohmic" case) but at any rate is not exponentially small in V, and the system will behave as a normal metal (conductor), possibly with a highly nonlinear conductance. Thus, depending on the structure of the energy levels as a function of flux, the system may behave as an insulator, a normal metal or a superconductor. It may seem a trivial point, but is worth some emphasis, that to prove that a particular system is a superconductor it is not enough to demonstrate that it is not a normal metal (which could be done, for example, by demonstrating that there is a finite (volume-independent) gap in the elementary excitation spectrum): it is essential to show also that it is not an insulator!

Let us now obtain a sufficient condition for a given system to be an insulator (case (1a)). In this case it must be possible to modify the ground state Ψ_0' adiabatically to cope with the varying SVBC (eqn. (3.9)) with no cost in energy (or more precisely a cost which vanishes exponentially with increasing volume, etc.). It is clear that a sufficient condition for this to be the case is that any path which takes us from $(\theta_i,\theta_j,\theta_k\ldots)$ to $\theta_j + 2l\pi$, $\theta_j + 2m\pi$, $\theta_n + 2n\pi\ldots)$ crosses at least one region in which the many-body wave function is exponentially small. For, in this case, it is possible to put the necessary "kinks" in the wave function entirely in such regions, i.e. take $\Psi'(\Phi) = \Psi \exp if$ where f is constant everywhere except in these "special" regions; then, according to the argument developed in Lecture 1, the potential energy is unaffected and the kinetic energy is affected only up to exponentially small terms. Note that it is <u>not enough</u> that all paths going from $(\theta_1\theta_2\ldots\theta_j\ldots\theta_N)$ to $(\theta_1\theta_2\ldots\theta_j+2\pi\ldots\theta_N)$ cross such a "special" region: there may be paths going from $(\theta_1\theta_2\ldots\theta_j\theta_k\theta_\ell\ldots\theta_N)$ to $(\theta_1\theta_2\ldots\theta_j+2\pi, \theta_k+2\pi, \theta_\ell+2\pi\ldots\theta_N)$ which fail to do so, and in that case we have no guarantee that we can deform the wave function adiabatically with no energy cost in the way described. (In words, the wave function may become exponentially small if we try to take <u>one</u> particle around the ring by itself, but not if we take two or more round together.) However, it is clear that if n is the minimum number of particles which have to be taken round in this way, then the "flux quantum" is h/ne, i.e. we can say that $F(\Phi + jh/ne) = F(\Phi)$ for all integral j. The "sufficient condition" described here may be seen to be satisfied in the case, for example, of an ordinary insulating crystalline solid (this is most easily seen by using the Wannier representation) or an "Anderson insulator" in which the relevant part of the many-electron wave function is a Slater determinant of states localized on particular impurities. We will call many-body wave functions which do not satisfy this sufficient condition "topologically extended."

One may ask what is the situation if the many-body wave function, rather than vanishing exponentially in a particular finite region of space, has isolated nodal planes on which it is zero? To answer this question we need to distinguish between the nodal surfaces which automatically occur in Fermi systems because of the necessity to antisymmetrize the wave function, and other ("non-symmetry-dictated") nodal surfaces (NSDNS). If all of the relevant paths cross at least one NSDNS, we can put in the necessary "kink" there, and the system must have $F(\Phi) \leq F(0)$, i.e. it must correspond either to case (1) or to case (3) above. An amusing application of this principle is to a set of an even number N of spinless fermions, with arbitrary one- and two-particle interactions, on an effectively one-dimensional ring at $T = 0$. ("Effectively one-dimensional" means that any dependence of the many-body wave function on the (r,z) coordinates is negligible for the purposes of the ensuing argument.) Let Ψ_0 be the ground state many-body wave function in zero flux: clearly this can be taken real. Now consider a path which takes a single particle, say number 1, around the ring once. Since we must recover the original amplitude at the end of the operation, it must cross an even number of nodal surfaces. On the other hand, only N–1 of these – an odd number – are symmetry-dictated (the planes $\theta_1 = \theta_j$, j=2...N). Consequently we must cross at least one NSDNS on each such path. We cannot, unfortunately, conclude* without further argument that $F(\Phi)$ cannot have a minimum at $\Phi = 0$, because the corresponding argument for two (or any even number of) particles taken around the ring fails: but we can conclude that if it indeed has such a minimum (i.e. if the system is "superconducting") then the flux quantum is an even submultiple of h/e. It is easy to see that the "complementary" argument for an odd number N of particles (starting from the (real) ground state wave function for $\Phi = h/2e$) does not run into this complication, and for this case the conclusion

* This conclusion was incorrectly drawn by the author in earlier discussions of this problem.

$$F(\Phi) \leq F(h/2e) \qquad \text{(N odd)} \tag{3.13}$$

is rigorous.

So far I have discussed only that characteristic of superconductors which I have called "AB diamagnetism". To conclude this lecture, I would like to revert more briefly to the other main characteristic, namely the metastability of supercurrents. As we have seen, a necessary and sufficient condition is that there should exist states corresponding to a finite supercurrent (or angular momentum) separated from the ground state by free energy barriers large compared to the thermal energy kT (and that the rate of quantum tunnelling through these barriers should be negligible). Whether such barriers indeed exist is a matter for detailed calculation using the favored model of the particular system in question: it should be emphasized that the phenomenon of NCRI (or "AB diamagnetism") does not itself automatically guarantee this, cf. Lecture 4.

In the literature on superconductivity and superfluidity one often finds reference to an alleged alternative necessary and sufficient condition for a system to be superfluid (superconducting), namely the famous "Landau criterion": a system is alleged to be superfluid for velocities $< v$ if and only if its elementary excitations satisfy the energy-momentum inequality

$$\varepsilon(p) \geq vp. \tag{3.14}$$

Despite its ubiquity in the literature, the Landau criterion is in my opinion almost totally misleading. There is to the best of my knowledge only one situation in which it is genuinely relevant, namely when a small object such as an ion moves at velocity v through the (stationary) system. Then, as may easily be seen fromthe laws of conservation of energy and momentum, eqn. (3.14) is indeed the condition for the object to suffer no friction. However, it is a very long step indeed, involving a number of implicit assumptions, from there to the statement that the system is "superfluid" in the sense of sustaining metastable currents in a ring geometry (let alone that it shows NCRI!), and indeed it is easy to find counter examples both to its necessity and to its sufficiency. Cases where the system is superfluid (superconducting) despite failure of the Landau condition for any finite v include the so-called gapless superconductors and the A phase of ^{3}He (with the so-called $\boldsymbol{\ell}$-vector pinned by, say, dipole forces): cases where it is normal despite the fact that the Landau criterion holds include He I (the normal phase of liquid ^{4}He above the λ-temperature*) and amorphous solids (where, unlike in crystalline solids, the spectrum of the phonons does not bend back to zero at the "reciprocal lattice vector", indeed none such exists). The latter case is quite intriguing: a ring of ordinary window glass certainly does not show NCRI, and one's intuition says that even at $T = 0$ it will not remain in a (metastable) state of rotation for arbitrarily long times, but the actual microscopic mechanism by which it comes to rest is not so immediately obvious and perhaps deserves further investigation. In any case, it is a mystery to me why the Landau criterion is still widely believed to be of any central relevance to the question of superfluidity.

Lecture 4: Superfluidity of ^{4}He

Having discussed in the last three lectures the general phenomenology of superfluidity and superconductivity and the general properties the many-body wave function must possess in order for these phenomena to occur, we now turn to the question of how these properties are realized in specific systems. In this lecture we discuss liquid ^{4}He. As usual, we shall consider an annulus of thickness small compared to its radius R. Since the system is neutral, it is somewhat more convenient to work not explicitly in terms of the "vector potential" $\underset{\sim}{A}(\underset{\sim}{r})$ as in the last lecture, but rather in terms of the equivalent rotation $\underset{\sim}{\omega}$, which is related to it by $\underset{\sim}{A}(\underset{\sim}{r}) \rightleftarrows (m/e)\, \underset{\sim}{\omega} \times \underset{\sim}{r}$. Then, as we saw in Lectures 3 and 1, the condition for superfluidity in the sense of NCRI is that when the standard replacement $\underset{\sim}{p} \to \underset{\sim}{p} - e\underset{\sim}{A}$ is made the free energy is a nontrivial function of $\underset{\sim}{A}$, i.e. $\underset{\sim}{\omega}$ or, what is equivalent, that the effective Hamiltonian $H_0 - \underset{\sim}{\omega} \bullet \underset{\sim}{L} + 1/2\, I_{cl}\omega^2$ should induce a lab. frame angular momentum $\underset{\sim}{L} = -\partial F/\partial\underset{\sim}{\omega} + I_{cl}\underset{\sim}{\omega}$ which is less then the classical value $I_{cl}\underset{\sim}{\omega}$. As we already saw, two necessary (but not sufficient) conditions for this are that the system should obey quantum rather than classical <u>dynamics</u>, and that the relevant wave functions should be "topologically extended."

Let us consider the very simplest models which might explain the superfluidity of a system such as ^{4}He. First, consider a <u>free gas of distinguishable</u>

* Since this is intrinsically a finite temperature case, one might perhaps argue that one should generalize the concept of "elementary excitation" to include processes in which an existing phonon is destroyed and another created, allowing the Landau criterion to fail; but this seems very forced.

particles obeying quantum mechanics at T = 0. The single-particle energy eigenfunctions and energies, and the associated angular momenta, are

$$\psi_\ell(\theta) = (2\pi)^{-1/2} \exp i\ell\theta, \quad \varepsilon_\ell = \ell^2\hbar^2/2mR^2, \quad 1 = \ell\hbar. \tag{4.1}$$

If n_ℓ is the number of particles in state ℓ, the expectation value of the effective Hamiltonian $H_0 - \underset{\sim}{\omega}\bullet\underset{\sim}{L}$ (we can omit the term in ω^2 which is independent of the distribution) is given by the expression

$$\langle H_0 - \underset{\sim}{\omega}\bullet\underset{\sim}{L}\rangle = \sum_\ell n_\ell(\ell^2\hbar^2/2mR^2 - \hbar\omega\ell). \tag{4.2}$$

It is immediately clear that the minimum value is obtained by taking $n_\ell = N$, the total number of particles, for that value of ℓ, say ℓ_m, which minimizes the quantity in brackets, and $n_\ell = 0$ for all other ℓ. Now for $\omega < \hbar^2/2mR^2$ we have $\ell_m = 0$: so $n_\ell = N\delta_{\ell,0}$, the total angular momentum, $\Sigma_\ell\hbar\ell n_\ell$, is clearly zero and we get the phenomenon of NCRI. More generally, it is clear that if we plot the behavior of $\langle\underset{\sim}{L}\rangle$ as a function of $\underset{\sim}{\omega}$ we get the behavior shown in fig. 3, with plateaux $L = N\ell\hbar$ disposed symmetrically about the values $\omega = (\ell+1/2)\hbar/mR^2$. If we represent this in the language of a dependence of the free energy on fictitious flux as in Lecture 3, we see that it corresponds qualitatively to the curves shown in fig. 1a, with a flux "quantum" h/e (and cusps at the maxima).

At first sight, therefore, this would seem an adequate model for superfluidity. However, already at finite but small temperatures things are quite different. At temperature T the distribution of the distinguishable particles for the effective Hamiltonian $H_0 - \underset{\sim}{\omega}\bullet\underset{\sim}{L}$ is simply the Maxwell one, i.e.

$$\langle n_\ell\rangle = Z^{-1} \exp - \beta\tilde{\varepsilon}_\ell, \quad \tilde{\varepsilon}_\ell \equiv \ell^2\hbar^2/mR^2 - \hbar\omega\ell \tag{4.3}$$

and it is a straightforward exercise to verify that for $kT >> \hbar^2/mR^2$ the expectation value of angular momentum $\langle\underset{\sim}{L}\rangle = \Sigma_\ell\ell\hbar\langle n_\ell\rangle$ is practically indistinguishable from its classical value $I_{cl}\underset{\sim}{\omega}$ (If the sum is replaced by an integral, the two are identical, cf. the argument in Lecture 1.). This agrees with our intuition that when kT is large compared to characteristic energy spacings quantum mechanics reduces to all intents and purposes to classical mechanics: we have already seen that no classical system can show NCRI.

The case of a free Fermi gas in one dimension is quite similar, except that the relevant energy spacing is now that near the Fermi "points", i.e. $n\hbar^2/mR^2$

rather than $\hbar^2/mR^2$ where n is the maximum quantum number corresponding to an occupied state. This is of order $k_F R$, $k_F \sim a^{-1}$ where a is the interparticle spacing, so the effect is to replace $\hbar^2/mR^2$ by a quantity of order $\hbar^2/mRa$. Now for typical "macroscopic" values of R (say 1 cm), m the mass of a helium atom and a ~ 3Å we have $\hbar^2/mR^2 \sim 10^{-15}$K and $\hbar^2/mRa \sim 10^{-7}$K, both well below currently attainable temperatures. Taking into account the transverse dimensions turns out, if anything, to worsen the situation. Thus neither the free "classical" (i.e. distinguishable) gas nor the free Fermi gas can be relevant to superfluidity as observed on a typical laboratory scale. (However, for electrons in mesoscopic rings the conditions $kT \lesssim \hbar^2/mR^2$ and, a fortiori, $kT \lesssim \hbar^2/mRa$ are nowadays attainable, and such systems have indeed been recently shown to show a phenomenon which in our language is essentially NCRI. To discuss this question properly, however, one needs to consider the effects of finite transverse dimensions, impurities, phonons etc. It is conceivable that in the next decade the relevant low-temperature regime will be obtained in another normal Fermi system which does not suffer from some of these complications, namely dilute solutions of ^{3}He in ^{4}He.)

Free Bose gas.

Since we believe that ^{4}He atoms are indeed bosons, this seems a more promising line. Moreover, the observed λ-temperature of ^{4}He, below which the liquid is superfluid, is 2.17K at saturated vapor pressures, not very different from the Bose condensation temperature (~ 3K) of a free Bose gas of the relevant mass and density. We will show that a free Bose gas does indeed show the property of NCRI.

The thermal equilibrium distribution of a gas of noninteracting bosons over the single-particle states with (effective) energies ε_ℓ is given by the standard formula

$$\langle n_\ell(T)\rangle = (\exp \beta(\varepsilon_\ell - \mu) - 1)^{-1} \tag{4.4}$$

where the chemical potential μ cannot be positive.* In a three-dimensional system, as T decreases, μ approaches zero from below at a finite temperature $T_0 \cong 3.5\, \hbar^2/ma^2k_B$ (a = average interparticle spacing). For $T < T_0$ a finite fraction N_0/N of the total number of particles N occupies the lowest single-particle state, while the rest are distributed over the excited states so that

* We choose the zero of energy so that the lowest single-particle state has energy zero.

$\langle n_\ell \rangle = (\exp \beta(\tilde{\varepsilon}_\ell - 1)^{-1}$. In effect, N_o of the particles (the so-called "condensed" ones) are at zero temperature, the rest at the actual temperature T.

We can now repeat the argument we gave for the distinguishable case at T = 0. The total angular momentum is just a sum of two terms; the first is the contribution of the "condensed" particles, which is (N_o/N) times the "zero-T" contribution calculated above, the second that of the rest, namely $1-(N_o/N)$ times the "finite-T" result, which, since we are interested in temperatures such that $kT >> \hbar^2/mR^2$, is just $I_{cl}\underset{\sim}{\omega}$. The resultant graph of angular momentum versus angular velocity is that given in fig. 4, and it is easy to see that this model indeed exhibits the phenomenon of NCRI, with a "superfluid fraction" simply equal to the "condensed fraction":

$$\rho_s(T)/\rho = N_o(T)/N. \tag{4.5}$$

The free Bose gas is, however, not an adequate model for the superfluidity of ^4He, because it does not possess the other defining characteristic of a superfluid, namely the ability to sustain metastable supercurrents. To see this, we note that under the relevant conditions, with the walls at rest, the relevant Hamiltonian is just the original lab. frame one and the total energy is simply given by

$$E = \sum_\ell n_\ell \varepsilon_\ell \quad , \qquad \varepsilon_\ell = \ell^2 \hbar^2/2mR^2. \tag{4.6}$$

Suppose that we start from a state in which (for example) a large number (~ N) of particles are in a state ℓ ($\neq 0$) carrying a circulating current. Then there is nothing to prevent the atoms dropping back "one by one" from the finite-ℓ state to the zero-ℓ one. This process can take place just as for an electron in a single atom: in that case we can, crudely speaking, describe the process by a single-particle wave function which interpolates smoothly from (say) a p-state at time zero to the ground s-state at time ∞:

$$\psi(t) = \alpha(t)\psi_p + \beta(t)\psi_s\,, \quad |\alpha(t)|^2 + |\beta(t)|^2 = 1\,, \ \alpha(0) = 0, \ \alpha(\infty) = 1. \tag{4.7}$$

The expectation value of the energy is given by $\langle E(t) \rangle = E_p |\alpha(t)|^2 + E_s |\beta(t)|^2$ and decreases monotonically in time (provided $\alpha(t)$ does). Note that there is no difficulty with the topology of the wave function: the p-state has "winding number" 1 (i.e. the phase goes through 2π as we go once around the nucleus), the s-state 0, so to get from one to the other the wave function must at some time go through zero amplitude at some point on the orbit. It may be verified that this indeed happens when $|\alpha(t)| = |\beta(t)|$.

(Note: the above description is of course too naive, since in reality the electron should not even be described, during the radiation process which takes it from the p- to the s-state, by a pure state. Rather, one must work in terms of the coupled wave function of the electron plus radiation field. Nevertheless, the above crude description brings out the essential topological point.)

Thus, if our model is to explain not only NCRI but also the stability of supercurrents, it is essential to take into account also the inter-particle interactions – which one might have anyway suspected to be quite important on a priori grounds.

Bose condensation in an interacting system

By "Bose condensation" in an interacting system we mean that we can find a basis of one-particle states $\phi_\nu(r)$ (which may or may not be the single-particle eigenfunctions of the noninteracting gas) such that in equilibrium below some temperature T_0 we have

$$n_{\nu_0} \sim N, \quad n_\nu \sim 1 \text{ for } \nu \neq \nu_0 \tag{4.8}$$

(Note that we assume that only one single-particle state is macroscopically occupied: cf. below.). Generalizing slightly, we assume that in a nonequilibrium situation statement (4.8) is still true but both $\phi_\nu(r)$ and n_{ν_0} may be a function of time.

Does Bose condensation actually occur in an interacting system? Obviously there is no general theorem to this effect (otherwise there would be no crystalline phase of ^{4}He!). However, it can be shown that if perturbation theory starting from the free Bose gas converges, Bose condensation does indeed occur at sufficiently low temperatures: but in general $N_0 \neq N$ even at zero tempreatures, and the transition temperature T_0 is different from that of the free gas. As far as I know there are no rigorous theorems about the value of T_0 in an interacting three-dimensional system. However, it is possible to use an inequality due to Hohenberg (much better known for its application to show that condensation cannot occur at all in 1D, or in 2D at $T \neq 0$) to show that for temperatures far above T_0^{free} the condensate fraction, if it exists at all, must be very small. The basic inequality, for a translation-invariant system in unit volume, reads

$$N_k \geq (\frac{N_o}{N})\frac{mk_BT}{\hbar^2k^2} - \frac{1}{2} \qquad (k \neq 0) \tag{4.9}$$

and a little algebra then yields the result (since $\sum_{k\neq 0} n_k \equiv N - N_o$)

$$\frac{N_o}{N} \leq \frac{1}{1+\alpha(T/T_o^{free})^{3/2}} \tag{4.10}$$

where α is a numerical constant of order 1.

Superfluid hydrodynamics of a Bose-condensed system

On the basis of the ansatz (4.8), generalized as indicated to nonequilibrium situations, we can define the order parameter $\Psi(\underset{\sim}{r},t)$ of the system by

$$\Psi(\underset{\sim}{r},t) \equiv \sqrt{n_{\nu_o}(t)}\ \phi_{\nu_o}(\underset{\sim}{r},t). \tag{4.11}$$

This may be related to the concept of off-diagonal long-range order (ODLRO): Consider the quantity $\langle\psi^\dagger(\underset{\sim}{r},t)\psi(\underset{\sim}{r}',t)\rangle$ for large $|\underset{\sim}{r}-\underset{\sim}{r}'|$, where $\psi,\psi^\dagger$ are the usual Bose field operators. In the basis in which n_ν is diagonal this quantity is

$$\sum_\nu n_\nu(t)\ \phi_\nu^*(\underset{\sim}{r},t)\ \phi_\nu(\underset{\sim}{r}',t). \tag{4.12}$$

Now, since the various ϕ_ν are orthogonal, and the n_ν for $\nu \neq \nu_o$ are each of order 1, for sufficiently large $|\underset{\sim}{r}-\underset{\sim}{r}'|$ their contribution will cancel and we are left only with the nonvanishing contribution of the state ν_o:

$$\langle\psi^\dagger(r,t)\psi(r',t)\rangle_{|r-r'|\to\infty} \to n_{\nu_o}\ \phi_{\nu_o}^*(r,t)\phi_{\nu_o}(r',t)$$

$$\equiv \Psi^*(r,t)\Psi(r',t). \tag{4.13}$$

Eqn. (4.13) is sometimes used as the basis of an alternative definition of the order parameter.

Next let us define the superfluid velocity $\underset{\sim}{v}_s(\underset{\sim}{r},t)$. For orientation consider first a single particle. We have for the probability density $\rho(r,t)$ and particle current density $\underset{\sim}{j}(r,t)$ the expressions

$$\rho(\underset{\sim}{r},t) \equiv \psi^*(r,t)\psi(r,t) \tag{4.14a}$$

$$\underset{\sim}{j}(\underset{\sim}{r},t) \equiv -\frac{i\hbar}{2m}(\psi^*(r,t)(\underset{\sim}{\nabla}\psi(r,t) - \psi(r,t)(\underset{\sim}{\nabla}\psi^*(r,t)). \tag{4.14b}$$

Let us write the complex amplitude $\psi(r,t)$ as $A(r,t) \exp i\phi(r,t)$, where A and ϕ are real. Then we see that the single-particle "velocity", $\underset{\sim}{v}(r,t) \equiv j(\underset{\sim}{r},t)/\rho(\underset{\sim}{r},t)$, is given by the expression

$$\underset{\sim}{v}(\underset{\sim}{r},t) = \frac{\hbar}{m} \underset{\sim}{\nabla}\phi(r,t). \tag{4.15}$$

By analogy with this expression, we write for the Bose superfluid

$$\Psi(\underset{\sim}{r},t) \equiv A(\underset{\sim}{r},t) \exp i\phi(\underset{\sim}{r},t) \tag{4.16}$$

$$\underset{\sim}{v}_s(\underset{\sim}{r},t) \equiv \frac{\hbar}{m} \underset{\sim}{\nabla}\phi(r,t). \tag{4.17}$$

That is, $\underset{\sim}{v}_s$ is, apart from a constant, simply the gradient of the phase of the single-particle wave function into which Bose condensation occurs. Note that the definition of $\underset{\sim}{v}_s$ is an entirely microscopic one and does <u>not</u> assume hydrodynamic equilibrium.

Three immediate consequences follow from the definition (4.17). First, everywhere in the liquid we have

$$\text{curl } \underset{\sim}{v}_s(\underset{\sim}{r},t) = 0. \tag{4.18}$$

(It should be noted that the "mean velocity" $\underset{\sim}{v}(\underset{\sim}{r},t)$ of the liquid as a whole, defined as $\underset{\sim}{j}(r,t)/\rho(r,t)$ where ρ and j are the <u>total</u> density and current, does <u>not</u> in general satisfy eqn. (4.18).) Secondly, in a multiply connected geometry, if $\oint$ indicates the integral around a contour contained entirely in the liquid but surrounding one or more holes (where $\underset{\sim}{v}_s$ is not defined), the fact that the phase $\phi(\underset{\sim}{r},t)$ must be single-valued modulo 2π implies the result

$$\oint \underset{\sim}{v}_s \cdot d\underset{\sim}{l} = nh/m \tag{4.19}$$

which is consistent with the discussion in Lecture 1. Finally, it follows from eqn. (4.17) that the rate of change of $\underset{\sim}{v}_s$ is given by

$$\left(\frac{d\underset{\sim}{v}_s}{dt}\right) = \underset{\sim}{\nabla}\Omega, \quad \Omega \equiv \frac{\hbar}{m}\frac{\partial\phi(r,t)}{\partial t}. \tag{4.20}$$

Eqn. (4.20) is general, but is not very useful until we have an expression for $\partial\phi/\partial t$.

We now consider a "bulk" situation in which the system is, at least locally, in or close to thermodynamic equilibrium, with the walls at rest, except that the superfluid velocity $\underset{\sim}{v}_s(\underset{\sim}{r})$ has some prescribed (time-independent) value (subject, of course, to eqns. (4.18)–(4.19)). An example would be the metastable current-carrying state in an annulus corresponding to $n \neq 0$ in (4.19). Assuming on symmetry grounds that the current $\underset{\sim}{j}(\underset{\sim}{r})$ is zero for zero $\underset{\sim}{v}_s$, and expanding to lowest order in $\underset{\sim}{v}_s$, we can write

$$\underset{\sim}{j}(\underset{\sim}{r}) = \int K_s(\underset{\sim}{r},\underset{\sim}{r}')\,\underset{\sim}{v}_s(r')\,d\underset{\sim}{r}' \tag{4.21}$$

where the kernel $K_s(\underset{\sim}{r},\underset{\sim}{r}')$ is uniquely defined as a function of the thermodynamic parameters and $(\underset{\sim}{r}-\underset{\sim}{r}')$. In the usual cases of interest $\underset{\sim}{v}_s(\underset{\sim}{r})$ is essentially constant over the range of K_s, so we can approximate (4.21) by the local equation

$$\underset{\sim}{j}(\underset{\sim}{r}) = \rho_s v_s(\underset{\sim}{r}) \qquad \rho_s(r) \equiv \int K_s(\underset{\sim}{r},\underset{\sim}{r}')\,d\underset{\sim}{r}' \tag{4.22}$$

thus defining the superfluid density ρ_s. Note that ρ_s, in distinction to $\underset{\sim}{v}_s$, is an essentially hydrodynamic concept. The "normal density" ρ_n is defined simply as $\rho - \rho_s$. Finally, consider a bulk situation in which thermodynamic equilibrium is attained subject to a prescribed $\underset{\sim}{v}_n(r)$ and conservation of total current $\underset{\sim}{j}$ (e.g. the situation in the middle of a wide tube). Under these conditions we define a normal velocity $\underset{\sim}{v}_n$ by the formula

$$\underset{\sim}{j}(\underset{\sim}{r}) = \rho_s \underset{\sim}{v}_s(\underset{\sim}{r}) + \rho_n \underset{\sim}{v}_n(\underset{\sim}{r}). \tag{4.23}$$

$\underset{\sim}{v}_n$, again a hydrodynamic concept, can be regarded as the velocity of the imaginary "walls" with which the system would be in equilibrium subject to the prescribed value of $\underset{\sim}{v}_s$: in other contexts, e.g. the "rotating-wall" experiment in an annulus, it has just this significance.

To complete our superfluid hydrodynamics we need an expression for the Ω of eqn. (4.20) i.e. for $\partial\phi/\partial t$. A quick, if somewhat nonrigorous, way of obtaining this is the following: We start the commutation relations of the Bose field operators

$$[\psi(r),\psi^{\dagger}(r')] = \delta(\underset{\sim}{r}-\underset{\sim}{r}') \qquad (4.24)$$

and write the operator relation $\psi(r) = \exp i\phi(r)\, \rho^{1/2}(r)$, with ρ and ϕ taken as operators. Then, a commutation relation which is compatible with (4.24) is

$$[\rho(r),\phi(r')] = i\delta(\underset{\sim}{r}-\underset{\sim}{r}'). \qquad (4.25)$$

We average this over a finite volume, defining $N = \int \rho(r)dr$, $\phi = V^{-1}\int\phi(r)dr$. This gives

$$[N,\phi] = i, \qquad (4.26)$$

i.e. the total number of particles in the volume and the "phase" are conjugate variables. A more careful consideration reveals that the "phase" occurring in (4.26) is indeed effectively the phase of the condensate wave function, i.e. the variable we called $\phi(r,t)$ above (suitably averaged). It then follows from (4.26) that

$$\frac{\partial\phi}{\partial t} = \frac{-1}{\hbar}\left(\frac{\partial E}{\partial N}\right) \qquad (4.27)$$

where $\partial E/\partial N$ should be interpreted as "at constant $\langle n_\nu\rangle$ for $\nu \neq \nu_0$", i.e. in effect, at constant entropy. Thus, in conditions close to hydrodynamic equilibrium, $\partial E/\partial N$ is just the chemical potential μ and we finally get from (4.20)

$$\dot{\underset{\sim}{v}}_s = -\frac{1}{m}\underset{\sim}{\nabla}\mu. \qquad (4.28)$$

Eqn. (4.28), along with eqns. (4.28), (4.23) and the usual equation $\dot{\underset{\sim}{J}} = -(\rho/m)\underset{\sim}{\nabla}p$ for the acceleration of the total current, form the basis of superfluid hydrodynamics.

One amusing immediate application is to prove that for a translation-invariant system described by the "minimal" hydrodynamic variables, the normal density is zero at zero temperature, i.e. $\rho_s = \rho$ (even though in an interacting system $N_0 \neq N$). We use the above equation for $\dot{\underset{\sim}{J}}$ and compare it with the time derivative of (4.23). Then, using (4.28), we obtain

$$\rho_n\underset{\sim}{v}_n = \frac{1}{m}(\rho\underset{\sim}{\nabla}p - \underset{\sim}{\nabla}\mu). \qquad (4.29)$$

The RHS is zero at T = 0 from the Gibbs-Duhem relation, and since $\underset{\sim}{v}_n$, if it is meaningful, cannot be automatically constant, we obtain the advertised result. Note that it fails for a system with "non-minimal" hydrodynamic variables such as the hypothetical "supersolid", since the Gibbs-Duhem relation then takes a more complicated form.

We finally return briefly to the important question of the metastability of current-carrying states. Suppose we quench through the Bose condensation temperature T_λ while rotating the annular container at angular velocity $\underset{\sim}{\omega}$. The effective Hamiltonian is then $H_0 - \underset{\sim}{\omega} \bullet \underset{\sim}{L}$ and we expect, as argued above for the free Bose gas, condensation into the single-particle state with the lowest value of $\ell^2\hbar^2/2mR^2 - \ell\hbar\omega$; this corresponds to $\Psi(\theta) \sim \exp i\ell\theta$, $\underset{\sim}{v}_s = \ell\hbar/mR$, $\underset{\sim}{j} = \rho_s \underset{\sim}{v}_s = \rho_s \ell\hbar/mR$ (assuming that we have stopped the container and allowed the normal component to come to rest). So far, everything is the same for the free Bose gas and the interacting one. The crucial difference, however, is that the latter, in distinction to the former, has a <u>finite compressibility</u>. Now we have seen that to go from an order parameter proportional to $e^{i\ell\theta}$ to the one (~ const.) characterizing the ground state we need at some stage to introduce a node into the condensate wave function $\phi_{v_0}(\underset{\sim}{r})$; and from normalization this implies we must build it up elsewhere above its original value. Because of the finite compressibility, <u>this requires finite extra energy</u> and provides the required free energy barrier which ensures the metastability of the current-carrying state. In practice, the "cheapest" way to change the topology is to generate vortex rings or to move pre-existing vortices in the liquid; this leads to an observable decay of the supercurrent under certain special conditions, e.g. very close to T_c where ρ_s and hence the free energy barrier is small.

A note on the Landau criterion (cf. Lecture 3): although the interacting Bose liquid (whose spectrum for small p is of the phonon type, $\varepsilon(p) = cp$) satisfies the Landau criterion while the free Bose ($\varepsilon(p) = p^2/2m$) does not, I believe it is quite misleading to regard this as the <u>reason</u> for the metastability of supercurrents in the former as opposed to the latter. Rather, both the metastability and the form of the energy spectrum are consequences of the same fundamental characteristic, the finite compressibility of the interacting system.

Lecture 5: Essentials of BCS Theory of Superconductivity

As we all know, the phenomenon of superconductivity in, at least, the "old" (pre-1986) superconductors is generally believed to be explained with impressive quantitative accuracy by one variant or another of the original BCS theory, i.e. the theory of formation of Cooper pairs. Since an account of this theory and the relevant calculations can be found in innumerable textbooks, in this lecture I shall be quite sketchy, omitting most of the calculational details, and just try to highlight a few points which I believe are not universally appreciated.

Cooper pairing is a phenomenon which is believed to be more or less universal in Fermi liquids which have an attractive interaction between (quasi) particles near the Fermi surface. In this lecture and the next I deal with the simplest case, where the distinction, in the normal phase, between bare particles and quasiparticles is negligible and the most attractive interaction is in a relative s-state. One can approach the problem in (at least) two ways: (a) by writing down a "model" Hamiltonian and finding the exact ground state (the original BCS method),and (b) by writing down the exact Hamiltonian and making an ansatz for the wave function. I choose the latter, which is the analog of the Hartree-Fock method for the normal state.

We consider an arbitrary system of spin-1/2 fermions* with Hamiltonian

$$H = \sum_{k\sigma} \xi_k a^+_{k\sigma} a_{k\sigma} + \frac{1}{2} \sum_{\substack{k_1k_2k_3k_4 \\ \sigma\sigma'}} \langle k_1k_2 | V | k_3k_4 \rangle a^+_{k_1\sigma} a^+_{k_2\sigma'} a_{k_3\sigma'} a_{k_4\sigma} \qquad (5.1)$$

where $\xi_k \equiv \hbar^2 k^2/2m$ and the potential matrix element $\langle k_1k_2 | V | k_3k_4 \rangle$ is for the moment arbitrary. The fundamental ansatz for the many-body wave function $\Psi_N(r_1r_2..r_N:\sigma_1\sigma_2...\sigma_N)$ which embodies the hypothesis of Cooper pairing is

$$\Psi_N = \mathcal{N}^{1/2} \mathcal{A}\, \phi(\underset{\sim}{r}_1,\underset{\sim}{r}_2,\sigma_1\sigma_2)\, \phi(\underset{\sim}{r}_3,\underset{\sim}{r}_3,\sigma_3\sigma_4)...\phi(\underset{\sim}{r}_{N-1},\underset{\sim}{r}_N,\sigma_{N-1}\sigma_N) \qquad (5.2)$$

where the two-particle function $\phi(r_1r_2\sigma_1\sigma_2)$ is arbitrary except that it must (a) be antisymmetric under the exchange $\underset{\sim}{r}_1\sigma_1 \leftrightarrows \underset{\sim}{r}_2\sigma_2$ and (b) correspond to a finite value* of $\langle (r_1-r_2)^2 \rangle$ (i.e. fall off sufficiently fast for $|\underset{\sim}{r}_1-\underset{\sim}{r}_2| \to \infty$); $\mathcal{N}$ is a normalization constant and A antisymmetrizes between all pairs of particles

---

* For convenience the total number N is chosen even.

* It should be noted that the normal ground state of a non-interacting Fermi gas may be written in the form (5.2) with ϕ satisfying (a) but not (b).

(e.g. 1 and 3) which are not in the same bracket in (5.2). Any system whose ground state is of the general form (5.2) will be said to undergo Cooper pairing.

The case of a translationally invariant superconductor with pairing in the conventional $\ell = 0$ spin singlet state corresponds to the choice

$$\phi(r_1,r_2:\sigma_1\sigma_2) = 2^{-1/2}(\uparrow_1\downarrow_2-\downarrow_1\uparrow_2)\,\phi(|\underset{\sim}{r}_1-\underset{\sim}{r}_2|) \tag{5.3}$$

in an obvious notation. Then if $\chi(\underset{\sim}{k}) \equiv \chi(|\underset{\sim}{k}|)$ is the Fourier transform of $\phi(|\underset{\sim}{r}_1-\underset{\sim}{r}_2|)$, the many-body state Ψ (eqn. (5.2)) is written in second-quantized notation

$$\Psi_N = \mathcal{N}^{1/2}\,(\sum_k \chi(k)\, a^+_{k\uparrow}\, a^+_{-k\downarrow})^{N/2}|\mathrm{vac}\rangle \equiv \mathcal{N}^{1/2}\,\widehat{\Omega}^{\,N/2}|\mathrm{vac}\rangle \tag{5.4}$$

(where $|\mathrm{vac}\rangle$ is the vacuum state). The state (5.4) is by construction an eigenstate of total particle number N.

As is well known, the method used by BCS to find the optimum choice of the function $\phi(\underset{\sim}{r}_1-\underset{\sim}{r}_2)$, i.e. of $\chi(\underset{\sim}{k})$, involves relaxing the constraint of fixed particle number N and replacing it by the condition that the expectation value $\langle H-\mu N\rangle$ should be a minimum, where μ is the chemical potential, which for the moment we regard as an externally controlled thermodynamic parameter. This is equivalent to minimizing the Hamiltonian (5.1) with the replacement

$$\xi_k \rightarrow \xi_k - \mu \equiv \varepsilon_k. \tag{5.5}$$

We note explicitly a point which may seem trivial but becomes important below, namely that when μ is positive (as is usually assumed implicitly in BCS theory) the minimum value of $|\varepsilon_k|$ is zero, whereas if $\mu< 0$ it is $-\mu$. The ansatz which replaces (5.2) in the BCS method is (apart from normalization)

$$\Psi = \exp\,\widehat{\Omega}\,|\mathrm{vac}\rangle \tag{5.6}$$

which after a little algebra can be rewritten

$$\Psi = \prod_k (1 + \chi_k\, a^+_{k\uparrow}\, a^+_{-k\downarrow})|\mathrm{vac}\rangle. \tag{5.7}$$

In effect, we are writing the many-body wave function not as an antisymmetrized product of wave functions referring to pairs of <u>particles</u> (as in (5.2)) but rather as a product of wave functions referring to pairs of states $\underset{\sim}{k}\uparrow, -\underset{\sim}{k}\downarrow$ and to the possible degrees of occupation of these states. To each pair $\underset{\sim}{k}\uparrow, -\underset{\sim}{k}\downarrow$ there corresponds a 4-dimensional Hilbert space, which is spanned by the four basis vectors

$$|00\rangle_k,\quad |01\rangle_k,\quad |10\rangle_k,\quad |11\rangle_k \tag{5.8}$$

where, for example, $|01\rangle_k$ indicates the state in which the single-particle state $\underset{\sim}{k}\uparrow$ is empty while $-\underset{\sim}{k}\downarrow$ is occupied. In this notation we can write the normalized many-body wave function in the conventional form

$$\Psi = \prod_k \Phi_k\,,\qquad \Phi_k \equiv u_k|00\rangle + v_k|11\rangle \tag{5.9}$$

where the u's and v's, which are functions only of $|\underset{\sim}{k}|$ in this case, satisfy

$$|u_k|^2 + |v_k|^2 = 1\quad,\qquad v_k/u_k = \chi_k. \tag{5.10}$$

It is clear that u_k can be chosen real without loss of generality. Starting from a particular choice of Ψ, i.e. particular values of the v_k, we can now recover the particle-conserving wave function Ψ_N by a trick due to Anderson: we define

$$v_k(\phi) \equiv v_k e^{i\phi},\qquad \Phi_k(\phi) \equiv u_k|00\rangle + v_k(\phi)|11\rangle, \tag{5.11}$$

$$\Psi(\phi) \equiv \prod_k \Phi_k(\phi).$$

Then Ψ_N is nothing but the quantity

$$(2\pi)^{-1}\int_0^{2\pi} \exp(-iN\phi)\Psi(\phi)d\phi = \Psi_N. \tag{5.12}$$

It should be emphasized that the wave function Ψ (or the particle-conserving form (5.12) derived from it) describes a whole class of many-body wave functions, not just the single member of that class which describes the ground state. We will call the class of states so described "generalized BCS states." To find the form of $\chi(\underset{\sim}{k})$ (or u_k, v_k) which corresponds to the ground state one must of course minimize the expectation value of the Hamiltonian

(5.1) with the replacement (5.5) over the wave function (5.9). The procedure is straightforward and I just quote the results.

The expectation value of the kinetic energy K (with the $-\mu N$ term included) is given by

$$\langle K\rangle = \sum_{k\sigma} \varepsilon_k \langle n_{k\sigma}\rangle = \sum_{k} 2\varepsilon_k |v_k|^2. \tag{5.13}$$

Were this the only contribution, the ground state would of course correspond to $v_k = \theta(-\varepsilon_k)$, which may be seen to be just the Fermi sea filled up to energy μ. In the potential energy three terms survive. The first two are just the standard Hartree and Fock terms:

$$\langle V\rangle_{\text{Hartree}} = \Big(\sum_{\substack{kk'\\ \sigma\sigma'}}\Big) \langle k'k | V | kk'\rangle \langle n_{k\sigma}\rangle\langle n_{k'\sigma'}\rangle \tag{5.14a}$$

$$\langle V\rangle_{\text{Fock}} = -\Big(\sum_{\substack{kk'\\ \sigma}}\Big) \langle kk' | V | kk'\rangle \langle n_{k\sigma}\rangle\langle n_{k'\sigma}\rangle. \tag{5.14b}$$

A point which is not always sufficiently emphasized in discussions of the BCS theory is that in general these terms are not small; in fact, it turns out that they may well be much larger than the "BCS" terms (below). However, the crucial point is this: The solution of the problem neglecting the terms (5.14) shows that the value of $\langle n_k\rangle$ is distorted from its value ($\theta(-\varepsilon_k)$) in the normal state only over a range $\sim \Delta$ (the zero-temperature energy gap, see below), and moreover in such a way that the sum of $\langle n_k\rangle$ over this range is unchanged (cf. eqn. (6.19) below). Thus, provided that the matrix elements in (5.14) are slowly varying over a range $\sim \Delta$, the expectation values of the Hartree-Fock terms are virtually unchanged from their (possibly large) values in the normal states, and therefore do not contribute to the minimization process. The proviso is crucial, and it is important to bear in mind that it may not be satisfied to an adequate approximation in all superconductors (cf. Lecture 9). For the moment we shall assume that the Hartree-Fock terms may indeed be neglected.

The third class of terms which survive in $\langle V\rangle$ have no normal-state analog:

$$\langle V\rangle_{\text{pair}} = \Big(\sum_{kk'}\Big) \langle k',-k' | V | k,-k\rangle\, u_k^* v_k u_{k'} v_{k'}^*. \tag{5.15}$$

From now on we shall for convenience write

$$\langle k',-k'|V|k,-k\rangle \equiv V_{kk'}, \qquad u_k^* v_k \equiv F_k \qquad (\equiv \langle a_{-k\downarrow} a_{k\uparrow}\rangle). \tag{5.16}$$

As we shall see, the quantity F_k plays a crucial role in BCS theory. The expectation value (5.15) is then written

$$\langle V\rangle_{pair} = \sum_{kk'} V_{kk'}\, F_k F_{k'}^*. \tag{5.17}$$

Note there is neither a factor of 1/2 nor a spin sum.

It is interesting to compare the expression (5.17) with the formula for the potential energy of a diatomic molecule. In terms of the Fourier transform ψ_k of the relative wave function $\psi(\underset{\sim}{r}_1-\underset{\sim}{r}_2)$, the latter reads (with the same definition of $V_{kk'}$ as in (5.16)),

$$\langle V\rangle_{mol} = (\sum_{kk'})V_{kk'}\, \psi_k \psi_{k'}^*. \tag{5.18}$$

We see therefore that F_k, or more precisely its Fourier transform $F(\underset{\sim}{r})$, plays the same role in the Cooper-pair problem as the relative wave function $\psi(\underset{\sim}{r})$ does for the diatomic molecule. We will see below that the correspondence is more far-reaching than the simple comparison of (5.17) and (5.18) would suggest.

We now minimize the sum of (5.13) and (5.17) subject to the constraint $|u_k|^2 + |v_k|^2 = 1$; probably the most transparent way of carrying out the calculation is to introduce the Anderson "pseudospin" representation, in which one writes $v_k = \sin\theta_k/2$, $u_k = \cos\theta_k/2$ and regards θ_k as the angle of a fictitious "spin" associated with the pair state $(\underset{\sim}{k},-\underset{\sim}{k})$. The result of the minimization procedure may be written in the form

$$F_k = \Delta_k/2E_k \tag{5.19}$$

where the quantity Δ_k is defined by

$$\Delta_k \equiv -\sum_k V_{kk'} F_{k'} \tag{5.20}$$

and

$$E_k \equiv (\varepsilon_k^2 + |\Delta_k|^2)^{1/2}. \tag{5.21}$$

The quantity Δ_k defined by (5.20) is sometimes called the "pair field" or "off-diagonal field" (in the Anderson pseudospin representation, it is the magnetic "field" in the x-direction, while that in the z-direction is ε_k). Combining eqns. (5.19)–(5.20), one can write the result in either of two equivalent forms:

$$\Delta_k = -\sum_k V_{kk'}\, \Delta_{k'}/2E_{k'} \tag{5.22}$$

or

$$2E_k F_k = -\sum_k V_{kk'} F_{k'} \tag{5.23}$$

with E_k defined by (5.21). Eqn. (5.22) is the standard form of the BCS gap equation, while (5.23) brings out the close analogy to the two-particle problem, for which the relevant Schrödinger equation is*

$$(2\varepsilon_k - E)\psi_k = -\sum_k V_{kk'} \psi_{k'}. \tag{5.24}$$

Eqn. (5.22) is rather general; in particular, it does not depend on any particular assumptions about the form of the pair potential $V_{kk'}$. (It does, however, assume implicitly that the Hartree-Fock terms may be safely neglected, cf. above.) We will postpone a detailed study of the solution until we have generalized it to finite temperature.

We conclude this lecture with a remark about the excitation spectrum. It is intuitively clear from the nature of the ground state wave function (5.9) that one way of creating an excited state is to write

$$\Psi_{exc} = \Big(\prod_{k \neq k_o} \Phi_k\Big)\, \Phi'_{k_o}$$

where Φ'_k is a state in the 4-dimensional Hilbert space of the pair $(k\uparrow, -k\downarrow)$ which is orthogonal to the "ground-pair" state Φ_k. An obvious guess for two of the remaining energy eigenstates is the "broken-pair" states $|01\rangle$ and $|10\rangle$, and if this is right then the unique remaining state which is orthogonal to the three we have identified is the linear combination of $|00\rangle$ and $|11\rangle$ which is orthogonal to Φ_k ("excited-pair" state). It is straightforward to show that these

* Note that the E in (5.24) occurs, in a variational calculation, as a result of the normalization condition on ψ. There is no analog in (5.23), since the quantity $\Sigma_k\, |F_k|^2$ does not have to satisfy a normalization condition (see below).

ansätze for Φ_k' indeed correspond to extrema of $\langle H-\mu N\rangle$, with energy E_k (relative to the ground-pair state) for the first two and $2E_k$ for the third. Thus, the complete spectrum of energy eigenfunctions and eigenvalues for the 4-dimensional Hilbert space $|k,-k\rangle$ is:

$$\Phi_{GP} = u_k|00\rangle + v_k|11\rangle \qquad E = 0 \quad \text{(by convention)} \tag{5.25a}$$

$$\Phi_{BP} = |10\rangle \text{ or } |01\rangle \qquad E = E_k \tag{5.25b}$$

$$\Phi_{EP} = v_k^*|00\rangle - u_k^*|11\rangle \qquad E = 2E_k. \tag{5.25c}$$

Thus, the minimum energy of excitation of states of this class is min $E_k \equiv$ min $(\varepsilon_k^2 + |\Delta_k|^2)^{1/2}$. If $\mu > 0$ and Δ_k is not a wildly varying function of $\underset{\sim}{k}$ near the Fermi surface, the minimum will be equal to $\Delta(0)$, the value of $|\Delta_k|$ evaluated at the Fermi energy ($\varepsilon_k = 0$). Hence $\Delta(0)$ is often known as the "energy gap". In addition, there may be (indeed must be, in a consistent theory) collective excitations in which many pair states are simultaneously excited; however, in a real superconductor the most important of these, the density fluctuation mode, has its frequency pushed up by the Coulomb force to the plasma frequency ω_p, which is much larger than any of the characteristic frequencies ($\sim \Delta/\hbar$, etc.) relevant to Cooper pairing, and has much the same character as in the normal phase.

Needless to say, it is possible and sometimes convenient to describe the various states (5.25) in the language of Bogolinkov quasiparticles. If we define quasiparticle operators in the standard way by

$$\alpha_{k\sigma} \equiv u_k a_{k\sigma} - \sigma v_k a^{+}_{-k,-\sigma} \tag{5.26}$$

etc., then it is easy to verify that apart from phase factors

$$\alpha_{k\uparrow}|GP\rangle = \alpha_{-k\downarrow}|GP\rangle = 0 \tag{5.27a}$$

$$\alpha^{+}_{k\uparrow}|GP\rangle = |BP\rangle_1 , \quad \alpha^{+}_{-k\downarrow} = |BP\rangle_2 \tag{5.27b}$$

$$\alpha^{+}_{k\uparrow}\alpha^{+}_{-k\downarrow}|GP\rangle = |EP\rangle \tag{5.27c}$$

so the "excited-pair" state can be regarded as a "doubly broken" ground-pair state.

Lecture 6: Finite-T BCS Theory: Properties of the Paired State

The obvious generalization to finite temperature of the ansatz (5.7) for the ground state is to write the density matrix $\hat{\rho}$ of the many-body system in the form

$$\hat{\rho} = \prod_k \hat{\rho}_k \tag{6.1}$$

where $\hat{\rho}_k$ is a density matrix in the 4-dimensional Hilbert space of the pair of states $(\underset{\sim}{k}\uparrow, -\underset{\sim}{k}\downarrow)$. We saw that at zero temperature the energy eigenstates in this space are given by (5.25), and it is natural to assume that the situation is similar at finite temperature, with the parameters u_k, v_k and the excitation energies functions of T. If so, then since the GP, BP and EP states are distinguishable Boltzamann statistics applies, and the probabilities that the various states are realized are given by

$$P_{GP}:P_{BP}:P_{EP} = 1:\ \exp -\beta E_k:\ \exp -2\beta E_k \qquad (\beta \equiv 1/k_B T). \tag{6.2}$$

With the total probability normalized to 1, this gives the important result*

$$P_{GP} - P_{EP} = \tanh \frac{1}{2} \beta E_k. \tag{6.3}$$

The optimum choice of the parameters $u_k(T)$, $v_k(T)$ is given by minimizing the free energy $\Omega \equiv \langle H - \mu N \rangle - TS$. The procedure is a straightforward generalization of that at T = 0, with the entropy term given by

$$S = \sum_k S_k, \quad S_k = -k_\beta \sum_n P_n \ln P_n \tag{6.4}$$

where the P_n's are given by (5.2). Using the normalization condition $|u_k(T)|^2 + |v_k^2(T)| = 1$, and also the fact that to the extent that the so-called "particle-hole symmetry" around the Fermi surface is valid we have

---

* It is important to remember that there are <u>two</u> "broken-pair" states.

$$\sum_k \varepsilon_k f(E_k) = 0 \tag{6.5}$$

we obtain for Ω the expression

$$\Omega = \sum_k \{\varepsilon_k[1-(2|v_k|^2-1)\tanh \beta E_k/2] - \frac{E_k}{e^{\beta E_k}+1} - 2\beta^{-1}\ln(1+e^{-\beta E_k})\}$$

$$+ \sum_{kk'} V_{kk'} F_k F^*_{k'} \tag{6.6}$$

where the quantity $F_k(T)$, which is the quantum-mechanical <u>and thermal</u> average of $a_{k\uparrow}a_{-k\downarrow}$, is given by

$$F_k(T) \equiv u_k(T)\, v_k(T) \tanh \frac{1}{2}\beta E_k(T). \tag{6.7}$$

We write Ω in the rather messy-looking form (6.6) because it will be useful in the discussion of Ginzburg-Landau theory. Minimization of (6.6) subject to the normalization condition on u_k and v_k gives after a little algebra the finite-temperature BCS gap equation for the quantity

$$\Delta_k(T) \equiv \sum_k V_{kk'} F_{k'}(T) \tag{6.8}$$

namely

$$\Delta_k(T) = \sum_k V_{kk'} \frac{\Delta_{k'}(T)}{2E_{k'}(T)} \tanh \frac{1}{2}\beta E_{k'}(T) \tag{6.9}$$

with

$$E_k(T) \equiv (\varepsilon_k^2 + |\Delta_k(T)|^2)^{1/2}. \tag{6.10}$$

Again, the simplest way of viewing this result physically is probably via the Anderson pseudospin representation: Each "spin" sits, as at T = 0, in a self-consistent field due to the other "spins", but the expectation value of $\langle \underset{\sim}{\sigma} \cdot \underset{\sim}{\hat{H}} \rangle$ is now reduced by the thermal factors to $\tanh \frac{1}{2}\beta E_k$.*

Let us now discuss briefly the nature of the solution to (6.9), treating the T = 0 limit as a special case. We note that (6.9) is a quite general result within

* Note not $e^{\beta E_k} - e^{-\beta E_k}$, because spin "zero" corresponding to broken-pair states, is now allowed.

the BCS ansatz (6.1) for the nature of the density matrix: it does not assume that the potential has the BCS model form, namely

$$V_{kk'} = -V_o\theta(\varepsilon_c - |\varepsilon_k|)\theta(\varepsilon_c - |\varepsilon_{k'}|) \tag{6.11}$$

where ε_c is a phenomenological cutoff energy (taken in the original BCS paper as the Debye energy). However, we will now show that provided that the true potential $V_{kk'}$ regarded as a function of ε_k and $\varepsilon_{k'}$ is negligibly varying over an energy range of the order of the maximum value of $|\Delta_k|$, then its replacement by the form (6.11) is justified, with the caveat the constant V_o must itself be taken to be a function of the cutoff ε_c. To demonstrate this we write the gap equation, eqn. (6.9), in the symbolic matrix form

$$\Delta = -\widehat{V}\,\widehat{Q}\Delta, \quad (\widehat{V})_{kk'} \equiv V_{kk'}, (\widehat{Q})_{kk'} \equiv \delta_{kk'}(\tanh \beta E_{k'}/2)2E_{k'}. \tag{6.12}$$

We introduce projection operators $\widehat{P}_1$ and $\widehat{P}_2 \equiv \widehat{1} - \widehat{P}_1$ which project off values of $\underset{\sim}{k}$ with $|\varepsilon_k| > \varepsilon_c$ and $|\varepsilon_k| < \varepsilon_c$ respectively. Then eqn. (6.12) can be rewritten formally,

$$\Delta = -\widehat{V}(\widehat{P}_1 + \widehat{P}_2)\widehat{Q}\Delta \tag{6.13}$$

and this can be rearranged to read

$$\Delta = -\widehat{t}\widehat{P}_2\widehat{Q}\Delta \tag{6.14}$$

where the matrix $\widehat{t}$ is defined by

$$\widehat{t} \equiv \frac{\widehat{V}}{1+\widehat{V}\widehat{P}_1\widehat{Q}} \tag{6.15}$$

i.e. it is a "t-matrix", defined within the shell $|\varepsilon_k| < \varepsilon_c$, which takes into account repeated "scatterings" into the states outside this shell. The crucial point, now, is that by hypothesis we can choose ε_c small enough that $V_{kk'}$ is constant as a function of energy within this shell, but at the same time much larger than either $\Delta_k(T)$ or T (since, as we shall see, the relevant values of T are $\lesssim \Delta(0)$). Then the factor $\widehat{P}_1\widehat{Q}$, i.e. $(\tanh \beta E_k/2)/2E_k$ taken for $|\varepsilon_k| > \varepsilon_c$, may be approximated with negligible error by $1/2|\varepsilon_k|$, and hence the matrix $\widehat{t}$ depends neither on Δ_k nor on temperature; moreover, since $V_{kk'}$ is by hypothesis constant as a function of ε_k and $\varepsilon_{k'}$, $t_{kk'}$ is also*, so it can at best be a function of angle. However, for the case of single isotropic BCS pairing described by eqn. (5.3) Δ_k is independent of angle, and it then follows from the form of (6.9) that only the zeroth harmonic of $t_{kk'}$ is relevant. Hence, finally, eqn. (6.15) reduced to (6.9) with (6.11), with $(-V_o)$ the isotropic average of $t_{kk'}$. Note that V_o so defined depends (logarithmically) on ε_c, in just such a way that the final value of the physically relevant quantity, the gap, is independent of the choice of ε_c.

Inserting the form (6.11) into (6.19), we see that the gap Δ_k is independent of $\underset{\sim}{k}$ within the shell: $\Delta_k \equiv \Delta(T)$. Replacing the sum by an integral in the standard way, we find that $\Delta(T)$ is given by the implicit equation

$$\frac{1}{N(0)V_o} = \int_0^{\varepsilon_c} \frac{\tanh\,[1/2\,\beta(\varepsilon^2 + |\Delta|^2)^{1/2}]}{(\varepsilon^2 + |\Delta|^2)^{1/2}}\, d\varepsilon \qquad (6.16)$$

with N(0) the density of states of <u>one</u> spin population at the Fermi surface. It is clear that this equation has a solution only for $V_o > 0$, and that then $\Delta(T)$ is a decreasing function of temperature and must tend to zero at some temperature T_c (corresponding to $\beta = \beta_c$). At T_c eqn. (6.16) is satisfied with $\Delta = 0$. Hence we eliminate V_o from this equation and write

$$\int_0^{\varepsilon_c} \frac{\tanh\,[1/2\,\beta(\varepsilon^2 + |\Delta(T)|^2)^{1/2}]}{(\varepsilon^2 + |\Delta(T)|^2)^{1/2}}\, d\varepsilon = \int_0^{\varepsilon_c} \frac{\tanh\,(1/2\,\beta\varepsilon)}{\varepsilon}\, d\varepsilon. \qquad (6.17)$$

In the limit $\varepsilon_c \gg \Delta, kT_c$, (which we must assume for self-consistency, cf. above) it follows from (6.17) that $\Delta(T)$ can be written in the form

$$\Delta(T) = kT_c\, f(T/T_c) \qquad (6.18)$$

where f(x) is a universal function which turns out to be well approximated by $1.76\,(1-x^4)^{1/2}$. Thus, once T_c is fixed, $\Delta(T)$ and hence, as we shall see, <u>essentially all interesting properties* of the superconducting state are uniquely determined</u>, quite irrespective of the detailed form of the original

---

* Here we have made use of the assumption that $V_{kk'}$ is not appreciably a function of $\underset{\sim}{k}$ even when k' is outside the shell.

* This does not apply to a few properties which are sensitive to the short-range behavior of the "pair wave function" $F(\underset{\sim}{r})$ (see below), such as the dipole energy in the case of anisotropic pairing.

potential $V_{kk'}$. This explains the amazing quantitative success of the original BCS theory based on the model choice of potential (6.11).

Let's now discuss briefly some of the properties of the superconducting state. First, the single-particle distribution function is given by

$$\langle n_k\rangle(T) = \frac{1}{2}\left(1 - \frac{\varepsilon_k}{(\varepsilon_k^2 + |\Delta(T)|^2)^{1/2}} \tanh\left[\frac{1}{2}\beta(\varepsilon_k^2 + |\Delta|^2)^{1/2}\right]\right) \qquad (6.19)$$

and therefore, even at zero temperature, is qualitatively similar to its finite-temperature form in the normal state, except that it has a much longer "tail". A more interesting quantity is the function (we assume as usual translation-invariance)

$$F(\underset{\sim}{r}) \equiv \langle \psi_\uparrow(r_1)\psi_\downarrow(r_2)\rangle_{\underset{\sim}{r}_1-\underset{\sim}{r}_2=\underset{\sim}{r}} = \sum_k F_k e^{i\underset{\sim}{k}\cdot\underset{\sim}{r}} \qquad (6.20)$$

which, as we have seen, plays the role of the "wave function of the Cooper pairs". In fact, the contribution of the BCS pairing terms to any (spin-independent) two-particle expectation value of the form $\Sigma_{ij}\, Q(\underset{\sim}{r}_i - \underset{\sim}{r}_j)$ is given by the expression

$$\langle Q\rangle_{BCS} \equiv \int dr\, Q(r)\, |F(\underset{\sim}{r})|^2,$$

of which the result (5.18) for the potential energy is a special case. Since F_k is given in terms of Δ by eqns. (6.8) and (6.9), we have

$$F(\underset{\sim}{r}) = \sum_k \frac{\Delta_k}{2E_k} \tanh\left(\frac{1}{2}\beta E_k\right) e^{i\underset{\sim}{k}\cdot\underset{\sim}{r}}. \qquad (6.21)$$

Note that the sum in (6.21) goes over all k, both within and without the BCS "shell" $|\varepsilon_k| < \varepsilon_c$, and hence we cannot in general approximate Δ_k by a constant. However, from a comparison of (5.23) (or its finite-T analog) with the Schrödinger equation (5.24) we can infer that for $|\varepsilon_k| >> \Delta$, kT, F_k is simply proportional to the relative wave function of two particles colliding in free space at the Fermi energy. The same then follows for the real-space pair wave function F(r) for $r << \hbar v_F/\Delta(0)$. In particular, when r is large compared to the range of the interparticle potential but small compared to $\hbar v_F/\Delta(0)$, the pair wave function F(r) is simply, apart from a possible phase shift, the wave function of two free particles with equal and opposite momentum at the Fermi surface, i.e. $F(r) \sim (\sin k_F r)/k_F r$. When, however, r becomes of the order of

$\hbar v_F/\Delta(0)$, the pair wave function begins to fall off exponentially, as $e^{-r/\xi}$ where ξ is a weak function of T but is always of order $\hbar v_F/\Delta(0)$. Thus the "radius" of a Cooper pair is of the order of $\hbar v_F/\Delta(0)$, or equivalently of the "Pippard coherence length" ξ which is conventionally defined as $\hbar v_F/\pi\Delta(0)$. Note also that the quantity

$$n_c(T) \equiv \int |F(r)|^2 \, dr$$

which can be regarded as in some sense the "density of Cooper pairs" is of order $(\Delta(0)/\varepsilon_F)n$ at $T = 0$ and of order $(\Delta^2(T)/kT_c{\bullet}\varepsilon_F)n$ as $T \to T_c$, hence much smaller than the total density of electrons n.

Once we are given the form of the density matrix, eqn. (6.1), the calculation of the thermodynamic and response properties of the system as a function of temperature is a straightforward problem which is treated in many textbooks, so I will not go into it in detail here. However, there is one general point which is worth mentioning: When the system, in the superconducting state, is subjected to an external perturbation, there are two different ways in which it can respond to it, which are illustrated most simply for the case of a spatially homogeneous perturbation. In this case we should expect from symmetry that the perturbed density matrix is still of the general form (6.1), with the $\hat{\rho}_k$ in general changed from their original values. Now if we consider a definite pair of states $(\underset{\sim}{k}\uparrow, -\underset{\sim}{k}\downarrow)$ there are two ways of changing $\hat{\rho}_k$: (1) we can change the u_k's and v_k's, or (2) we can change the relative probabilities for the GP, BP and EP states. In some cases only one of these possibilities is relevant. For example, if we apply a weak magnetic field $\underset{\sim}{H}$, the energies of the GP and EP states are unaffected, since the Zeeman terms for the up- and down-spin electrons cancel. Thus to this order one gains no advantage by changing the u's and v's. However, the energies of the $|1,0\rangle$ and $|0,1\rangle$ BP states are shifted by $-\mu H$ and $+\mu H$ respectively where μ is the Bohr magneton, and therefore in thermal equilibrium their probability relative to one another is shifted by $\exp{-2\beta\mu H}$. As a result, the spin susceptibility χ relative to its normal-state value has the well-known Yosida form

$$\chi(T)/\chi_n = \int_0^\infty \frac{\beta}{2} \operatorname{sech}^2 \frac{1}{2}\beta\varepsilon \, d\varepsilon \equiv Y(T) \qquad (6.22)$$

and tends to zero as $T \to 0$. A similar argument applies to the normal density, which we recall can be defined in terms of the current $\underset{\sim}{J}$ generated by a vector potential $\underset{\sim}{A}$ under the constraint that the Cooper pairs remain at rest: the latter condition ensures that the pairs contribute nothing, and the term $-\underset{\sim}{J}\cdot\underset{\sim}{A}$ in the Hamiltonian splits the two BP states $|10\rangle$ ($J = +\hbar k/m$) and $|01\rangle$ ($J = -\hbar k/m$) in just the same way as the magnetic field does. We find the standard result

$$\rho_n(T)/\rho = Y(T) \tag{6.23}$$

and note for future reference that $\rho_s \equiv \rho - \rho_n$ tends to zero as $\Delta^2(T)$ in the limit $T \to T_c$.

On the other hand the procedure of changing the chemical potential μ, which defines the compressibility, causes a change in both the distribution function and the parameters u_k and v_k of the pair wave function. This is because for a given pair state $(\underset{\sim}{k}\uparrow, -\underset{\sim}{k}\downarrow)$ the equilibrium values of the u_k's and v_k's are functions of μ: in fact, the standard result (obtainable from the minimization of Ω with respect to the v_k) is

$$u_k^2 = \frac{1}{2}(1 + \varepsilon_k/E_k)\,, \quad v_k^2 = \frac{1}{2}(1 - \varepsilon_k/E_k) \tag{6.24}$$

where the "energy" ε_k <u>is measured from the chemical potential</u>. Hence a change in μ changes the u_k's and v_k's for any given $\underset{\sim}{k}$: at zero temperature this is the <u>only</u> way in which we can produce a finite compressibility. At finite T both this mechanism and the change of the occupation factors P_{GP}, etc., contribute, so that to order $(\Delta/\varepsilon_F)^2$ the compressibility is temperature-independent and equal to its normal-state value. We may regard the two mechanisms intuitively as expressing respectively the contribution of the "superfluid component" (change of parameters of pair wave function) and of the "normal component" (change of the occupation probabilities). However, it should be emphasized that the actual magnitude of the contribution of the two components depends on the property being considered, so that for example the "superfluid density" as conventionally defined in terms of the current response is not proportional to the contribution of the "superfluid component" to the compressibility. The separation is however conceptually useful when discussing certain types of nonequilibrium phenomenon, in which the system may have time to adjust the parameters v_k, etc., but not to change the thermal occupation probabilities.

Lecture 7: Derivation of Ginzburg-Landau Theory from BCS Theory

As you know, the Ginzburg-Landau theory of superconductivity is a phenomenological theory which antedated by several years the BCS theory. It is a development of the more general Landau-Lifshitz theory of second-order phase transitions, and works in terms of a complex scalar "order parameter" $\psi(\underset{\sim}{r})$, which GL interpreted, correctly as it turns out, as something like a Schrödinger wave function. The fundamental ingredient in GL theory is an expression for the free energy which is postulated to be valid for small ψ and T near the transition temperature T_c: omitting a term $F_o(T)$ which does not involve ψ or the electromagnetic vector potential $\underset{\sim}{A}(\underset{\sim}{r})$, this is:

$$F_{GL} = \int d\underset{\sim}{r}\, \{\alpha(T)|\psi(r)|^2 + \frac{1}{2}\beta(T)|\psi(\underset{\sim}{r})|^4 + \gamma(T)|(-i\hbar\underset{\sim}{\nabla} - e^*A(\underset{\sim}{r})\psi(\underset{\sim}{r})|^2$$

$$+\frac{1}{2}\mu_o^{-1}(\underset{\sim}{\nabla}\times\underset{\sim}{A}(\underset{\sim}{r})^2\}. \tag{7.1}$$

The coefficients α, β, γ are assumed to have the temperature-dependence

$$\alpha(T) \propto (T-T_c), \quad \beta(T) \propto \text{const}, \quad \gamma(T) \propto \text{const}. \tag{7.2}$$

Note that the normalization of $\psi(\underset{\sim}{r})$ is arbitrary: GL in their original work fixed it so that $\gamma(T) = 1/2m$, but it is convenient for our purposes to leave it more general. The Ginzburg-Landau equations for $\psi(\underset{\sim}{r})$ and $\underset{\sim}{A}(\underset{\sim}{r})$ follow by minimizing F with respect to these variables.

Very soon after the development of the microscopic BCS theory, Gor'kov in a classic paper showed that under appropriate circumstances the GL equations could be derived from it, and related the coefficients appearing in (7.1) to the gap $\Delta(T)$ of BCS theory. His derivation, which uses the technique of temperature-dependent Green's functions, is elegant but may not make it easy to grasp intuitively what is going on. In addition, by its nature it is confined to the case of thermodynamic equilibrium and does not automatically suggest ways of generalizing it. For these reasons I sketch in this lecture a more elementary, if somewhat less rigorous, derivation, which gives directly the free energy (7.1) and identifies the coefficients α, β, γ and e^* appearing in it.

We first make a guess as to the correct identification of the "order parameter" $\psi(\underset{\sim}{r})$ of GL theory. Let us generalize eqn. (6.20) to take into account a possible breaking of translation invariance by defining

$$F(\underset{\sim}{r}_1,\underset{\sim}{r}_2) \equiv \langle\psi\uparrow(\underset{\sim}{r}_1)\psi\downarrow(\underset{\sim}{r}_2)\rangle \tag{7.3}$$

and write F as a function of the center-of-mass variable $\underset{\sim}{R} \equiv (\underset{\sim}{r}_1+\underset{\sim}{r}_2)/2$ and relative coordinate $\underset{\sim}{\rho} \equiv \underset{\sim}{r}_1-\underset{\sim}{r}_2$. (Thus, in the present notation $\underset{\sim}{r} \to \underset{\sim}{\rho}$ in eqn. (6.20) and related equations). Then it is a reasonable guess that, apart perhaps from normalization, we have

$$\psi_{GL}(\underset{\sim}{R}) = F(\underset{\sim}{R},\underset{\sim}{\rho} = 0) \equiv \langle\psi\uparrow(\underset{\sim}{R})\psi\downarrow(\underset{\sim}{R})\rangle \tag{7.4}$$

i.e. ψ_{GL} is nothing but the Cooper-pair "wave function" at the relative origin, regarded as a function of the <u>center-of-mass</u> variable (as distinct from the discussion following (6.20), where it was assumed to be constant in this variable and studied as a function of the <u>relative</u> coordinate). From now on, since hopefully no confusion will arise, we shall revert to the notation $\underset{\sim}{r}$ for the COM coordinate in accordance with (7.1). We note that in the case $\psi(\underset{\sim}{r})$ = const. ψ_{GL} is given by the simple expression

$$\psi_{GL} = \sum_k F_k \tag{7.5}$$

where F_k is defined as the thermal and quantum-mechanical average of $\langle a_{k}\uparrow a_{-k}\downarrow\rangle$, i.e. by eqn. (6.7):

$$F_k \equiv u_k v_k \tanh \frac{1}{2}\beta E_k(T). \tag{7.6}$$

For the moment we make no assumptions about the actual form of u_k and v_k.

In the following I shall work for simplicity, as did Gor'kov, with the simple BCS model defined by eqn. (6.11). (I believe it should be fairly straightforward to generalize the argument to the more realistic case, but at the time of writing have not done so.) We first consider the case where $\psi(\underset{\sim}{r})$ is constant in space <u>but not necessarily given by its equilibrium value</u> ψ_o (which is given by eqn. (6.21) with $\underset{\sim}{r}$ (≡ relative coordinate!) = 0, where Δ is the equilibrium BCS gap), and express the free energy Ω as a function of ψ. (This is analogous to expressing the free energy of a ferromagnetic material as a function of the magnetization $\underset{\sim}{M}$, without assuming that the latter necessarily has its equilibrium value.) For clarity we write

$$\Omega \equiv \Lambda + \langle V\rangle, \qquad \Lambda \equiv \langle K-\mu N\rangle - TS \tag{7.7}$$

where K and V are respectively the kinetic and potential energies. The expression for $\langle V \rangle$ as a function of ψ is trivial: from (6.6), (6.11) and (7.5) we have simply

$$\langle V \rangle = -V_0 |\psi|^2. \tag{7.8}$$

We now come to the only tricky part of the argument, the dependence of the quantity Λ on ψ. To obtain the minimum of Λ for given ψ, we add the usual Lagrange multiplier term $-\lambda\psi$ and minimize the expression $\Lambda - \lambda\psi$. But it is clear that this is nothing other than the problem of minimizing the total free energy (eqn. (6.6)) with the self-consistently determined gap Δ replaced by λ. Consequently, the solution must be given by

$$\psi(\lambda{:}T) = \sum_k \frac{\lambda}{2E_k(\lambda)} \tanh \beta E_k(\lambda)/2, \quad E_k(\lambda) \equiv (E_k^2 + |\lambda|^2)^{1/2}. \tag{7.9}$$

Moreover, since the minimum corresponds to $\partial\Lambda/\partial\psi = \lambda(\psi{:}T)$, we have apart from a ψ-independent term

$$\Lambda(\psi{:}T) = 2 \int_0^{\psi} \lambda^*(\psi'{:}T)d\psi', \tag{7.10}$$

– a result which can be verified by direct expansion of the term in brackets in (6.6). Now, we know that the equilibrium superconducting-state free energy, i.e. $\Omega(\psi,T)$ minimized with respect to ψ, becomes equal to the normal-state free energy (which is zero in our convention) at the temperature $T = T_c$, so at this point the term in $|\psi|^2$ in Λ must be exactly cancelled by the potential term $-V_0|\psi|^2$, i.e. the coefficient must be V_0. Thus, we can write quite generally

$$\Omega(\psi,T) = 2\{ \int_0^{\psi} \lambda^*(\psi'{:}T)\, d\psi' - \int_0^{\psi} \lambda^*(\psi'{:}T)d\psi' \,|_{T=T_c,\psi\to 0} \,. \tag{7.11}$$

The potential V_0 has fallen right out of the problem! It should be emphasized that the expression (7.11), with $\lambda(\psi)$ given implicitly by eqn. (7.9), is actually valid for all values of T and ψ, not just for small ψ and T close to T_c. However, simple algebraic expressions are obtained only in this limit (the so-called Ginzburg-Landau regime). To obtain them, we expand $\psi(\lambda)$ (eqn. (7.9)) in the form

$$\psi(\lambda) = \lambda\{A(T_c) - (T_c - T)\, A'(T_c) + B(T_c)|\lambda|^2 + \ldots\} \tag{7.12}$$

invert to find $\psi(\lambda)$ up to order ψ^3, substitute in (7.11) and finally, for convenience redefine ψ by setting $\psi_{new} \equiv A^{-1}(T_c)\psi_{old} \equiv V_o \, \Sigma_k F_k$ (so that in equilibrium the new ψ is simply the energy gap Δ, and more generally is the "off-diagonal field"). In this way we obtain

$$\Omega(\psi) = -A'(T_c)(T_c - T)|\lambda|^2 - \frac{1}{2}B(T_c)|\lambda|^4 \tag{7.13}$$

where the coefficients A', B are explicitly given by

$$A'(T_c) = \frac{d}{dt}\sum_k (2\varepsilon_k)^{-1} \tanh\frac{1}{2}(\beta\varepsilon_k/2)\,|_{T=T_c} = -\frac{1}{2}\frac{1}{T_c}\left(\frac{dn}{d\varepsilon}\right) \tag{7.14a}$$

$$B(T_c = -\left\{\frac{d}{d|\lambda|^2}\sum_k \frac{1}{2(\varepsilon_k^2 + |\lambda|^2)^{1/2}} \tanh\left[\frac{1}{2}\beta_c(\varepsilon_k^2 + |\lambda|^2)^{1/2}\right]\right\}_{T=T_c,\lambda=0}$$

$$= -\frac{1}{2}\left(\frac{dn}{d\varepsilon}\right)\frac{7\zeta(3)}{16\pi^2}\frac{1}{(k_B T_c)^2} \tag{7.14b}$$

where the second expressions are valid in the limit $\beta_c\varepsilon_c >> 1$ which is assumed in using the BCS model (see Lecture 6). Comparing (7.13) with the phenomenological expression (7.1) with Ω identified with the F used there, we see that the "bulk" terms indeed have the GL *form*, with coefficients given, for our chosen normalization, by

$$\alpha(T) = \frac{1}{2}\left(\frac{dn}{d\varepsilon}\right)\left(\frac{T}{T_c} - 1\right), \qquad \beta = \frac{1}{2}\left(\frac{dn}{d\varepsilon}\right)\frac{7\zeta(3)}{16\pi^2}\frac{1}{(k_B T_c)^2} \tag{7.15}$$

in accordance with the assumption (7.2). Needless to say, minimization of the bulk terms with respect to ψ leads to a value of ψ which agrees with the $T \to T_c$ limit of the BCS expression for $\Delta(T)$.

We now turn more briefly to the remaining terms in (7.1). We first consider a situation where $\psi(\underset{\sim}{r})$ is slowly varying in space but the electromagnetic vector potential $\underset{\sim}{A}(\underset{\sim}{r})$ is zero. In the general case we have to consider separately spatial variations in the magnitude and phase of $\psi(\underset{\sim}{r})$, and there is no guarantee that the coefficients are related in such a way that the bending energy can be written in the simple form const. $|\underset{\sim}{\nabla}\psi|^2$ as in (7.1). However, since the absolute phase of the pair wave function, hence of the order parameter $\psi(\underset{\sim}{r})$, has no physical significance, the bending terms must be invariant under the replacement $\psi(r) \to \psi(r)e^{i\alpha}$ (α = const. in space), and if they are analytic functions of Re ψ, Im ψ in the limit $\psi \to 0$ (an assumption, but an eminently reasonable one), then in this limit the form in (7.1) is the only one which can satisfy this constraint. Once terms of higher order in ψ then the second are allowed, we can no longer make this argument.* However, if we do restrict ourselves to terms of second order in ψ, we can infer the coefficient $\gamma(T)$ by considering the special form of $\psi(\underset{\sim}{r})$ corresponding to a variation of phase only: i.e. the form

$$\psi(\underset{\sim}{r}) = \psi_0 \, e^{i\underset{\sim}{k}\cdot\underset{\sim}{r}} \tag{7.16}$$

where we have assumed for simplicity that the magnitude of ψ_0 is the equilibrium one. It is clear that the form (7.16) describes a situation in which the Cooper pairs are moving with center-of-mass momentum $\hbar\underset{\sim}{q}$. Since the system is in thermodynamic equilibrium subject to this constraint, the associated "bending" energy should be given (cf. Lecture 1) by the expression $1/2\, \rho_s v_s^2$, where $\underset{\sim}{v}_s$ is the "superfluid velocity" discussed in Lectures 1-3. How is this related to $\underset{\sim}{q}$? Following the argument of Lecture 4, and taking into account that the "mass" associated with the center-of-mass motion of the Cooper pairs is twice the electron mass m, we conclude that the correct expression for $\underset{\sim}{v}_s$ is

$$\underset{\sim}{v}_s(\underset{\sim}{r}) = \frac{\hbar}{2m} \underset{\sim}{\nabla}\phi(r) \qquad (\psi(r) \equiv |\psi(r)| \exp i\phi(r)) \tag{7.17}$$

and hence in the situation considered we have $\underset{\sim}{v}_s(r) = \hbar\underset{\sim}{q}/2m$. Thus, inserting the form (7.16) in (7.1) and equating the bending term to $1/2\, \rho_s v_s^2$, we obtain the relation

$$\gamma(T) = \frac{\hbar^2}{8m^2} \frac{\rho_s(T)}{|\psi_0(T)|^2}. \tag{7.18}$$

This expression is of course valid independently of the normalization of ψ. Finally, we obtain an expression for $\rho_s(T)$ up to second order in $|\Delta(T)|^2$ by expanding the expression (6.23) for ρ_n up to this order, and in accordance with our normalization convention set $\psi_0(T)$ in (7.18) equal to $\Delta(T)$. This finally gives the result (n = total density of electrons)

---

* Indeed, it may be explicitly shown by considering the effect of "Fermi-liquid" corrections (Lecture 9) that the form (7.1) cannot hold in general.

$$\gamma(T) = \frac{n\hbar^2}{4m} \frac{7\zeta(3)}{8\pi^2 k_B^2 T_c^2} \tag{7.19}$$

in agreement with (7.2). Needless to say, to obtain the original form postulated by GL, $\gamma(T) = 1/2m$, one would simply have to change the normalization of ψ; then, of course, the expressions for α and β are multiplied by appropriate constants.

The generalization to the case of a non-zero electromagnetic vector potential $\underset{\sim}{A}(\underset{\sim}{r})$ is rather obvious. In the presence of a finite $\underset{\sim}{A}$, the velocity of the Cooper pairs is related to their canonical momentum p_c by $\underset{\sim}{v} = 1/m_p(\underset{\sim}{p}_c - e_p\underset{\sim}{A})$, where m_p and e_p are respectively the mass 2m and charge 2e of the pair, and therefore the appropriate definition of $\underset{\sim}{v}_s$ is

$$\underset{\sim}{v}_s(\underset{\sim}{r}) = \frac{\hbar}{2m}(\underset{\sim}{\nabla}\phi(r) - 2e\underset{\sim}{A}(r)).$$

We thus recover the general gradient term in (7.1), with γ given by (7.18) and e^* identified as the pair charge 2e.

Finally, the last term in (7.1) is simply the electromagnetic field energy $\int 1/2\, \mu_0^{-1} B^2(r)\, d\underset{\sim}{r}$. Thus, starting from the microscopic BCS theory and specializing to the "GL" regime of small ψ and T near T_c, we have recovered the phenomenological free energy (7.1), with the identifications

$$\alpha(T) = \frac{1}{2}\left(\frac{dn}{d\varepsilon}\right)\left(\frac{T}{T_c} - 1\right), \qquad \beta = \frac{1}{2}\left(\frac{dn}{d\varepsilon}\right)\frac{7\zeta(3)}{16\pi^2}\frac{1}{(k_B T_c)^2}$$

$$\gamma = \frac{n\hbar^2}{4m} \bullet \frac{7\zeta(3)}{8\pi^2(k_B T_c)^2}, \quad e^* = 2e$$

and ψ identified with the "off-diagonal field" $V_o \sum_k F_k$.

Lecture 8: Aspects of Anisotropic Superfluidity in ^{3}He

The phase diagram of liquid ^{3}He below 3 mK in zero magnetic field is shown in fig. 5. The "N" (normal) phase is well described by Landau's theory of a normal Fermi liquid, while the A and B phases, discovered in 1972, are by now almost universally believed to be "anisotropic superfluid" phases which involve a generalized type of BCS pairing and are qualitatively similar to the BCS phase of ordinary superconductors.

In superconducting metals, BCS pairing is conceived, at least in the simplest models, as taking place between the electrons themselves. However, in a degenerate strongly interacting system, such as liquid ^{3}He, which is described by Landau's Fermi liquid theory, it turns out that in the normal phase the elementary excitations are not real ^{3}He atoms but rather Landau "quasiparticles", which one can conceive intuitively as something like an atom accompanied by its "dressing cloud" of other atoms, the whole complex behaving like a single fermion. Since the energy scale involved in the formation of quasiparticles is large compared to the energy scale ($\sim kT_c$) for formation of the superfluid state, it is natural to regard the Cooper pairing as taking place between already formed Landau quasiparticles, and the variables $\underset{\sim}{r}$, σ which occur in (8.4) below should strictly speaking be regarded* as characterizing a quasiparticle rather than a real particle. We will introduce other effects which follow from using a Landau Fermi-liquid description as they arise: see also next lecture.

Since we have seen that a necessary condition for the formation of Cooper pairs is an attraction between (quasi) particles near the Fermi surface, an obvious question is the origin of this attraction in liquid ^{3}He. It is generally believed that there are two main ingredients. One is the attractive part of the standard van der Waals interaction between neutral atoms; this, to a good approximation at least, is simply a fixed function of the relative separation. Of course, at short distances ($\lesssim$ 2Å) the interaction is strongly repulsive due to the overlap of the atomic cores. This has an important consequence: Since Cooper pairs (unlike diatomic molecules in free space) cannot be formed out of arbitrarily low-momentum states but must have a magnitude of relative momentum close to $2k_F$, it is advantageous to choose the relative angular momentum ℓ so that the first maximum of the (unperturbed) relative wave function lies not at the origin as it would for an s-wave state, but rather near the minimum r_m of the van der Waals potential, i.e. crudely speaking, so that $k_F r_m \sim \ell$. This argument suggests that the optimum value of ℓ would be 2 or, possibly, 1, but anyway not zero as in simple BCS theory. A second contribution to the interaction is more interesting: The Landau theory of Fermi liquids allows for various kinds of interactions between quasiparticles in the normal

---

* Of course the "coordinate" of a quasiparticle can only be meaningfully defined on a scale large compared to its "size" (a few Å). Since this is small compared to the Cooper pair radius, this is not a particular difficulty.

state, in particular to a sort of spin-dependent interaction which may be mimicked by a short-range "molecular field" $\underset{\sim}{H}$ such that, approximately,

$$H(r) \sim -\zeta_o \underset{\sim}{S}(r). \qquad (8.1)$$

where $\underset{\sim}{S}(r)$ is the spin density. The <u>direct</u> effect of this field (equivalent to an interaction $\sim \zeta_o \underset{\sim}{\sigma} \bullet \underset{\sim}{\sigma}' \, \delta(\underset{\sim}{r})$ between quasiparticles) is not particularly interesting. More interesting is the <u>indirect</u> effect: a quasiparticle spin at point $\underset{\sim}{r}$ produces the field (8.1), which then polarizes the liquid at a nearby point $\underset{\sim}{r}'$:

$$\underset{\sim}{S}(\underset{\sim}{r}',t') \sim \chi(\underset{\sim}{r}'-\underset{\sim}{r},t'-t) \, H(\underset{\sim}{r}). \qquad (8.2)$$

The spin polarization $S(\underset{\sim}{r}')$ in turn produces a molecular field $-\zeta_o S(r')$, with which a quasiparticle at point $\underset{\sim}{r}'$ can then interact. In this way, we produce an effective spin-dependent interaction V_{eff} between quasiparticles at points $\underset{\sim}{r}$ and $\underset{\sim}{r}'$ given schematically* by

$$V_{eff}(\underset{\sim}{r}-\underset{\sim}{r}',t-t') = -\zeta_o^2 \, \underset{\sim}{\sigma} \bullet \underset{\sim}{\sigma}' \, \chi(\underset{\sim}{r}-\underset{\sim}{r}',t-t'). \qquad (8.3)$$

Irrespective of the sign of ζ_o (which is actually negative for ^3He) this interaction is attractive for a total spin S of 1 and repulsive for $S = 0$, and therefore tends to favor triplet pairing. It is also, crudely speaking, proportional to the spin susceptibility of the liquid, a feature which gives rise to some interesting effects which we will mention in the next lecture. To generalize the BCS theory to the case of pairing in a state of finite relative angular momentum ℓ and, possibly $S \neq 0$ we simply write down once again the fundamental ansatz (5.2) for the many-body wave function Ψ_N, namely

$$\Psi_N = \mathcal{N}^{-1/2} \mathcal{A} \, \phi(\underset{\sim}{r}_1\underset{\sim}{r}_2\sigma_1\sigma_2) \, \phi(\underset{\sim}{r}_3\underset{\sim}{r}_4\sigma_3\sigma_4) \ldots \phi(\underset{\sim}{r}_{N-1}\underset{\sim}{r}_N\sigma_{N-1}\sigma_N) \qquad (8.4)$$

but do not now assume that the two-particle function ϕ has the simple form (5.3); rather, assuming only that the center of mass is at rest, we write it in the more general form

---

* The correct result is actually (8.3) symmetrized by $r, t \rightleftarrows r', t'$.

$$\phi(\underset{\sim}{r}_1\underset{\sim}{r}_2{:}\sigma_1\sigma_2) = \phi(\underset{\sim}{r}_1-\underset{\sim}{r}_2{:}\sigma_1\sigma_2) \tag{8.5}$$

without for the moment making any further restrictions.

The Pauli principle implies that if the function ϕ has even spatial parity it corresponds to total spin $S = 0$, whereas odd parity implies $S = 1$. In principle a superposition of these two types of function is possible, but it seems to be of little practical interest, so we shall assume that one case or the other is realized for ^3He. Let us start with the case of even parity and $S = 0$. Then the form of ϕ is again (5.3), except that the spatial part $\phi(\underset{\sim}{r}_1-\underset{\sim}{r}_2)$ may now be a function of the direction of the relative separation $\underset{\sim}{r}_1-\underset{\sim}{r}_2$ as well as its magnitude. The generalization of the arguments of Lectures 5 and 6 to this case is straightforward, indeed almost trivial, and we recover a gap equation which is formally identical to (6.9), namely

$$\Delta_k = -\sum_{k'} V(\underset{\sim}{k},\underset{\sim}{k}') \frac{\Delta_{k'}}{2E_{k'}} \tanh \frac{1}{2}\beta E_{k'} \tag{8.6}$$

the only difference being that it is not now assumed that Δ_k is independent of the direction of $\underset{\sim}{k}$. Just as in the isotropic case considered in Lecture 6, we can carry out a renormalization procedure to reduce the actual problem to a model problem characterized by a BCS-like interaction*

$$V(\underset{\sim}{k},\underset{\sim}{k}') = -V(\hat{\underset{\sim}{k}}\bullet\hat{\underset{\sim}{k}}')\,\theta(\varepsilon_c-|\varepsilon_k|)\,\theta(\varepsilon_c-|\varepsilon_{k'}|) \tag{8.7}$$

where however $V(\hat{\underset{\sim}{k}}\bullet\hat{\underset{\sim}{k}}')$ is not a constant but must be expanded in Legendre polynomials:

$$V(\hat{\underset{\sim}{k}}\bullet\hat{\underset{\sim}{k}}') = -\sum_{\substack{\ell\\=\text{even}}} V_\ell P_\ell \hat{\underset{\sim}{k}}\bullet\hat{\underset{\sim}{k}}'). \tag{8.8}$$

It is straightforward to show that eqn. (8.6), with the form (8.7) of $V(\hat{\underset{\sim}{k}}\bullet\hat{\underset{\sim}{k}}')$, has a nontrivial solution below a temperature T_c defined by

$$k_BT_c = 1.14\,\varepsilon_c \exp\{-1/N(0)V_\ell\} \tag{8.9}$$

---

* We assume isotropy of the original interaction $V(\underset{\sim}{r}-\underset{\sim}{r}')$, which is correct for ^3He provided small effects like the dipole forces are neglected.

where V_ℓ is the largest positive member of the expansion coefficients in (8.8) (i.e., the one corresponding to the maximum attraction). (The transition temperature for the isotropic case of Lectures 5-6 is the special case $\ell = 0$ of (8.9).) Just below T_c the gap Δ, which is a function, for the model (8.7), only of the direction of $\hat{\underset{\sim}{k}}$, is some combination of spherical harmonics $Y_{\ell m}$ corresponding to this "most attractive" ℓ-value:

$$\Delta(\hat{k}:T) = \sum_m a_m(T) Y_{\ell m}(\hat{k}). \tag{8.10}$$

At lower temperatures the exact solution may have an admixture of other spherical harmonics of the same parity, but unless another V_ℓ of the same parity is very close to the "most attractive" one this is very small; in practice, in the case of ^3He, it is usually neglected. Even with this approximation, however, the various $Y_{\ell m}$ for the chosen value of ℓ are not degenerate, because of the nonlinear dependence of the free energy on $\Delta(\hat{\underset{\sim}{k}})$, so that the $a_m(T)$ in (8.10) are actually uniquely fixed up to an overall rotation (e.g. for $\ell = 2$ the spherical harmonic P_{22} does better than P_{20}, and a particular linear combination does better than either). One very important point, however, is that no linear combination of spherical harmonics corresponding to a value of ℓ other than zero can make the gap isotropic, nor even nodeless: $\Delta(\hat{\underset{\sim}{k}})$ must have, at least, point nodes somewhere on the Fermi surface. This has the important consequence that the minimum excitation energy of a Bogoliubov quasiparticle is now zero rather than a finite value Δ as in the isotropic BCS case, so that the density of states is now finite for all energies, and hence that properties of the system which have contributions exclusively from the normal quasiparticles (such as the specific heat or, for singlet pairing, the spin susceptibility) now tend to zero as $T \to 0$ as a power law rather than exponentially as in the isotropic case (the exact power being determined by the property considered and the nature of the nodes (point, line etc.)).

As a result of the anisotropy of the energy gap $|\Delta(\hat{\underset{\sim}{k}})|$, the condensation energy at $T = 0$ of a superfluid with angular momentum ($\ell \neq 0$) and with a given value of V_ℓ is slightly less then that of an isotropic superfluid with a V_0 equal to V_ℓ. It is therefore intuitively appealing to try to think of a more sophisticated type of solution than that given by (8.4), which would permit the energy gap to be isotropic even for $\ell \neq 0$ pairing. In 1961, Gor'kov and Galitskii obtained, by a decoupling of the equation of motion for the two-particle Green's functions of the system which they could show was asymptotically exact in the thermodynamic limit, a state of the many-body system which had just this property and indeed had lower energy than the "standard" solution described above. Unfortunately, it was very soon shown to most people's satisfaction that this solution of the Green's function equations, "exact" though it is, does not correspond to any solution of Schrödinger's equation! This should be an awful warning to the generation who imbibed Green's function methods with their mother's milk: the Green's functions of a system, at least those of any order less than the total number of particles N, are averaged properties and hence do not necessarily embody all the relevant information and constraints.

The above analysis of the $\ell \neq 0$ spin singlet case is generally believed not to be immediately applicable to the superfluid A and B phases of liquid ^{3}He, which are thought to involve pairing with $\ell = 1$ and $S = 1$. We therefore now consider the case of odd-parity, spin triplet pairing. It is obvious that in this case the function $\phi(r_1-r_2,\sigma_1\sigma_2)$ can be expanded, in an obvious notation, in terms of the three Zeeman substates:

$$\phi(\underset{\sim}{r}_1-\underset{\sim}{r}_2) = \phi_{\uparrow\uparrow}(\underset{\sim}{r}_1-\underset{\sim}{r}_2{:}\sigma_1\sigma_2)\,|\uparrow\uparrow\rangle + \phi_{\uparrow\downarrow}(r_1-r_2)\frac{1}{\sqrt{2}}(|\uparrow\downarrow + \downarrow\uparrow\rangle$$

$$+\ \phi_{\downarrow\downarrow}(\underset{\sim}{r}_1-\underset{\sim}{r}_2)\,|\downarrow\downarrow\rangle \tag{8.11}$$

where the functions $\phi_{\alpha\beta}$ are in general complex. If we consider any function of the form (8.11) for given $(\underset{\sim}{r}_1-\underset{\sim}{r}_2)$, there is a general theorem that we can always choose new spin axes so that $\phi_{\uparrow\downarrow} = 0$. However, it may not be possible to choose in this way the same set of axes for all values of $\underset{\sim}{r}_1-\underset{\sim}{r}_2$. If it is possible, we speak of an "equal-spin-pairing" (ESP) state (and call the set of axes so defined, which is not necessarily unique, the "ESP" axes); the A phase of ^{3}He is usually believed to be such a state. A special subclass of ESP states, which is again believed to include ^{3}He–A, and which we shall call "factorizable" (not a standard designation), is characterized by the feature that the ratio of $\phi_{\uparrow\uparrow}$ and $\phi_{\downarrow\downarrow}$ is independent of $\underset{\sim}{r}_1-\underset{\sim}{r}_2$ and is a complex number of modulus unity, so that we can write

$$\phi(\underset{\sim}{r}_1-\underset{\sim}{r}_2{:}\sigma_1\sigma_2) = \phi(\underset{\sim}{r}_1-\underset{\sim}{r}_2)\,(|\uparrow\uparrow\rangle + e^{i\chi}\,|\downarrow\downarrow\rangle). \tag{8.12}$$

For such a state another choice of axes (with z-axis rotated by $\pi/2$) is possible and often convenient, such that only the $S_z = 0$ Zeeman component is represented, i.e. such that $\phi(\underset{\sim}{r}_1-\underset{\sim}{r}_2{:}\sigma_1\sigma_2) = \phi(r_1-r_2) \times 1/\sqrt{2}\,(|\uparrow\downarrow + \downarrow\uparrow\rangle$. It is clear that factorizable states can be discussed almost exactly like spin singlet states: the number-nonconserving representation is formally identical to (5.9), i.e.

$$\Psi = \prod_k (u_k + v_k a^+_{k\uparrow} a^+_{-k\downarrow} | vac\rangle. \tag{8.13}$$

the only difference being that the parameter v_k is now odd in v_k, so as to preserve the correct Fermi statistics. The gap equation (8.6) still holds (with $V(\hat{k}\cdot\hat{k}')$ however now of odd parity), and all our remarks about the anisotropy of the gap, etc., go through as before.

The many-body wave function (8.11), with ϕ given by the form (8.12), has the characteristic feature that all the N particles are condensed, in pairs, into a linear combination of the $S_z = +1$ and $S_z = 1$ states (in the "ESP" axes). This may also be seen by transforming the number-nonconserving version (8.13) back into the ESP axes, where it reads

$$\Psi = \prod_k \{u_k + v_k(a^+_{k\uparrow} a^+_{-k\downarrow} + e^{i\Delta\phi}\, a^+_{k\downarrow} a^+_{-k\downarrow})\} | vac\rangle. \tag{8.14}$$

Thus, whether or not we require particle number to be conserved, the wave function is not an eigenstate of total spin projection S_z. Just as in the case of particle number (see Lecture 5) we can obtain such an eigenstate, if we wish, by multiplying $\Psi \equiv \Psi(\Delta\phi)$ by exp $iS_z \bullet \Delta\phi$ and integrating over the phase $\Delta\phi$: let us call the resulting state Ψ_{S_z}. Now in the case of particle number, the physically correct choice for an isolated system (where N is rigorously conserved) is Ψ_N, eqn. (5.2) rather than (5.9) (though no physical expectation values actually depend on this choice). What is the situation for S_z? An important difference, here, is that in practice there will almost always be some interactions in the system, however small, which violate conservation of S_z and can give rise to physical effects; the most important such interaction in ^{3}He is the tiny nuclear dipole-dipole interaction. Such interactions, it turns out, give in general a term in the expectation value of the energy which depends on $\Delta\phi$ and moreover is extensive: e.g. for ^{3}He, which has a pair function $\phi(\underset{\sim}{r}_1 - \underset{\sim}{r}_2)$ which, were it to describe a diatomic molecule, would correspond to relative angular momentum $\hbar$ along some axis (say for definiteness the x-axis), the expectation value of the total nuclear dipole energy is equal to $-g_D \cos \Delta\phi$, where $g_D \propto N$. On the other hand, there is an energy which depends on S_z, namely the usual polarization energy $S_z^2/2\chi$, where χ is the Pauli spin susceptibility of the superfluid liquid along the ESP axes, which turns out to be (very nearly) equal to that of the normal liquid, and is, of course, itself extensive. Thus, the total relevant energy is of the form

$$E(S_z,\Delta\phi) = \frac{S_z^2}{2\chi} - g_D \cos \Delta\phi. \tag{8.15}$$

Now it is straightforward to show (cf. Lecture 4) that S_z and $\Delta\phi$ satisfy the commutation relation

$$[S_z,\Delta\phi] = -i \tag{8.16}$$

and hence the mean-square fluctuations satisfy

$$\langle S_z^2\rangle \bullet \langle(\Delta\phi)^2\rangle \geq 1/4. \tag{8.17}$$

It then follows that $E(S_z,\Delta\phi)$ is minimized, for a large enough value of the product $g_D\chi \sim N^2$, when $\langle S_z^2\rangle \sim N$ and $(\Delta\phi)^2 \sim 1/N$. Hence in the thermodynamic limit $N \to \infty$ $\Delta\phi$ is precisely fixed and the wave function is indeed $\Psi(\Delta\phi)$ rather than Ψ_{S_z}. For a small enough value of N the reverse would be the case; however if we put in the numbers appropriate to ^{3}He, it appears that the crossover takes place at values of N so small (volume $\sim 10^6$Å^3) that it is unlikely to be of great experimental interest.

The relations (8.15)–(8.16) have another interesting consequence; they describe a simple quantum pendulum, and since with a reasonable value of N we are in the extreme semiclassical limit, the motion of $\Delta\phi$ when excited is exactly that of a classical pendulum, namely harmonic oscillation for small displacements, succeeded by nonlinear libration and rotation at larger amplitudes. This behavior is seen experimentally (though one usually actually measures the dynamics of the conjugate variable $S_z(t)$ rather than $\Delta\phi$ itself), the small oscillations being precisely the well-known longitudinal magnetic resonance oscillation of ^{3}He–A (a similar but rather more complicated resonance occurs in the B phase).

We now turn to the most general case, that of spin triplet pairing in a non-ESP state. For simplicity we shall however impose the constraint of unitarity, which for present purposes means that for any given direction $\hat{\underline{k}}$ we can choose axes so that the pair function ϕ can be written in the form occurring in (8.13), involving only the $S_z = 0$ Zeeman substate. Both the A and B phases of superfluid ^{3}He are believed to satisfy this condition. It then follows that the number-nonconserving wave function can be written formally in the form (8.13), provided that the spin axes are chosen differently for different $\underset{\sim}{k}$. In a representation in which the spin axes (labelled by α, β) are uniformly chosen, this wave function becomes

$$\Psi = \prod_k (u_k + v_{k\alpha\beta} a^+_{k\alpha} a^+_{-k\beta}) |\mathrm{vac}\rangle \tag{8.18}$$

(so that in general all Zeeman components are represented in the Cooper pair, cf. eqn. (8.11)). The matrix $\hat{v}_k$ whose elements are $v_{k\alpha\beta}$ is related to the original v_k by

$$\hat{v}_k = \begin{pmatrix} \sin\theta e^{i\phi} & \cos\theta \\ \cos\theta & \sin\theta e^{-i\phi} \end{pmatrix} v_k. \tag{8.19}$$

The matrix $\hat{v}_k \hat{v}_k^+$ is $|v_k|^2$ times the unit matrix in any representation. Considering now the case of zero temperature, let us evaluate the expectation value of the energy as a function of the elements $v_{k,\alpha\beta}$ of the matrix $\hat{v}_k$. The kinetic energy may be evaluated separately for each pair state ($\underset{\sim}{k}$,$-\underset{\sim}{k}$) using the axes in which only the $S_z = 0$ Zeeman state is populated, and is given by the representation-independent expression

$$\langle K - \mu N \rangle = 2 \sum_k \varepsilon_k |v_k|^2 \equiv \sum_k \varepsilon_k \,\mathrm{Tr}\, v_k v_k^+. \tag{8.20}$$

The potential energy is evaluated by factorizing the two-particle expectation values into sums of products of the quantities $F_{k\alpha\beta} \equiv \langle a^*_{k\alpha} a^+_{-k\beta} \rangle = u_k v_{k\alpha\beta}$:

$$\langle V \rangle = \sum_{kk'} u^*_k u_k v_{k\alpha\beta} v^*_{k'\beta\alpha}. \tag{8.21}$$

We minimize the sum of (8.20) and (8.21) subject to the constraint $|u_k|^2 + |v_k|^2 = 1$. The result is most easily expressed by introducing a "partial" gap (matrix) $\Delta_{k,\alpha\beta}$; this is then found to satisfy the generalized BCS gap equation

$$\hat{\Delta}_k = -\sum_{k'} V(\underset{\sim}{k}\underset{\sim}{k}') \frac{\hat{\Delta}_{k'}}{2E_{k'}} \tag{8.22}$$

$$E_k \equiv (\varepsilon_k^2 + |\Delta_k|^2)^{1/2}, \qquad |\Delta_k|^2 \equiv \frac{1}{2} \mathrm{Tr}\, \Delta_k \Delta_k^+. \tag{8.23}$$

Just as in the spin singlet case, the quantity E_k turns out to be the excitation energy of a Bogoliubov quasiparticle (which is spin-independent for a unitary state), so the quantity $|\Delta_{\hat{k}}|$ defined by (8.23) plays the role of the energy gap for direction $\hat{\underset{\sim}{k}}$. An important difference from the "factorizable" case discussed above is that for the special case $\ell = 1$ (which is believed to be relevant for ^{3}He) it is possible to get an isotropic energy gap, even though the individual matrix elements $\Delta_{k\alpha\beta}$ must, from (8.22), be (approximately) spherical harmonics of order ℓ. As an example, we could take

$$\hat{\Delta}(\hat{\underset{\sim}{k}}) = \Delta_o \begin{pmatrix} \sin\theta e^{-i\phi} & \cos\theta \\ \cos\theta & \sin\theta e^{i\phi} \end{pmatrix} \tag{8.24}$$

where θ, ϕ now represent the angle of $\hat{k}$ on the Fermi surface. This state, which is believed, up to a relative rotation of spin and orbital coordinates, to describe the B phase of ^{3}He, corresponds intuitively to Cooper pairs with total (spin and orbital) angular momentum J = 0 and has an isotropic gap, $|\Delta(\hat{\underset{\sim}{n}})|^2 = |\Delta_o|^2$. It is usually known as the Balian-Werthamer state.

The generalization to finite temperature is almost trivial, since the entropy term, and the thermal occupation factors, can be evaluated individually for each $\underset{\sim}{k}$ and are exactly the same functions of E_k as in the spin singlet case. Consequently the generalization to finite temperatures of eqn. (8.22) consists, as in the singlet case, simply of multiplying the quantity $\hat{\Delta}_{k'}/2E_k$ on the RHS by the thermal factor tanh $\beta E_{k'}/2$.

There is no space to do more than mention here a number of interesting properties of an anisotropic superfluid such as ^{3}He which are qualitatively different from these of the a simple isotropic BCS one:

(1) There exist a large number of collective excitations which correspond to a vibration of the internal structure of the Cooper pairs and hence have no analogy in the isotropic case (where this structure is trivial).

(2) It is possible, in principle at least, to use the coherent behavior of the Cooper pairs to "amplify" tiny effects which at the level of a single atom are unbelievably tiny. For example, ^{3}He–A, where the ground state wave function is invariant under neither inversion (P) nor time reversal (T) may display a tiny magnetic moment in zero field due to the fact that in the rotation of a diatomic molecule, or Cooper pair, the electronic and nuclear orbital magnetic moments do not exactly cancel; while ^{3}He–B, which breaks P but not T, may show minute effects from that part of the weak interaction (the dominant term) which has this property.

(3) The first-order phase transition between ^{3}He–A and ^{3}He–B is the only known (laboratory) example of a transition between two different superfluid phases, and poses many unique problems.

(4) Finally, it turns out that while the B phase of ^{3}He is probably just as "good" a superfluid as ^{4}He, the A phase, while it is believed to show NCRI, would not be "superfluid" in the sense of carrying metastable persistent currents were it not for an accident involving the tiny nuclear dipole interaction: the topological considerations which guarantee the metastability of current-carrying states in ^{4}He or ^{3}He–B do not persist for the more complicated pair wave function of the Cooper pairs in ^{3}He–A.

Lecture 9: Beyond BCS

Lectures 5-8 have treated the original BCS theory and its generalization to the case of Cooper pairing with nonzero spin and/or orbital angular momentum. In this lecture I shall try to summarize some of the known ways of taking into account effects which are neglected in this simple treatment. It should be strongly emphasized that all of the generalizations mentioned in this lecture have in common with the original BCS theory that the many-body wave function is conceived to be, schematically, of the form (5.2), namely

$$\Psi = \mathcal{N}^{-1/2}\mathcal{A}\phi(\underset{\sim}{r}_1\underset{\sim}{r}_2\sigma_1\sigma_2)....\phi(\underset{\sim}{r}_{N-1}\underset{\sim}{r}_N\sigma_{N-1}\sigma_N) \tag{9.1}$$

(or at least be "topologically isomorphic" to it); the differences lie (a) in the detailed form assumed for ϕ, (b) in the energetics involved in optimizing the wave function, and/or (c) in the calculation of the response to external fields. The first four effects we consider involve only (b) and (c); ϕ is still assumed to have the form (8.5).

1. Transition to diatomic molecules.

An important assumption implicit in the original BCS theory is that the energy gap Δ is small compared to the Fermi energy ε_F; or, what turns out to be essentially the same thing, that the Cooper-pair radius ξ_o is large compared to the interparticle spacing, a. Yet the ansatz (5.2) could, obviously, formally describe also a Bose-condensed gas of tightly bound diatomic molecules; in this case we expect $\xi_o << a$.

One can study the crossover between the two types of limiting behavior at $T = 0$ using a simple model of a dilute Fermi gas interacting via a potential whose parameters are close to the critical values at which an s-wave bound state of two particles in free space appears. Thus the s-wave scattering length a_s is large compared to the range r_0 of the potential and may have either sign. Under these conditions one expects intuitively that all the detailed properties of the interaction potential are buried in the single parameter a_s(so that r_0 is irrelevant) and the problem is characterized by a single dimensionless parameter, which we may conveniently take as $a/a_s \equiv \zeta$. The limit $\zeta \to +\infty$ then corresponds to the case of Bose condensation of diatomic molecules, while $\zeta \to -\infty$ corresponds to the standard Cooper-pair limit.

To study the crossover behavior at $T = 0$ all we need do is to construct the ansatz (9.1) for the many-body wave function and go through exactly the same procedure as in Lecture 5, with one difference: Whereas in the standard BCS theory it is implicitly assumed (and can be shown a posteriori to follow from the BCS solution) that in the superconducting state the chemical potential μ is approximately the same function of particle density n as in the normal phase, in the more general problem this is no longer true: μ must be fixed self-consistently from the relation ($\xi_k \equiv \hbar^2k^2/2m$)

$$\langle N\rangle \equiv \sum_{k\sigma} \langle n_{k\sigma}\rangle = \sum_{k} \left(1 - (\xi_k - \mu)/E_k\right)$$

$$(E_k \equiv (\xi_k - \mu)^2 + |\Delta_k|^2)^{1/2} \tag{9.2}$$

along with the BCS gap equation

$$\Delta_k = -\sum_{k'} V_{kk'}\Delta_{k'}/2E_{k'}. \tag{9.3}$$

A renormalization of the gap equation similar to that employed in Lecture 6, along with (9.2), gives us two equations for the two unknowns Δ ($\equiv \Delta(\underset{\sim}{k} = 0)$ and μ in terms of the known quantities a and a_s: as expected, both Δ and μ can be written in the form $\varepsilon_F f(\zeta)$, where ε_F is the Fermi energy of the noninteracting gap with the given value of a. The equations turn out to have no singularity for any value of ζ (including the value $\zeta = 0$ at which the 2-particle bound state appears) and to give us the physically expected results in the limits $\zeta \to \pm\infty$. The only point at which anything changes qualitatively is the point at which $\mu = 0$ (which turns out to correspond to $\zeta \equiv \zeta_o \sim 1$): for $\zeta < \zeta_o$ Δ represents a true "energy gap" (minimum quasiparticle excitation energy), while for $\zeta > \zeta_o$ the latter is given by $(|\mu|^2 + |\Delta|^2)^{1/2}$, reducing to $-\mu$ in the Bose-condensed limit as expected.

It should be mentioned that the calculations for the case of p-wave pairing (which can be made physically realistic by turning on a large enough magnetic field to effect complete polarization of the system) are considerably less trivial, particularly as regards the renormalization of the gap equation (9.3), since Δ_k is now proportional to $|\underset{\sim}{k}|$ for small $\underset{\sim}{k}$ (and hence the system becomes "gapless" for all directions of $\underset{\sim}{k}$ at the point $\mu = 0$). Also, the generalization of these considerations to finite temperature is far from trivial and still very much an open question. (In the Bose limit we expect that there is a wide range of temperature where "molecules" are formed but are not Bose-condensed, whereas the BCS solution allows no such regime. How do we describe the intermediate case?)

2. Landau Fermi-liquid effects.

An important difference between Landau's theory of a normal Fermi liquid and that of a free Fermi gas is that the former introduces interactions $f(\underset{\sim}{p}\underset{\sim}{p}'\sigma\sigma')$ between the quasiparticles, or equivalently a set of molecular fields generated by various kinds of "polarization" of the system (density, spin density, current etc.). The simplest example is the Weiss-type molecular field induced by a spin polarization $\underset{\sim}{S}$:

$$\underset{\sim}{H}_{mol} = -\left(\frac{dn}{d\varepsilon}\right)^{-1} F_0^a \underset{\sim}{S} \tag{9.4}$$

where F_0^a is a dimensionless phenomenological parameter of the Landau theory.* We have alrady seen (eqns. (8.1)–(8.3)) that the molecular field generates both a direct and an indirect effective spin-dependent interaction between the quasiparticles (cf. also next section). Apart from this, the Landau terms have no effect on the Cooper-pairing process (since the latter, in the BCS limit $\Delta \ll \varepsilon_F$, involves no polarization which could generate a molecular field); however, they do affect the response to external fields. If for definiteness we consider the spin susceptibility and assume isotropic singlet pairing, then we can take the molecular field into account by writing

---

* Related to the ζ_o of eqn. (8.1) by $F_0^a \equiv \zeta_o(dn/d\varepsilon)$.

$$\underset{\sim}{S} = \chi_o(T)\, \underset{\sim}{H}_{total} \equiv \chi_o(T)(\underset{\sim}{H}_{ext} + \underset{\sim}{H}_{mol}) \tag{9.5}$$

where $\chi_o(T)$ is the quantity calculated without account of molecular-field effects; this is calculated as indicated in Lecture 6 with the result (6.22), i.e., since $\chi_n^{(o)} = dn/d\varepsilon$,

$$\chi_o(T) = (dn/d\varepsilon)\, Y(T) \tag{9.6}$$

when Y(T) is the Yosida function (6.22). Combining eqns. (9.3)–(9.6), we obtain for the true susceptibility $\chi \equiv d\underset{\sim}{S}/d\underset{\sim}{H}_{ext}$ the result

$$\chi(T) = \frac{(dn/d\varepsilon)Y(T)}{1+F_o^a Y(T)} \tag{9.7}$$

so that already χ is no longer a universal function of T/T_c. The effects of other "molecular fields" can be treated in the same way.

3. Interaction-modification effects.

In the BCS theory of superconductivity in metals, the effective attraction between electrons is due to virtual polarization of the ionic lattice, an effect whose dynamics is almost totally insensitive to the onset of superconductivity in the electron system. As a result, the effective interaction $V_{eff}(\underset{\sim}{r}_1-\underset{\sim}{r}_2)$ is simply a fixed function which is unchanged in the superconducting state from its normal-state value. By contrast, in liquid ^{3}He, and perhaps in heavy-fermion and/or high-T_c superconductors, an important part of the effective interaction between quasiparticles arises from virtual polarization of the liquid itself. For example, we saw in Lecture 8 that the Weiss-type molecular field associated with spin polarization generates an effective potential given schematically by

$$V_{eff}(\underset{\sim}{r}-\underset{\sim}{r}',t-t') = -\zeta_o^2\, \chi(r'-r,t'-t)\ \underset{\sim}{\sigma}\bullet\underset{\sim}{\sigma}'. \tag{9.8}$$

For qualitative purposes (only!) we may replace $\chi(r'-r,t'-t)$ by $\chi\delta(t-t')$, where χ is the static spin susceptibility of the system. Now the first interesting point is that χ is itself a function of the state of the system and hence may change in the superconducting state. Consider for example the case of spin singlet pairing, for which the term (9.8) gives a repulsive contribution to the effective pairing interaction. Because χ drops off with decreasing T in the superconducting state, this repulsion is weakened and the net effect is therefore that the effective attraction <u>increases</u> with decreasing T. Consequently, we should expect the ratio $\Delta(0)/kT_c$ to be <u>larger</u> than its BCS value (1.76).

An even more interesting effect occurs for spin triplet pairing. In the general case χ is now a <u>tensor</u> quantity χ_{ij}, and eqn. (9.8) must be generalized to

$$V_{eff}(\underset{\sim}{r}'-\underset{\sim}{r},t'-t) = -\zeta_0^2\, \chi_{ij}(r'-r,t'-t)\ \underset{\sim}{\sigma}_i \underset{\sim}{\sigma}'_j \sim -\chi_{ij}\underset{\sim}{\sigma}_i\underset{\sim}{\sigma}'_j. \tag{9.9}$$

Thus, the interaction is not only temperature-dependent but depends on the actual form of the gap matrix $\hat{\Delta}_k$ (eqn. (8.22)) since the latter determines χ_{ij}. It is believed to be mainly this effect which in ^3He stabilizes the A("ABM") phase at high pressures and temperatures relative to the B ("BW") phase, which in the BCS theory would always have a lower free energy.

<u>4. "Strong-coupling" effects.</u>

So far we have retained from the original BCS theory not only the form (8.5) of the pair function ϕ but also the implicit assumption that $V_{kk'}$ is effectively constant over an energy range >> Δ.* While the phrase "strong-coupling effects" does not seem to have a universally accepted conventional usage, it is often used in the traditional superconductivity literature to refer to cases such as Pb and Hg when the characteristic energy scale of the interaction (~ the phonon Debye energy) is not so large compared to Δ that the effects of corrections to this approximation are negligible.

In traditional superconductivity, the usual method of dealing with such effects is to write down and solve, usually numerically, the so-called Eliashberg equations for the normal and anomalous electron propagators. Because of the special feature of the electron-phonon interaction that the ratio of phonon to electron characteristic velocities is very small and the related theorem of Migdal, it turns out that the Eliashberg equations furnish a very accurate description of the superconducting state even in the presence of "strong-coupling" effects in the above sense, and the comparison of theory and experiment as regards (e.g.) tunnelling characteristics is extremely impressive. Unfortunately in the present context, the equations kill two birds with one stone: they deal simultaneously with the effects of the energy-dependence of the interaction (i.e. "strong-coupling" effects as defined above) and with the effects of emission of real phonons ("pair-breaking" effects, cf.

---

* In case (1) this is implicitly assured by the condition $|a_s| >> r_o$.

below). While in the case of the electron-phonon interaction these two effects are inextrically linked, it is not obvious that they must be so in the more general case, as in high-T_c materials, and consequently it would be desirable to study the effects of strong coupling separately from these of pair-breaking, which is somewhat difficult (though not impossible) in the Eliashberg formalism because of the necessity of solving the equations numerically. In addition, it is improbable that we have an analog of Migdal's theorem in the more general case, so that the conceptual basis of the extension of the results in Elishberg treatment to such a case is dubious. It is however perhaps worth noting that the result of such a treatment seems, for any reasonable form of electron-phonon interaction, to be to increase both the ratio $\Delta(0)/kT_c$ and the specific heat jump at T_c relative to the normal-state specific heat.

So far, we have always assumed eqn. (8.5), which is equivalent to the assumption that pairing takes place between states of opposite momentum $\underset{\sim}{k},-\underset{\sim}{k}$. For a translationally invariant system this is a natural assumption, since such states are eigenstates not only of the kinetic energy but of the Hartree-Fock Hamiltonian. However, once we relax the assumption of translation invariance (e.g. allow for a nontrivial one-particle potential $U(\underset{\sim}{r})$ the situation is more complicated. Suppose we choose a complete set of orthonormal one-particle eigenfunctions $\phi_n(\underset{\sim}{r}\sigma)$, which in general need not be eigenfunctions either of the kinetic energy or of the HF Hamiltonian. Then a possible generalization of (8.5) is

$$\Psi = \prod_n (u_n + v_n a_n^+ a_{\bar{n}}^+)|\text{vac}\rangle \tag{9.10}$$

where the $\bar{n}$ are a permutation of the n and the mapping $n \rightarrow \bar{n}$ is one-to-one. The generalized expressions for the one-particle and potential energies are*

$$\langle K - \mu N + U\rangle = \frac{1}{2}\sum_n (\bar{\varepsilon}_n + \bar{\varepsilon}_{\bar{n}})|v_n|^2 \tag{9.11}$$

where $\bar{\varepsilon}_n$ is the expectation value of the one-particle energy $-(\hbar^2/2m)\nabla^2 - \mu + U(\underset{\sim}{r}\sigma)$ in state ϕ_n, and

---

* We do not explicitly consider Hartree-Fock terms here, but they can be added to (9.10) without affecting the conclusions.

$$\langle V\rangle = \sum_{nn'} u_n^* v_n u_{n'} v_{n'}^* \int V(\underset{\sim}{r}-\underset{\sim}{r}')\, \phi_{\bar{n}}^*(r)\phi_n^*(\underset{\sim}{r}')\phi_{n'}(\underset{\sim}{r})\phi_{\bar{n}'}(\underset{\sim}{r}')\, dr dr'. \qquad (9.12)$$

For the purposes of the present discussion it is convenient to specialize to the case $V(\underset{\sim}{r}-\underset{\sim}{r}') \sim -V_o\delta(\underset{\sim}{r}-\underset{\sim}{r}')$ (similar considerations apply in the more general case). Then $\langle V\rangle$ takes the simple form

$$\langle V\rangle = -V_o \int d\underset{\sim}{r}\, |F(\underset{\sim}{r})|^2, \qquad F(\underset{\sim}{r})\,(\equiv F(\underset{\sim}{r},\underset{\sim}{r})) \equiv \sum_n u_n^* v_n \phi_n^*(\underset{\sim}{r})\,\phi_{\bar{n}}^*(r). \qquad (9.13)$$

It is clear that the potential energy is minimized (i.e. made most negative) when, crudely speaking, the ϕ_n are chosen so as to make $|F(\underset{\sim}{r})|$ a maximum, i.e. so that the contributions from the various terms $\phi_n^*(r)\phi_{\bar{n}}^*(r)$ contribute in phase. On the other hand, such a choice may not necessarily be the best from the point of view of the kinetic energy (9.11), (which fairly clearly is optimized by choosing the ϕ_n to be eigenstates of the 1-particle energy) and we must then work out the optimum trade-off. We now consider some specific cases.

(5) "Non-pair-breaking" perturbations. (interactions)

In certain cases we may be able to have our cake and eat it, that is, choose the ϕ_n to be one-particle (or Hartree-Fock) eigenstates and keep their contributions all in phase in (9.13). This is the case in particular when the perturbation is a one-particle potential $U(r\sigma)$ which preserves time-reversal invariance; the simplest case of this, though not the only one, is when $U(\underset{\sim}{r}\sigma) \equiv U(\underset{\sim}{r})$ is the spin-independent potential due to nonmagnetic impurities (or walls, etc.). In that case the complex conjugate $\chi_n^*(\underset{\sim}{r})$ of any energetic eigenstate $\chi_n(r)$ is itself an eigenstate with the same energy, so we can choose $\phi_{\bar{n}}(\underset{\sim}{r}) \equiv \phi_n^*(\underset{\sim}{r})$, i.e. pair in time-reversed states. It is clear that the different contributions in (9.13) now automatically stay in phase, so that this choice optimizes both $\langle K-\mu N+U\rangle$ and $\langle V\rangle$. In fact, we can now repeat almost word for word the discussion of Lectures 5 and 6, substituting n for the wave vector $\underset{\sim}{k}$, and the resulting expressions for $\Delta(T)$, the condensation energy etc., in terms of V_o and $N(0)$ are identical to those obtained there. Similar considerations apply to a good approximation even when $V(\underset{\sim}{r}-\underset{\sim}{r}')$ is not a δ-function, provided that its range is small compared to the Cooper-pair radius ξ_o. Thus the only way* in which the thermodynamics of a "dirty" superconductor with s-wave pairing† differs from that of the corresponding pure case ($U(\underset{\sim}{r}) = 0$) is through the effect of the

---

* There is actually a more subtle (but generally small) effect in the case where the Fermi surface of the pure metal is strongly anisotropic.

† The effect of dirt on pairing with nonzero angular momentum is radically different.

impurities on the density of states N(0) – not usually a very substantial effect. In particular T_c is not usually substantially affected by nonmagnetic inpurities. (The electrodynamics, however, is quite radically affected by "dirt".)

A particularly spectacular case of the above considerations is what is sometimes called a "localized superconductor", that is, a system in which the disorder is so strong that all the relevant one-particle states are localized, but which undergoes a Cooper-pairing transition. Contrary to naive intuition, although the normal (i.e. unpaired) state is insulating, the Cooper-paired state is not merely a conductor but a superconductor; it shows both the Meissner effect and metastability of supercurrents.

6. Static pair-breaking.

An interaction is called "pair-breaking" if it is such that it is impossible to choose the single-particle eigenfunctions $\phi_n(\underset{\sim}{r})$ so that they are eigenfunctions of the one-particle (or Hartree-Fock) energy while at the same time preserving the constructive interference in (9.13). We will see that interactions of the particles with another dynamical system (e.g. the phonons) is automatically pair-breaking, at least to the extent that real collision processes are involved, while interactions with c-number agencies ("static" perturbations) are pair-breaking (for s-wave pairing) only to the extent that they violate time reversal invariance. The best-known example of a static pair-breaking interaction is the effect of magnetic impurities.

In the case of static pair-breaking, two obvious suggestions are (a) to pair in eigenstates of the one-particle energy, thereby reducing F(r) and hence $|\langle V\rangle|$, or (b) to pair in time-reversed eigenstates, thereby increasing the one-particle energy since the ϕ_n are then in general not eigenstates of the latter. It turns out on detailed calculation that in the first case the value of $|\langle V\rangle|$ is exponentially suppressed, so that this is rarely advantageous. In case (b), we may estimate the increase of the one-particle energy crudely by the following argument: The one-particle states related by time-reversal (or one of each pair, at least) can be conceived as wave packets of the energy eigenstates with a "width" of order $\Gamma = \hbar/\tau_{p.b.}$ where $\tau_{p.b.}$ is the "pair-breaking time", i.e. v_F^{-1} times the mean free path against time-reversal-violating scattering processes. To pair time-reversed states we first have to reorganize the normal ground state so that it is these states rather than the one-particle eigenstates which form a T = 0 Fermi sea, and since the number of states involved is of order $(dn/d\varepsilon)\Gamma$, the energy cost of doing this is of order $(dn/d\varepsilon)\Gamma^2$. Now, it is a standard result of the original BCS theory that the condensation energy of the "pure" (U=0) superfluid ground state (i.e. its energy relative to the normal ground state, i.e. the free-particle Fermi sea) is of order $(dn/d\varepsilon)|\Delta(0)|^2$. Hence we expect that it becomes energetically disadvantageous to form a superfluid state (i.e. $T_c \to 0$) as soon as $\hbar/\tau_{p.b.} \gtrsim \Delta(0)$, and a more quantitative calculation confirms this prediction. As is well known, this system has many interesting features, including a "gapless" region for Γ just below its critical value, in which the minimum quasiparticle excitation energy is zero even though the quantity Δ is still finite and the system still superfluid (thus violating the Landau condition!).

7. Dynamic pair-breaking.

By "dynamic pair-breaking" I mean pair-breaking due to an interaction with another quantum-mechanical system such as the phonons. It is clear that in this case the true energy eigenstates of the "universe" involve correlated states of the electrons and phonons, i.e. they are of the general form $\Sigma_n c_n \phi_n(\underset{\sim}{r}) \chi_n(\underset{\sim}{R})$ where the $\chi_n(\underset{\sim}{R})$ (which do not necessarily form an orthogonal set) are phonon wave functions. If we take these correlated functions as the basis for pairing, then because the χ_n associated with different ϕ_n are at least partly mutually orthogonal, we will get a "Franck-Condon" effect which drastically reduces $F(\underset{\sim}{r})$ (now defined as $\langle \psi_\uparrow^\dagger(\underset{\sim}{r}) \psi_\downarrow(\underset{\sim}{r}) \rangle$ averaged over the phonon variables) and hence $|\langle V \rangle|$.* So, similarly to case (6), we should expect it to be advantageous to use the single-electron ("phonon-undressed") states as the basis for pairing; as above, this costs us an energy $\sim (dn/d\varepsilon)\Gamma^2$. Unlike the case of static pair-breaking, however, Γ is now a quite sensitive function of energy (it usually varies as $(\varepsilon - \varepsilon_F)^3$ or T^3); on dimensional grounds we should expect that superconductivity is destroyed when $\Gamma(T_c) \gtrsim \Delta(0)$, a condition which is nowhere near met in the old-fashioned superconductors (even in Hg or Pb), but may be encountered in the high-temperature ones, see next lecture. Unfortunately, most of our current quantitative understanding of dynamic pair-breaking is in the context of the Eliashberg solution for the specific electron-phonon system, where as pointed out above it gets mixed up with "strong-coupling" effects and from where it may not be transferable to the general ("non-Migdal") case.

* This simple argument is actually not totally convincing as it stands, and may need some amplification and/or modification.

8. Fulde-Ferrell state.

I finally want to mention briefly an unusual case in which, although the static interaction violates time-reversal invariance, it is still possible to pair in such a way as to preserve a large F(r) without a catastrophic cost in one-particle energy. This is the case of a translation-invariant system which forms pairs in a singlet s-wave state, in the presence of Zeeman coupling (but no "Ampère" ($\underset{\sim}{j} \bullet \underset{\sim}{A}$) coupling) to a static uniform magnetic field, H. A physical example might be a dilute solution of ^{3}He in ^{4}He, if T_c is high enough for this system to be experimentally accessible. The special feature which distinguishes this case is that the original H = 0 eigenstates remain eigenstates, but have their energies systematically shifted:

$$\xi_k \rightarrow \xi_k - \mu_B \underset{\sim}{\sigma} \bullet \underset{\sim}{H}.$$

Thus, the originally paired states ($\underset{\sim}{k}\uparrow, -\underset{\sim}{k}\downarrow$) are no longer degenerate in energy. We then apparently have two choices: (a) pair as in the original $\underset{\sim}{H} = 0$ state and sacrifice the magnetic polarization energy (BCS solution) or (b) adjust the Fermi energies to be equal for the two spin populations, so that the Fermi momenta are not, whereupon at first sight it is impossible to get a large F(r) and one must give up the idea of pairing (normal ground state). Clearly for $\mu H >> \Delta(0)$ solution (b) is preferable.

However, there is a way of starting from the normal (polarized) ground state and still getting a reasonable $F(\underset{\sim}{r})$. In their classic paper Fulde and Ferrell explored this question using the standard periodic boundary conditions and hence plane-wave eigenstates; however, the physics may be clearer if we work with a spherical sample and impose realistic (hard-wall) boundary conditions. Then the one-particle eigenstates are characterized by a radial quantum number n and angular quantum numbers ℓ and m, and the BCS ground state would, as is well known in a nuclear physics context, correspond to pairing of n, ℓ, m with n, ℓ, –m, (its time reverse). However, we can also consider the possibility of pairing n, ℓ, m with n + Δn, ℓ, –m where the energy corresponding to Δn is 2μH. This gives an oscillating form of the function F(r)

$$F(r) \sim A \cos[(2\mu H/v_F)r] \tag{9.14}$$

where r is the radial coordinate. Provided the range of V(r–r') is small compared to the "wavelength" $v_F/2\mu H$ (which may itself be small compared to

$\xi_0 \sim \hbar v_F/\Delta_0 \gtrsim \hbar v_F/\mu H$), this gives us the same potential energy as a BCS state with $F(r) = A$ except for a factor of 1/2 ($\equiv \cos^2\theta$). This results in a factor of 2 in the exponent of $\Delta(0)$ and T_c, so with this ansatz the condensation energy, though finite, is very much smaller than that of the BCS state. If we had used "periodic" boundary conditions in the sphere, so that one-particle states which vary as exp inr were available, the cosine in (9.14) could be replaced by a complex exponential, the factor of 1/2 would disappear and we would recover the results (for T_c and $\Delta(0)$) of Fulde and Ferrell. It is still not clear to me whether this crucial factor of 1/2 represents a real effect (i.e. it is essential, in any geometry, to apply hard-wall rather than periodic boundary conditions) or whether there is some fairly trivial way of reformulating the pairing ansatz, even with hard-wall boundary conditions, so as to remove it. It is possible that if ^3He–^4He mixtures indeed have an accessible T_c, we will some day obtain an experimental answer to this question.

Lecture 10: High-temperature superconductivity

There have by now been upwards of 20,000 papers on this subject, and it is clearly impossible to give any kind of comprehensive discussion in a one-hour lecture. I will therefore just try to remind you of the basic experimental facts and comment on a few of the obvious theoretical questions which these systems raise. For reference, the main classes of recently discovered materials which have transition temperatures higher than, or close to, the highest known before 1986 (~ 25K) are:

Formula	Popular name(s)	T_c(K) (approx.)*	Substitutions
$La_{2-x}Sr_xCuO_{4-\delta}$ ($0.05 \lesssim x \lesssim 0.32$)	LSCO, 214	40	Sr → Ba,Ca,Na
$YBa_2Cu_3O_{7-\delta}$ $\delta \lesssim 0.6$)	YBCO, 123	90	Y → any rare earth except Pr and possibly Ce,Tb,Pm
$Bi_2Sr_2Cu_{n-1}Cu_nO_{2n+4+\delta}$	BSCCO, 22,n-1,n	110	
$Tl_mBa_2Ca_{n-1}Cu_nO_{m+2n+2+\delta}$	TBCCO	125	Ba → Sr
$Ba_{1-x}K_xBiO_{3-\delta}$ ($0 < x \lesssim 0.5$)	BKB	30	
$Nd_{2-x}Ce_xCuO_4$ ($0.15 < x < 0.17$)	NCCO	25	Nd → most rare earths

Of these groups, there is by now a rather general belief that BKB and possibly NCCO are not "true" high-temperature superconductors, but are simply materials of the same general nature as the "old-fashioned" BCS superconductors which just happen, by a historical accident, to have been discovered since 1987 and to have a higher T_c than the pre-1987 materials. Conversely, with the advantage of hindsight we can see that among the pre-1987 materials there are a few which, while having T_c less than 25K, have much in common with the high-T_c materials, e.g. $BaPb_{1-x}Bi_xO_3$(T_c = 13K). Among such "low-temperature high-temperature superconductors" a particularly striking (recently discovered) member is "2201" (i.e. $Bi_2Sr_2CuO_{6+\delta}$), which has a transition temperature of 7K but in all other respects seems to behave exactly like a typical high-T_c material.

If we confine ourselves to the three main classes (LSCO, YBCO and the BSCCO and TBCCO groups) we find that the high-temperature superconductors have a number of features in common: All are <u>oxides</u>, and all contain

---

* T_c is usually a strong function of the stoichiometry (i.e. of x,δ): the figure quoted is the value at "optimum" stoichiometry, i.e. the maximum value attained.

CuO_2 planes* (or pairs or triplets thereof) which are separated by fairly large distances (10–15 Å); as a consequence, all have strongly anisotropic properties (the c-axis properties being different from the ab-plane one); all have electron densities (both overall and "in the planes") which are small compared to those of traditional superconductors; all have a very small Pippard coherence length (Cooper pair radius) ξ_0 and are therefore extreme type-II: and the LSCO and YBCO groups, at least, occur at stoichiometries close to an antiferromagnetic transition. Finally, all have highly unusual properties in the normal state (see below).

In trying to obtain a clear picture of the experimental properties of the high-temperature superconductors (hereafter HTS) one encounters a number of difficulties which are not usually met with the old-fashioned superconductors. First, because of the complicated stoichiometries it is by no means trivial to prepare single-phase materials, still less to obtain pure single crystals (this is particularly difficult with the LSCO group). Secondly, the O concentration, which has a dramatic effect on many of the properties, seems to be highly volatile and to depend sensitively on the details of the preparation process. Thirdly, even when one otherwise has a pure single crystal, the fact that the structures are generally orthohombic rather than tetragonal means that crystals prepared without special care are likely to be "twinned". This effect is particularly important in the YBCO group, where there is actually a rather strong anisotropy of the crystal as a whole because of the so-called "chains" of Cu and O atoms which lie off the CuO_2 planes and parallel to the b-axis: it is only quite recently that careful measurements on untwinned crystals have revealed that, contrary to expectation, these materials are quite strongly anisotropic even in the a-b plane itself.

Despite these difficulties, there are certain general qualitative conclusions and can draw, in particular from experiments involving substitution for the various elements involved. It seems clear that most of the "action", including that responsible for superconductivity, takes place in the CuO_2 planes and is sensitive primarily to the number of electrons which are available in these planes. The role of the off-plane atoms, be they Cu, O or a rare earth etc., is primarily to donate extra electrons or holes to the planes; for example, the "parent" compound of LSCO, La_2CuO_4, has one electron per primitive unit cell

* Conventionally, the orthohombic axes lying in the CuO_2 plane are known as a and b and the axis perpendicular to the planes as the c-axis.

and is at low temperatures a classical Mott antiferromagnetic insulator; doping with Sr, which lacks La's valence electron, creates holes which are believed to migrate to the CuO_2 planes. The irrelevance of the specific nature of the off-plane donor atoms to many of the principal properties of the HTS, in particular to superconductivity, is dramatically illustrated by the fact that when Gd is substituted for Y in YBCO, the Gd atoms undergo a magnetic ordering transition (at ~ 2K) and yet the transition temperature to the superconducting state is actually enhanced somewhat over that of YBCO! (By contrast, in traditional superconductors the introduction of magnetic ions is very deleterious to superconductivity (see Lecture 9, part (6)) and indeed until quite recently it was believed that superconductivity could not co-exist with magnetic ordering.)

Since it seems that it is the CuO_2 planes which play the principal role in (most of) the characteristic properties of the HTS, it is interesting to ask what we know a priori about the electronic states available there. Let us start, for concreteness, with the "parent" compound La_2CuO_4. The isolated Cu atom has 10 3d electrons and one 4s one, while La has one 5d and two 6s electrons; O has, of course, two holes in the p-shell. The general belief is that each La atom denotes 3 electrons and each Cu 2; those go to fill the 8 holes in the O p-shells, leaving the Cu atoms in the Cu^{2+} state, i.e. with 9 electrons (one hole) in the 3d shell (but see below). The material then has one hole per primitive unit cell (or, if we prefer, per Cu atom) and is observed, experimentally, to be an antiferromagnetic insulator. The five-fold degenerate d-level of the Cu atom is split by the crystal field in tetragonal symmetry into a 2-fold degenerate level (d_{xz}, d_{yz}) and three nondegenerate levels $d_{xy}, d_{3z^2-r^2}$ and $d_{x^2-y^2}$, where the notation is the standard one. Since the ordering of the levels given is believed to be that of increasing energy, the single unpaired electron (or hole) should reside in the $d_{x^2-y^2}$ level. However, we should expect that this level hybridizes quite strongly with the oxygen p_x and p_y levels. When we dope with Sr so as to provide extra holes in the CuO_2 planes, these will go into the appropriately hybridized states. It is then a fundamental question (see below) whether the resulting situation is better described by a traditional band picture, with 1 + x holes (or 1–x electrons) per unit cell, or rather by regarding the extra x holes as themselves occupying a "band".

We turn now to the experimental properties of the HTS. Here we find a strange paradox: the behavior of the normal state is highly anomalous, while that of the superconducting state is almost boringly "normal" (i.e. similar to traditional BCS superconductors: though see below). We discuss first the normal-state properties.

Perhaps the most striking single feature of the normal state of all the HTS* is the normal-state resistivity in the a-b plane, $\rho_{ab}(T)$. In sufficiently pure samples this seems to be very close to linear in temperature all the way from T_c (which, recall, is as low as 7K for 2201) up to the highest temperatures at which it is measured (600–1000K). The slope $B \equiv d\rho_{ab}/dT$ is ~ 0.4–0.5 μΩ cm/K for 123, 0.45 μΩ cm/K for 2212, 0.7 μΩ cm/K for 2201 and is estimated at 0.7–1 μΩ cm/K for 214. (In the case of both 123 and 2212 there is an in-plane anisotropy of about a factor of 2; the figures given above are averages.) Two striking features emerge. First, the resistivity at room temperature is about a factor of 100 greater than that of a typical metal such as Cu. Secondly, $\rho_{ab}(T)$ appears to be universal for the HTS up to a factor of about 2. It is at first sight tempting to think that this implies that the resistance $R_\square$ of a (square of) CuO_2 plane is universal, i.e. independent of the material in which the plane sits, but this would imply that for different materials ρ_{ab} is proportional to the (average) distance d between planes, which does not seem to be true:† in fact, if anything, the product $\rho_{ab}d^{-1}$ is less universal than ρ_{ab} itself.

Suppose we try to interpret ρ_{ab} by a simple Drude type of theory, so that $\sigma_{ab}(0) \equiv \rho_{ab}^{-1}$ is given by the expression

$$\sigma_{ab}(0) = n_{eff}e^2\tau/m^* \qquad (\equiv \varepsilon_0\omega_p^2\tau) \tag{10.1}$$

where n_{eff} is the number of mobile electrons per unit volume and m^* their effective mass. We can estimate n_{eff} in various ways; (a) by stoichiometric arguments, (b) from the observed value of the plasma frequency ω_p, (c) from the London penetration depth in the superconducting state. Although the absolute value of n_{eff} is obtained from these different measurements varies by a factor ~ 3, there is no evidence that ω_p is appreciably a function of T, and hence we find $\tau \propto T^{-1}$, or more quantitatively

$$\hbar/\tau = \text{const.}\ k_BT \tag{10.2}$$

* With the exception of so-called "60K 123" ($\delta \cong 0.4$) which shows a more complicated behavior, possibly because of oxygen ordering effects.

† Nor does it help to treat the double or triple planes in YBCO, BSSCO etc. as single ones.

where the constant is of order 1. We note that a similar behavior is found for ordinary metals at $T >> \theta_D$ (the Debye temperature), and in that case the constant is the dimensionless electron-phonon coupling constant λ, which is indeed of order 1. It is sometimes argued that since a linear form of $\rho(T)$ is observed to hold for some ordinary metals right down to $T \sim \theta_D/4$, it is possible to attribute the linear behavior in the HTS also to an electron-phonon mechanism (the anomalously high value of ρ_{ab} being, on this view, due to the paucity of carriers n_{eff}); however, it seems very difficult to maintain this view for 2201, where the range over which $\rho_{ab}(T)$ is linear extends down to $T \sim 0.02\ \theta_D$.

The c-axis resistivity $\rho_c(T)$ is much larger than ρ_{ab}, at room temperature by a factor $\sim 10^2$ for YBCO and 10^5 for BSSCO, and is therefore difficult to measure reliably. The majority belief is that ρ_c is approximately proportional to T^{-1}, hence inversely proportional to ρ_{ab} as a function of temperature, but this is controversial. In any case it is clear that the c-axis conductivity at all relevant temperatures is much smaller than the so-called "minimum metallic conductivity" ($\sim e^2/hd$) of Mott, and therefore it is usually believed that transport in the c-direction cannot be of the standard metallic (Bloch-wave) type.

After the ab-plane resistivity, the next most striking feature of the HTS is probably the Hall effect. In the very simplest (Drude) models of metallic conduction, the Hall coefficient R_H is inversely proportional to the number of carries: in fact we can define a "Hall number" n_H as $1/eR_H$. In the case of the HTS we find (for current and voltage both in the ab-plane, $\underset{\sim}{H}$ along the c-axis)

$$n_H(T) = A + BT \tag{10.3}$$

with the coefficient A nearly zero for YBCO (though finite for LSCO) and the coefficient B strongly correlated (as a function of doping) with T_c. On the other hand, as we have seen, the quantity $n_{eff}(T)$ as determined from the plasma frequency ω_p is practically constant as a function of temperature. The apparent discrepancy is actually not too surprising, since even in a simple metal n_H is equal to n_{eff} only under very stringent conditions. However, the qualitative difference as a function of temperature may suggest that something very anomalous is going on. The sign of n_H would correspond, in a Bloch-Wilson (band) picture, to hole rather than electron transport in the ab plane (in the c-direction, on the other hand, the sign is the opposite).

Many other properties of the normal phase have been measured (thermopower, infrared absorption, Raman scattering, ARPES, positron emission, specific heat, static magnetic susceptibility, NMR and NQR, tunnelling...) and in many cases are qualitatively different from those of an ordinary metal. Particularly noteworthy is that in tunnelling experiments the current I is proportional to V^2 rather than V, a highly unusual behavior (though not totally unknown).

A rather ubiquitous feature of the HTS is that as we "overdope", i.e. increase the hole concentration x past the point corresponding to the maximum in $T_c(x)$, the normal-state behavior becomes qualitatively closer to that of an ordinary metal: whereas underdoping makes it, if anything, even more anomalous. This puzzling feature is not currently well understood.

Turning now more briefly to the superconducting state, we note that the only major qualitative difference with ordinary superconductors seen so far lies in the electromagnetic properties, in particular in finite magnetic field. This is qualitatively understood as a consequence of the very short coherence length ξ_0 and consequent extreme type-II behavior, which among other things means that fluctuation affects which would be almost totally negligible in ordinary superconductors can play an important and even dominant role. As an example, the width of the transition to "zero" resistance, which for a traditional one-phase pure superconductor is usually unobservably small, can be large as $\sim T_c/3$ for BSCCO in a field of a few tesla. This is currently a very active area of research. Apart from this, the behavior of the HTS in the superconducting state appears to be qualitatively rather similar to that of the traditional BCS type, the main differences being that (a) certain experiments (Raman, specific heat, tunnelling...) suggest that there may be a finite electronic density of states even for energies small compared to kT_c (i.e., presumably, to $\Delta(0)$), (b) experiments which purport to measure the energy gap $\Delta(0)$ in the $T \rightarrow 0$ limit give values for $\Delta(0)/kT_c$ ranging from the BCS value (1.76) up to ~ 4, (c) many experiments near T_c suggest a puzzling absence of the "coherence" effects which are important in BCS superconductors (e.g. there is no "Hebel-Slichter peak" in the nuclear relaxation rate just below T_c).

We now consider some obvious theoretical questions, starting with the normal state. A first obvious question is: In thinking about the ab-plane properties, is it adequate to think of the different CuO_2 planes as essentially independent, i.e. as forming effectively a set of two-dimensional (2D) systems whose properties then simply add to give the "bulk" properties, or should we think of the material or rather a 3D system which happens to have a strong built-in anisotropy? There are strong arguments in favor of the former point of view: in particular, it appears to be almost impossible to interpret the c-axis properties in terms of coherent Bloch transport. Rather, one should almost certainly think of it as incoherent "hopping" transport strongly modulated by the in-plane fluctuations, and this would strongly suggest that the 3D aspects will be a small perturbation on essentially 2D behavior.

If this is so (and indeed, to some extent, if it is not) the next question is: Is the normal state a Fermi liquid in the sense of Landau, or something more exotic? Although the appropriate definition of a "Fermi-liquid" is not universally agreed, most people would probably call a system of fermions at $T = 0$ a "Fermi liquid" if and only if it satisfies the following three conditions:

(1) The charge and spin excitations are not separated,

(2) There exists a well-defined Fermi surface, i.e. a locus in $\underset{\sim}{k}$-space where the expectation value of particle number $n(\underset{\sim}{k})$ undergoes a finite jump,

(3) The inverse single-quasiparticle lifetime $\Gamma(\omega)$ tends to zero faster than ω as $\omega \equiv \varepsilon - \varepsilon_F$ tends to zero, where ε_F is by definition the quasiparticle energy at the Fermi surface.

Of course, to apply this definition of a Fermi liquid as it stands to the HTS one has to imagine one could somehow "shut off" the superconductivity, so as take the normal state right down to zero temperature; however, it is possible to generalize the conditions (1)–(3) so as to apply them at finite T. The question of whether any interacting 2D system can be a Fermi liquid, and in particular whether the electrons in the CuO_2 planes of the HTS are, is currently the subject of intense debate. It is interesting to note, in this context, that in 1D it can be shown that no Fermi system with finite interactions, whether repulsive or attractive, can form a Fermi liquid at $T = 0$, whereas in 3D there are believed to be many examples, such as the many metals which are believed to be normal in the limit $T \to 0$. The crucial question is: Is 2D more like 1D, like 3D or qualitatively different from both?

A possible alternative hypothesis to the standard Landau picture is what is called a "marginal Fermi liquid". This is a Fermi system in which the inverse lifetime $\Gamma(\omega)$ is precisely proportional to ω as $\omega \to 0$. It may be shown that in such a system the difference between the occupation numbers for energies $\varepsilon_F \pm \Delta\omega$ is proportional to $(\ln \Delta\omega)^{-1}$, so that condition (3) as well as condition (2) above is (marginally) violated. One reason (not the only one) why such a picture is thought by some to be relevant to the HTS is that if we calculate the band structure of noninteracting electrons in a square lattice of the CuO_2 type in the nearest-neighbor tight-binding approximation (NNTB), it turns out that for "half filling" (one electron per unit cell) the Fermi surface (line) is exactly a square. The peculiar geometry then implies that when interactions are introduced and the resulting lifetimes calculated by perturbation theory in the usual way, $\Gamma(\omega)$ indeed satisfies condition (2) (Essentially, this is because the relation between the energy and momentum δ-functions for real collision processes is different from the usual one, which gives $\Gamma(\omega) \sim \omega^2$ in any number of dimensions.). While in reality there will certainly be corrections to the NNTB approximation, this makes it plausible that the marginal Fermi-liquid picture might be at least qualitatively correct.

If one believes that the HTS in the normal phase are standard Landau Fermi liquids, the next question is the origin of the linear temperature-dependence of the ab-plane resistivity. In a Drude-type theory one may show that any boson collective excitation with a characteristic ("Debye") energy ε_D will, by providing a collision mechanism for the electrons, give a resistivity proportional to T for $kT >> \varepsilon_D$, just as in the well-known phonon case; moreover, there are at least "hand-waving" arguments to indicate that the constant which relates $\hbar/\tau$ to kT should indeed be of order 1 as observed. Various candidates for such a boson have been suggested (antiferromagnetic spin fluctuations, fluctuations associated with the proximity of a localization transition, etc.).

Even if one believes that the HTS in their normal phase are described by some version of standard Landau Fermi-liquid theory, one still has to consider two alternative versions: (a) The excitations which form the Fermi sea are the electrons themselves, so that the volume of the sea is proportional to $1-x$. One would then expect a Fermi surface approximating that described above for the NNTB approximation. (b) Rather, one should take as one's starting point the state which occurs at $x = 0$, which we know to be an antiferromagnetic Mott insulator. One then considers the nature of the excitations which result from the introduction of a single hole, using for instance the so-called "t–J" model. The excitations in question turn out, unsurprisingly, to obey Fermi statistics, but to have a dispersion relation quite different from the electrons in case (a). For finite x one then expects to obtain a Fermi sea of volume x rather than $1-x$, and which is of a quite different shape from that in case (a), involving in fact four different "pockets". It is surprising but true that the relevant experiments which purport to measure the Fermi surface (ARPES, positron annihilation, de Haas-van Alphen, etc.) or the effective number of carriers (Hall, plasma frequency) have so far failed to discriminate unambiguously between the hypotheses (a) and (b); some experiments appear to support (a), some (b) and a few neither. Of course, a substantial fraction of the theory community believes that something much more exotic is going on, and various more or less sophisticated theories starting from the so-called t–J model or something similar and involving concepts such as slave bosons, flux phases and anyons are currently under intensive investigation.

Before leaving the normal phase, we note that so far there is really no decent quantitative model for the c-axis transport properties; the only conclusion which can be reasonably drawn at present is the negative one that they almost certainly are not due to the coherent Bloch mechanism. My own prejudice is that the "detuning" of the coherent hopping by the large in-plane voltage fluctuations will turn out to play a major role.

We turn now to the superconducting phase. The first question one has to ask is: Is the mechanism of superconductivity essentially of the same nature as that in conventional superconductors, i.e. does it involve the formation of some kind of collective bound state of fermion pairs as described by eqn. (9.1) (Cooper pairing). or is it something much more exotic, such as the so-called "anyon" mechanism which is believed by many people (not, at least at the time of writing, including the present author) to provide a mechanism for superconductivity without the need for Cooper pairing or conventional ODLRO? My personal prejudice is that notwithstanding the differences noted above, the overall pattern of behavior of the HTS in the superconducting states is so close to that of the conventional variety that it would require a substantial coincidence (whose probability is, of course, difficult to estimate a priori) for them to be explainable equally well by an "exotic" model. It may be possible to resolve this question fairly unambiguously by experiment, since "exotic" theories typically predict qualitative features such as P- and T-violation which do not occur in at least the simpler types of pairing model.* At the time of writing the experimental situation is confused, with some groups reporting evidence for P- and T-violation and others reporting none.

If the mechanism is indeed some kind of Cooper pairing, is it essentially 2D or 3D in nature? Here I think there is already a fair degree of consensus: the same arguments that led us to reject Bloch waves as a mechanism for c-axis

---

* They are predicted for some types of "anisotropic superfluid" states such as the A phase of superfluid ^{3}He, cf. Lecture 8.

transport in the normal phase strongly suggest that the formation of Cooper pairs essentially takes place within the individual CuO_2 planes: the characteristically two-dimensional destruction of ODLRO by fluctuations (cf. Lecture 4) is then avoided, and superconductivity also in the c-direction ensured, by a residual Josephson-type coupling between neighboring planes.

If the general idea of Cooper pairing, and hence presumably of an "order parameter" $\Psi_{\alpha\beta}(r_1 r_2) \equiv \langle\psi_\alpha^\dagger(r_1)\psi_\beta^\dagger(r_2)\rangle$ is valid, then at the phenomenological level there should be at least some regime in which the system is well described by a Ginzburg-Landau type of theory, and we can then infer the parameters of this theory from experiment, e.g. on the so-called upper critical field H_{c2}. If now we assume (and it is a big "if"!) that the relation between the phenomenological parameters and microscopic quantities such as the Cooper pair radius ξ_o is the same as in the traditional BCS theory, then we find that this radius in the ab-plane is (for YBCO) about 12 Å; we may suppose that this is at least a reasonable order-of-magnitude estimate of ξ_o. Now let us use our knowledge of n_{eff} to calculate how many electrons lie within this radius. The answer is: only about 10! (By contrast, for a pure conventional superconductor the answer is ~ 10^9–10^{12}.) This means we are quantitatively very far from the usual BCS limit, in fact well into the "intermediate-coupling" regime considered in Lecture 9, subsection 1. One immediate consequence is that we should expect fluctuations to be vastly more important than in conventional superconductors, as indeed seems to be the case experimentally.

If the mechanism is Cooper pairing, is it "conventional" (i.e. with the simplest symmetry compatible with that of the lattice or "unconventional" (i.e. with a more complicated anisotropy, similar to the $\ell \neq 0$ pairing in ^3He)? The evidence here is apparently mutually contradictory; the apparent presence of low-lying electronic excitations (cf. above) would suggest nodes in the gap and hence an unconventional state, while the overall temprature-dependence of most quantities in the superconducting state would appear to be at least compatible with the hypothesis of conventional pairing. Additional evidence for this hypothesis, which unfortunately is for various reasons not conclusive, is the observation of a Josephson effect between the HTS and conventional superconductors.

What of the mechanism of attraction necessary for Cooper pairing? It seems unlikely (though not totally out of the question) that the conventional electron-phonon mechanism can produce transition temperatures as high as

125K. One therefore looks for an alternative collective boson mode, and it is natural to identify this mode with the one which is giving rise to the linear behavior of $\rho_{ab}(T)$ also in the normal phase, (if this is indeed the explanation of the latter). Such a mode might be an antiferromagnetic spin-fluctuation, a "nearly localizing" mode or perhaps something more exotic. Of course, it is possible that the whole assumption that one needs a collective boson is itself wrong.

Let me emphasize one quite general respect in which the HTS challenge theory, which is independent of the correct detailed model of the pairing interaction: We have already seen that the relation between the normal-state quasiparticle lifetime τ and the temperature, $\hbar/\tau \sim k_B T$, persists right down to T_c (in distinction to conventional superconductors, where τ^{-1} falls off much faster than T in a wide region above T_c). Hence, the energy "width" of the quasiparticles, which is almost certainly due primarily to a dynamic pair-breaking effect, is of order kT_c or Δ, to an order of magnitude, $\Delta(0)$. Although we have previous experience (e.g. in superconductors with magnetic impurities) of static pair-breaking effects which are this large, to the best of my knowledge there are no previous cases where dynamic pair-breaking effects produce a value of $\hbar/\tau(T_c)$ comparable to $k_B T_c$ (even for the so-called strong-coupling superconductors Hg and Pb, the ratio is only ~ 0.1). Moreover, there is no reason to believe that there is any analog of Migdal's theorem available in this case. It is a major conceptual challenge to the theory to provide a consistent quantitative treatment of such a situation.

Finally, of course, the $64K question is: Are there yet higher-temperature superconductors waiting to be found, perhaps even with T_c at room temperature or higher? Is the phenomenon of superconductivity indeed perhaps a very universal and ubiquitous one, with the conventional BCS superconductors only a very special case? In thinking about this it is as well to remember that the compounds made and tested since 1986 have very probably never existed previously (at least in bulk!), not just here on Earth but anywhere in the Universe! Here, as in ultralow-temperature physics, we are exploring physics that Nature never has, and it is this, to my mind, that makes condensed-state physics so unique and fascinating a discipline. On this inspirational note I close.

Suggestions for Supplementary Reading

Lectures 1–2. The original classic text on the phenomenology of superfluidity is F. London, Superfluids (2nd ed., New York, Dover, 1961). A more recent unified discussion of these topics is D. R. Tilley and J. Tilley, Superfluidity and Superconductivity, Adam Hilger, Bristol, 1986. The point of view taken here is similar to that of W. F. Vinen, in R. D. Parks (ed.), Superconductivity, Marcel Dekker, New York, 1969 (This useful collection is hereafter referred to simply as "Parks".). Two other relevant and exceptionally clear papers are F. Bloch, Phys. Rev. **B2**, 109 (1970) and P. W. Anderson, Revs. Mod. Phys. **38**, 298 (1966). The question of the London moment is recently discussed by B. Cabrera, Jpn. J. Appl. Phys. **26**, suppl. 3, 1961 (1987) and G. Baym, in *Frontiers and Borderlines in Many-Particle Physics*, ed. R. A. Broglia and J. R. Schrieffer, Italian Physical Society, Bologna, 1988.

Lecture 3. A more extended discussion of the topological arguments used here is given in A. J. Leggett, Physica Fennica **8**, 125 (1983).

Lecture 4. For a general discussion of NCRI-like phenomena in mesoscopic systems, see e.g. Y. Imry in Directions in Theoretical Physics, ed. G. Grinstein and G. Mazenko, World Scientific, New York, 1986. A number of the classic papers are superfluid ^{4}He are collected in Z. Galasiewicz, *Helium–4*, Pergamon, New York, 1971, and/or in D. Pines, *The Many-body Problem*, Benjamin, New York, 1961. For a proof that N_0 is finite for an interacting Bose liquid provided perturbation theory converges, see J. Gavoret and P. Nozières, Ann. Phys. (USA) **28**, 349 (1964). Hohenberg's theorem is proved in P. C. Hohenberg, Phys. Rev. **158**, 383 (1967). A detailed discussion and applications of two-fluid hydrodynamics can be found in I. M. Khalatnikov, *Theory of Superfluidity*, Benjamin, New York, 1965.

Lectures 5–6. There is a myriad of good books on the BCS theory of superconductivity. Three which approach the subject in a spirit similar to that of the lectures are A. D. C. Grassie, *The Superconducting State*, Sussex University Press, London, 1975; P. G. De Gennes, *Superconductivity of Metals and Alloys*, Benjamin, New York, 1966; and M. Tinkham, *Introduction to Superconductivity*, McGraw-Hill, New York, 1975. The question of the distinction between "superfluid" and "normal" response is discussed in much more detail, in the context of superfluid ^{3}He, in the lectures of D. Einzel at this School.

Lecture 7. The approach to the derivation of the Ginzburg-Landau theory given here is as far as I know not widely presented in textbooks (which generally prefer to follow the original Gor'kov derivation or something similar). It is presented (for the anisotropic case) in more detail in my Revs. Mod. Phys. paper cited under Lecture 8. Extensive applications of the GL theory are given by D. S. Fisher in his lectures at this School.

Lecture 8. A comprehensive and up-to-date account of the theory of superfluid ^{3}He is given by D. Vollhardt and P. Wölfle, *Superfluid* ^{3}He, Taylor and Francis, London, 1990. A review at the level and in the language of the lectures is A. J. Leggett, Revs. Mod. Phys. **47**, 331 (1975). For the Gor'kov-Galitskii conjecture, see L. P. Gor'kov and V. M. Galitskii, Zh. Eksp. Ter. Fiz. **40**, 1124 (1961), translation: Soviet Physics JETP **13**, 792 (1962); for counter-arguments see D. Hone, Phys. Rev. Letters **8**, 370 (1962); R. Balian et al., ibid, 372.

Lecture 9. A good reference for many of the topics of this lecture is the Parks collection: see especially the papers by Scalapino, McMillan and Rowell and Maki. A very clear discussion of the effect of time-reversal-conserving-and-violating perturbations is given by de Gennes, op. cit. The question of the transition to diatomic molecules is discussed for the s-wave case by A. J. Leggett, J. de Phys. (Paris) Colloq. C-7, 19 (1980), and for the p-wave case by M. G. McClure, thesis, University of Sussex, 1982. For the "localized superconductor", see M. Ma and P. A. Lee, Phys. Rev. **B32**, 5658 (1985).

Lecture 10. A good source for the experimental properties of the high-temperature superconductors is *Physical Properties of High-Temperature Superconductivity*, Vols. I and II, ed. D. M. Ginsberg, World Scientific, Singapore, 1989 and 1990. For a survey of current controversies see e.g., K. Bedell et al. (eds.), *High-Temperature Superconductivity*, Addison-Wesley, New York, 1990.

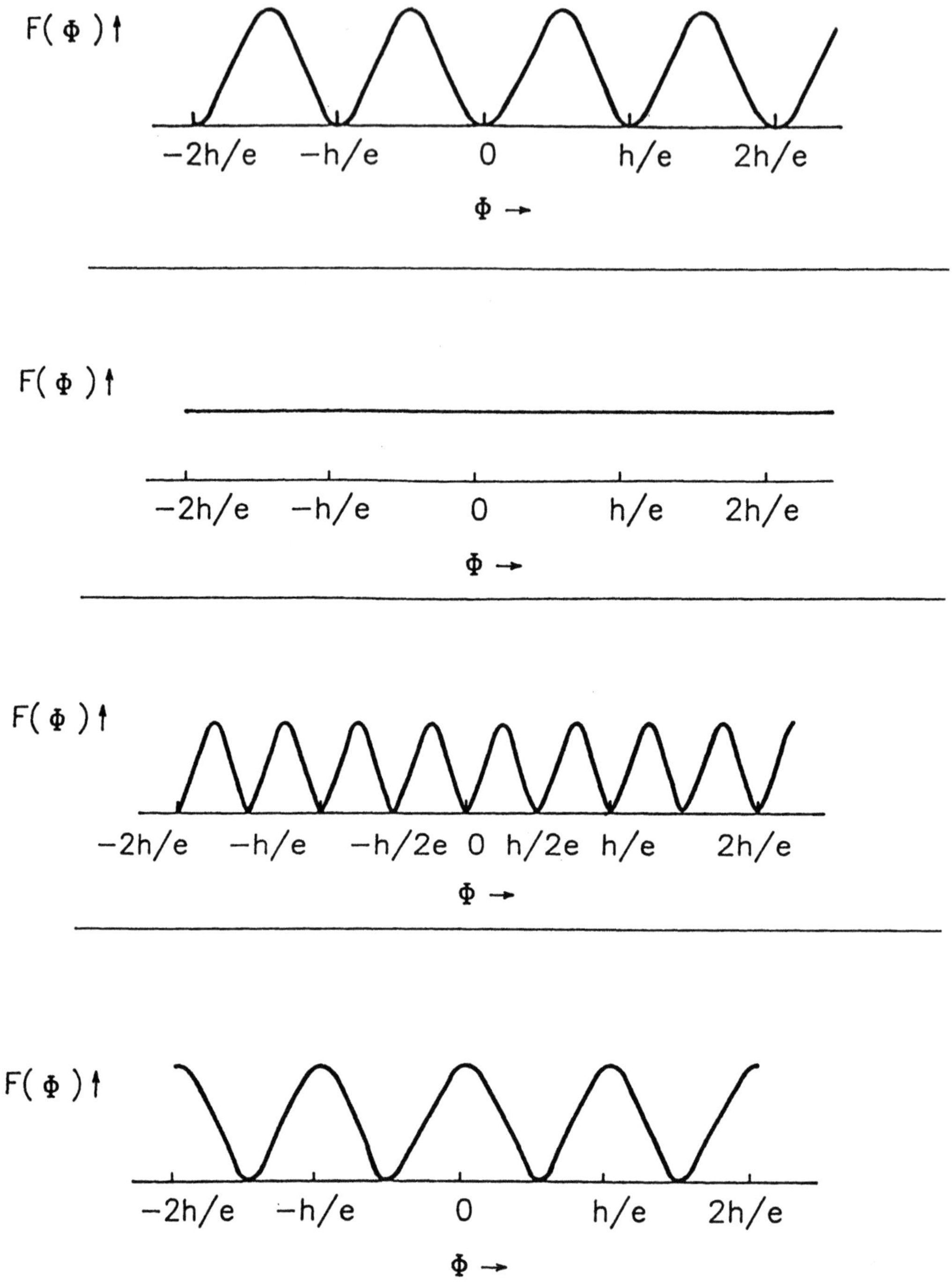

Figure 1(a–d): Possible forms of F(Φ) compatible with eqns. (3.6) and (3.10).

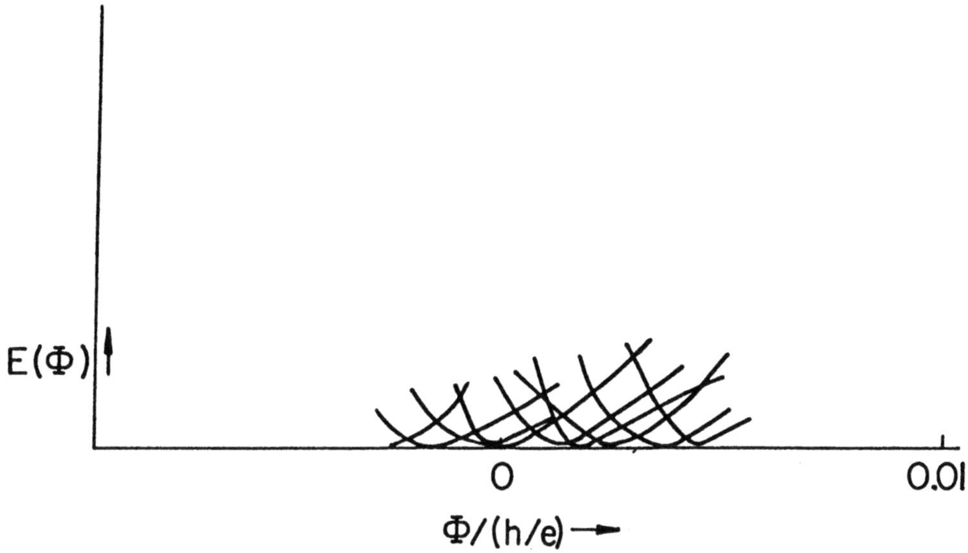

Figure 2: Energy–level diagram characteristic of a normal metal.

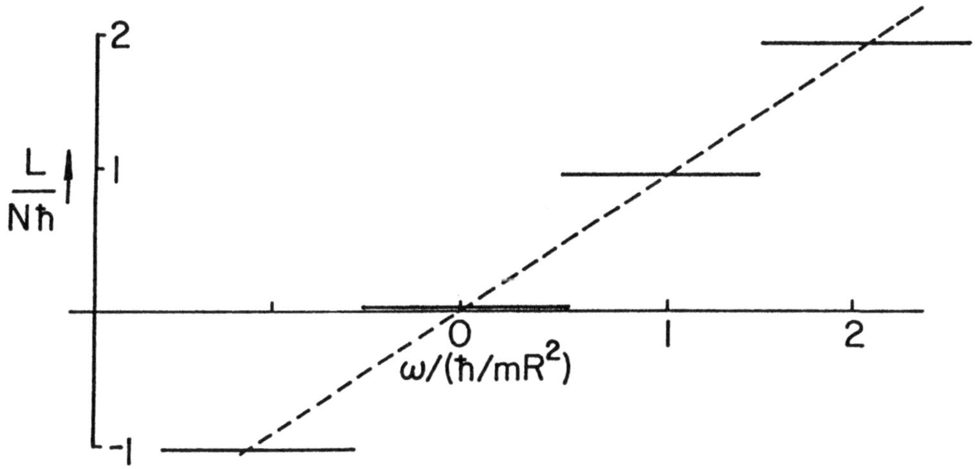

Figure 3: Angular momentum versus angular velocity for a free Bose gas (T = 0). The dashed line shows the behavior characteristic of a normal liquid.

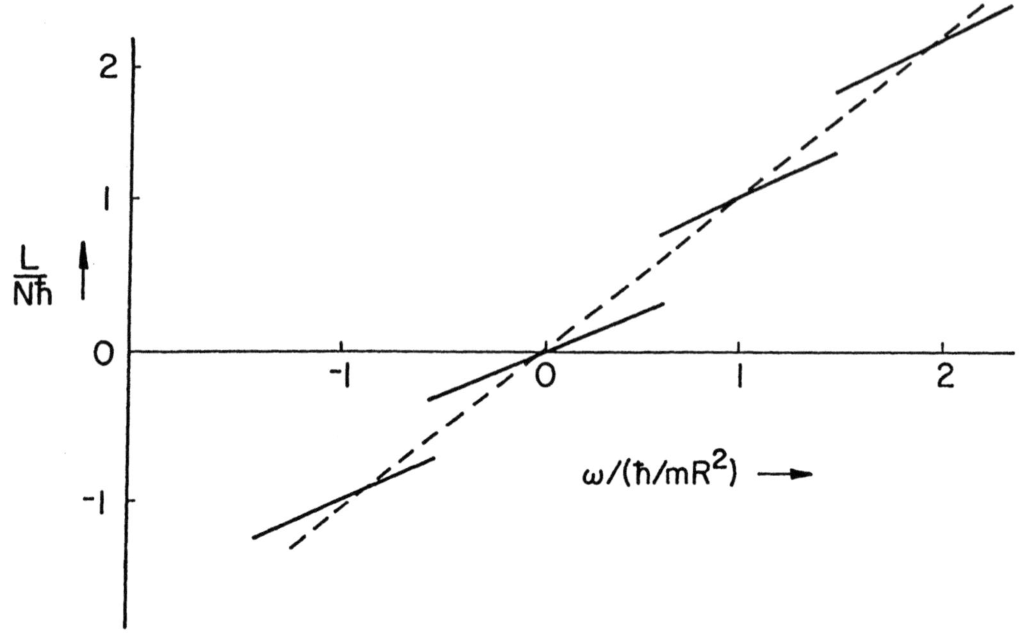

Figure 4: Angular momentum versus angular velocity for a free Bose gas ($T \neq 0$). The dashed line shows the behavior characteristic of a normal liquid.

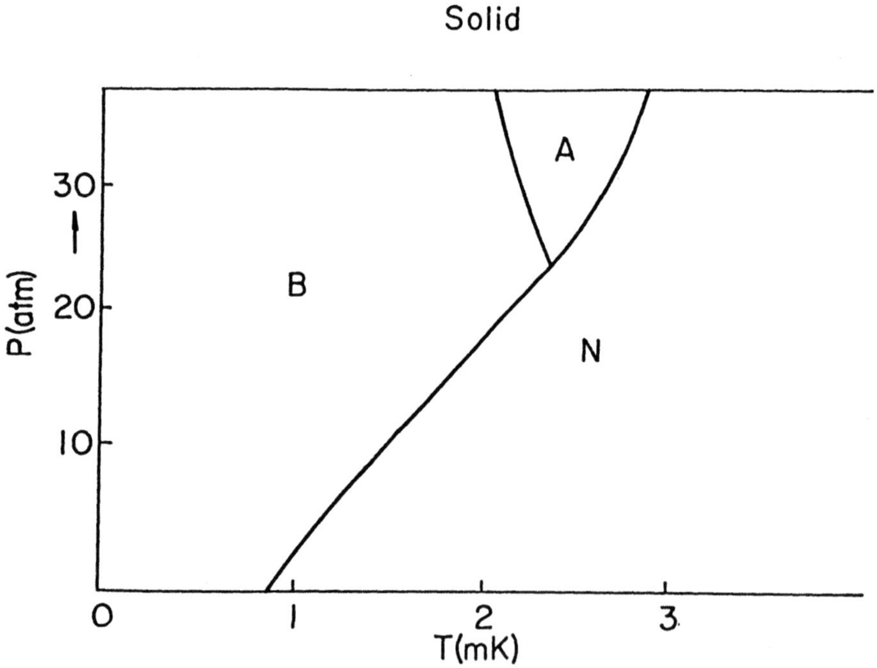

Figure 5: Phase diagram of ^3He below 3 mK in zero magnetic field.

SUPERFLUID ^{3}HE : SIZE EFFECTS, SURFACES, KAPITZA RESISTANCE AND THE ATTAINMENT OF ULTRA–LOW TEMPERATURES

R C RICHARDSON
Laboratory of Atomic and Solid State Physics,
Clark Hall, Cornell University, Ithaca NY 14853, U S A

LECTURE 1 : INTRODUCTION TO SUPERFLUIDS AND ADIABATIC DEMAGNETIZATION

In this, the first of five lectures, I am going to give an introduction to low temperature physics and discuss some of the phenomenology associated with this field, as Professor Tony Leggett has done in his first lecture, but from a different perspective. I then plan to review some of the simple underlying ideas relating to the quantum fluids ^{3}He, ^{4}He and electrons in superconductors. Finally I shall discuss methods of cooling samples to very low temperatures.

The central theme of low temperature physics is related to the third law of thermodynamics, viz. as the temperature T tends to zero, so the entropy S also tends to zero. For people in the field the interesting question is how matter rearranges itself to produce zero entropy. There are a large variety of ways and that is the fun of it. As a simple example, in liquid crystals the molecules are disordered at high temperatures. As the temperature is decreased, the thermal disorder changes at a phase transition and alignment of the molecules occurs.

Let us consider a system which is gaseous at room temperature. As the temperature is lowered, the gas will liquefy and, in most cases, then solidify. Some forms of matter remain in a fluid state down to the lowest temperatures. The problem of how order is achieved in these forms of matter constitutes a major area of research. The answer is to be found in a study of the superfluids, examples of which include ^{4}He, ^{3}He, electrons, neutrons (in neutron stars) and, in principle, spin–aligned atomic hydrogen.

The next idea, which I would like to discuss briefly, is that of coldness in a fluid. A fluid may be regarded as cold when the underlying particles comprising the fluid have become so *large* that their de Broglie wavelengths are comparable to the average distance between particles. The thermal de Broglie wavelength can be characterized as Planck's constant divided by the thermal momentum:

$$\lambda_{thermal} = \frac{h}{p_{thermal}} = \frac{h}{\sqrt{2mk_BT}}$$

When $\lambda_{thermal}$ exceeds the average inter–particle spacing, a crossover occurs between a regime of classical physics and quantum physics. Quantum properties will occur when

$$\lambda_{thermal} \geq (V/N)^{1/3}.$$

In the case of ^{4}He this is the condition in the *classical* quantum statistical picture for Bose condensation to occur. A detailed calculation shows that the crossover to the quantum regime takes place at ~3 K and experimentally the superfluid λ–transition occurs at 2.2 K, strongly supporting a connection between Bose condensation and the superfluid transition in ^{4}He. However the idea of this connection has not always been accepted. A large number of properties of superfluid ^{4}He do not depend in any explicit way on knowing whether or not there is a Bose condensate.

In the case of fermions, Fermi properties occur when the thermal wavelength of the particles is comparable to the inter–particle spacing. Examples include ^{3}He and the electrons in a normal metal. All fermions have a magnetic moment and therefore these systems show interesting magnetic properties. If spin–aligned atomic hydrogen becomes a superfluid, it too will have curious magnetic properties.

^{4}He, the classical Bose system, has no magnetic moment. A macroscopic occupation of the groundstate of ^{4}He starts to occur at the λ transition temperature. The groundstate occupation can be measured experimentally and even at absolute zero temperature it is not 100%. One may view each atom as being partially in a normal state and partially in the groundstate. The fact that all atoms share the groundstate leads to the remarkable properties of superfluid ^{4}He.

It is interesting to note that the thermal properties of liquid ^{4}He in bulk cannot be described by the quantum statistics of a non–interacting Bose gas. Because temperatures are measured and precisely controlled more easily in liquid ^{4}He than in almost any other material, measurements of the heat capacity at the λ transition have given impetus to studies of condensed matter theory, critical phenomena and scaling laws. ^{4}He has a many–body wavefunction describing the behaviour of all atoms collectively.

In ^{3}He quantum properties occur at approximately 1 K. For temperatures less than 1 K, ^{3}He is dominated by quantum statistics and behaves like a Fermi liquid. One of the basic rules of a collection of Fermi particles is that no two atoms can have the same set of quantum numbers. The characteristic Fermi velocity V_F is about 30 ms^{-1} at the Fermi energy. This is much lower than that for electrons in metals where $V_F \sim 10^5$ ms^{-1}. The fundamental properties of the ^{3}He fluid are nevertheless very similar to those of electrons in metals. For example, the heat capacity in ^{3}He is proportional to the absolute temperature for the same reasons that it is in a metal.

At the superconducting transition temperature, T_c, two important changes occur in the properties of electron systems. Firstly the heat capacity of the material changes, at least for normal superconductors where there is a second-order phase transition. Figure 1.1 (a) shows that, following linear behaviour with temperature, there is a sudden discontinuity in the heat capacity at T_c. Secondly the magnetic susceptibility of a superconductor changes suddenly at T_c. The susceptibility is often a difficult parameter to measure. When a superconductor is in a magnetic field and the material passes through T_c, a screening current is set up which makes it difficult for the magnetic field to penetrate the material. When the magnetic susceptibility can be measured in a homogeneous material, we find that below T_c it decreases and heads towards zero at the absolute zero of temperature (Figure 1.1 (b)). This behaviour is linked to the fraction of electrons which are in the paired state. At zero temperature all the electrons are paired and there is zero magnetic moment. The susceptibility below T_c therefore gives a measure of the fraction of normal electrons available in the system.

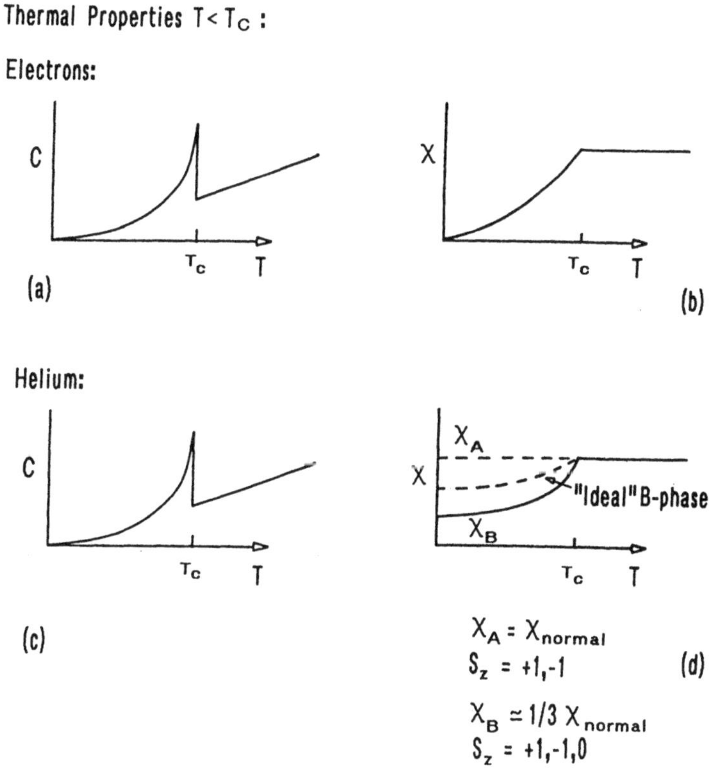

Figure 1.1 Specific heat C and magnetic susceptibility χ for electrons and ^3He.

Very similar results are obtained in the case of liquid ^{3}He. The heat capacity at the transition temperature for superfluid ^{3}He has a sharp discontinuity characteristic of a second order phase change (Figure 1.1 (c)). Scaling the heat capacity data for a typical superconductor with the value of T_c for ^{3}He, the graphs can be laid on top of each other. The results closely resemble the features predicted by the BCS theory of superconductivity. Figure 1.2 shows specific heat data for liquid ^{3}He obtained by Halperin et al. in early experiments performed at Cornell [3]. It was found that specific heat data for niobium could be superposed on these data.

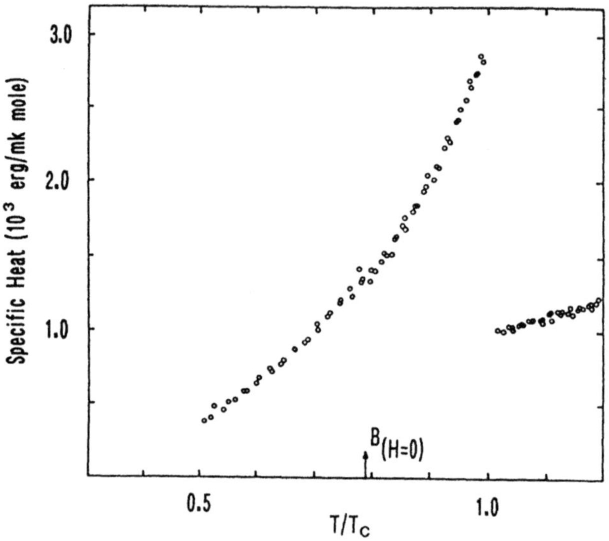

Figure 1.2 Specific heat data for liquid ^{3}He obtained in zero magnetic field at the melting pressure. The B transition occurs at $T/T_C = 0.791$. T_C is the temperature of the A transition (After V Halperin et al. [3]). The shape is very similar to that of specific heat curves for superconductors.

Recently workers on heavy fermion systems superposed the heat capacity data for a heavy fermion superconductor on those for liquid ^{3}He and concluded that the curves were similar and could therefore be interpreted as evidence for triplet pairing in these systems. This conclusion is clearly open to criticism in view of the fact that the ^{3}He heat capacity curves can also be superposed on singlet state superconductor heat capacity measurements.

The magnetic susceptibility, however, has a different behaviour (Figure 1.1 (d)). Instead of the magnetic susceptibility going to zero at zero temperature, it tends towards one third of the magnetic susceptibility that one would expect above the transition temperature (χ_{normal}). This may be explained by the fact that, in the case of triplet paired ^3He, where there are three spin components, $s_z = \pm 1$ and $s_z = 0$, only the susceptibility of the $s_z = 0$ component disappears as one goes to zero temperature. In the case of the ideal, non–interacting ^3He Fermi gas, a zero temperature susceptibility which is two thirds of the Pauli susceptibility may be predicted. The calculation for triplet pairing was completed by Tony Leggett roughly ten years before the discovery of superfluidity in ^3He [2]. The actual value obtained is a result of the change of the molecular field interaction by the pairing interaction. Calculations based on Fermi liquid theory can be used to account for the observations.

Figure 1.3 shows the experimental results obtained for superfluid ^3He at two different pressures. The relative susceptibility of superfluid ^3He is plotted as a function of reduced temperature. The small circles represent data, obtained by Osheroff [4], which confirm Professor Leggett's theory for the susceptibility of a triplet paired superfluid.

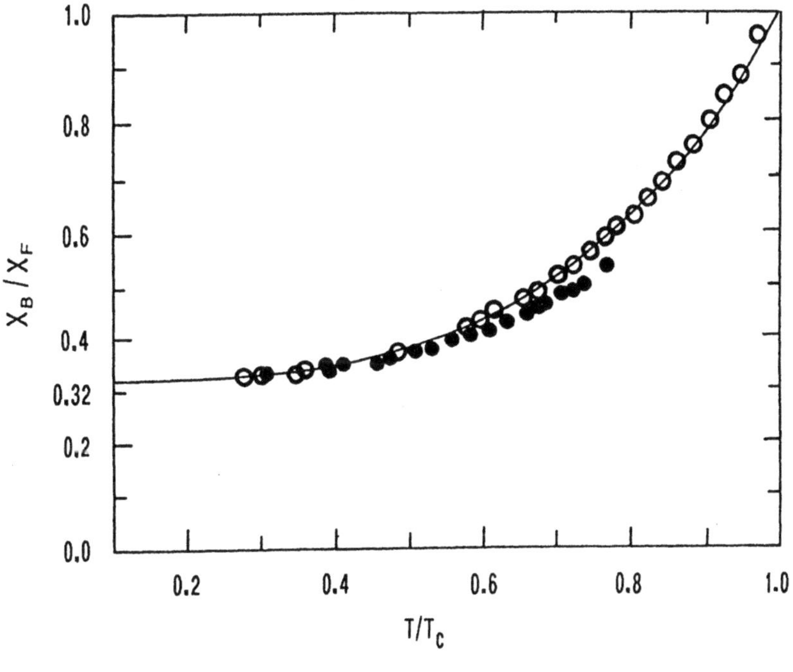

Figure 1.3 Relative susceptibility of superfluid ^3He as a function of reduced temperature. (After D D Osheroff [4]).

The following 3D ^3He phase diagram (Figure 1.4) is obtained by plotting the transition temperature as a function both of pressure and magnetic field [5]. On the right hand side of the diagram above the critical temperature the liquid behaves like a normal Fermi liquid, while at lower temperatures the liquid is a superfluid. A number of phases are demarcated. The low temperature low field phase is called the B–phase. Between a pressure of 21 bar and the melting pressure of 34 bar another phase called the A–phase is present even in zero magnetic field. In this phase the magnetic susceptibility is constant and the fluid has some remarkable magnetic resonance properties. As the magnetic field is increased, a third superfluid phase, A_1, develops. The pairing of spins is different in the three phases. In the A_1 phase the magnetic moments are aligned parallel to the magnetic field, so that there is only one substate $|\uparrow\uparrow\rangle$ ($S_z = -1$). In the A–phase the magnetic moments are paired in the up–up $|\uparrow\uparrow\rangle$ and down–down $|\downarrow\downarrow\rangle$ states ($s_z = \pm 1$). All three members of the spin triplet appear in the B phase: $|\uparrow\uparrow\rangle$, $|\downarrow\downarrow\rangle$ and $|\uparrow\downarrow + \downarrow\uparrow\rangle$.

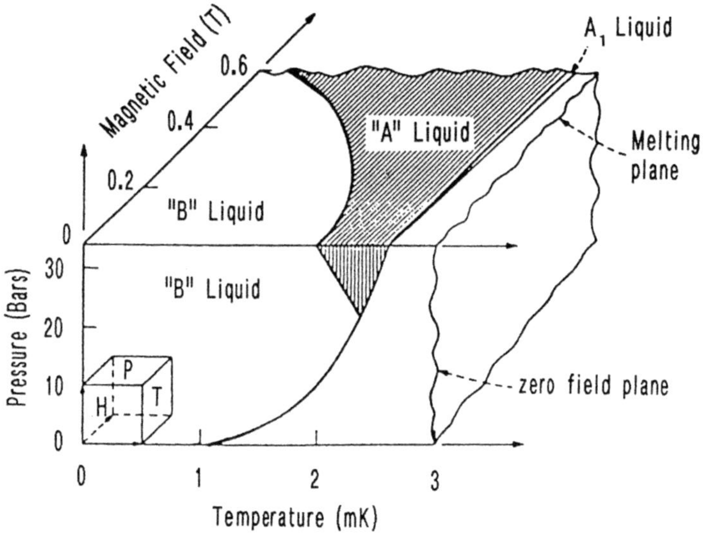

Figure 1.4 Phase diagram of ^3He showing the T—H plane at the melting pressure and the P—T plane in zero field. (After D N Paulson et al. [5]).

Table 1 compares superfluidity in ^3He with superconductivity for electron systems. For temperatures above the phase transition temperature the properties of both electrons and ^3He can be derived from a knowledge of the density of states at the Fermi energy E_F. In the past it was maintained that the ratio of the Fermi temperature T_F ($= E_F/k_B$) to the transition temperature was approximately 1000. The advent of high T_C superconductors has changed the situation.

Ambiguity arises over the Fermi temperature in the normal state of these materials but it appears that the ratio may approach 1 in high T_c superconductors and is certainly not larger than 10.

Table I: Comparison of certain properties of an electron fluid with the corresponding properties of liquid ^3He.

Comparison of ^3He with Electrons

Statistics:

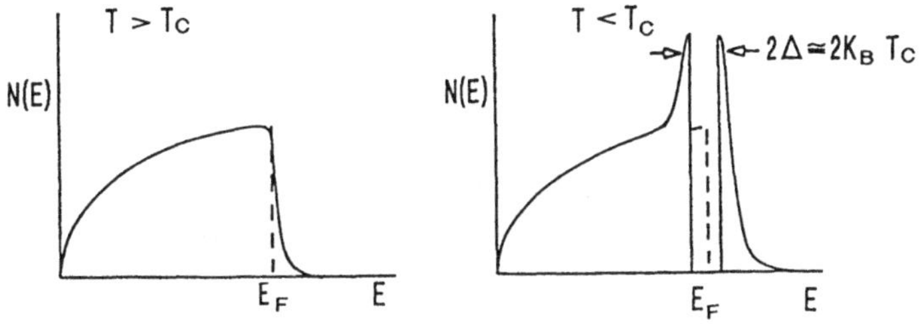

Property	Electrons	^3He
$T_F = E_F/k_B$	10^4K	2K
Tc	10^4K	2×10^{-3}K
Charge	2e	0
Wave function	$\psi(r)\,\phi(s)$=Odd	$\psi(r)\,\phi(s)$=Odd
Spin function	$\phi(s) = \lvert\uparrow\downarrow - \downarrow\uparrow\rangle$ =Odd	$\phi(s) = \lvert\uparrow\uparrow\rangle, \lvert\downarrow\downarrow\rangle$ =Even $\lvert\uparrow\downarrow + \downarrow\uparrow\rangle$
Space function	$\psi(r)$=Even $\ell=0$	$\psi(r)$=Odd $\ell=1$

In superconductors the particles are charged, while in the superfluid the ^3He atoms are not. In both systems the fundamental statistical rule is that the net wave function of the pairs has to have odd symmetry. In the case of all known superconductors the spin wave function is odd. The spatial wave function is even, so that the pairs form in a state with zero orbital angular momentum ($\ell = 0$). In the case of some of the heavy fermion superconductors it has been suggested that $\ell = 2$ but this has not been established. In ^3He the spin wave function is even, so that the spin angular momentum s = 1 and the spatial wave function is odd. Hence the state is formed with $\ell = 1$. This leads to a large number of interesting ramifications for superfluidity in ^3He. There is a preferred direction in space for the flow and the fluid behaviour has a set of properties, associated with long range alignment, called texture. It is suggested that these properties may have some relationship to the properties speculated for neutron stars.

Superfluids can be defined by their flow properties. The crudest criterion is that the fluids can leak through holes too small for a normal interacting fluid to pass through (Figure 1.5). A host of experiments have been performed in ^{3}He, ^{4}He and in rings of electrons with weak links or interruptions that control the interaction of the fluid on the one side with the fluid on the other side. The whole fluid can be described by a single wavefunction: $\psi \sim \rho^{\frac{1}{2}}\, e^{i\phi}$, where ρ is the superfluid density and ϕ specifies the phase.

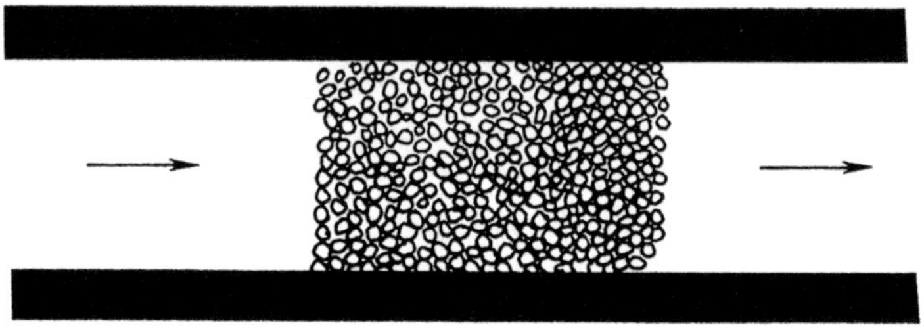

Figure 1.5 Schematic representation of flow through a superleak.

Figure 1.6 relates to an experiment performed by John Reppy and his group at Cornell a number of years ago [6]. The superfluid ^{3}He is trapped between two parallel plates and the plates are then set in torsional motion. In one case the magnetic field is parallel to the plates and in the

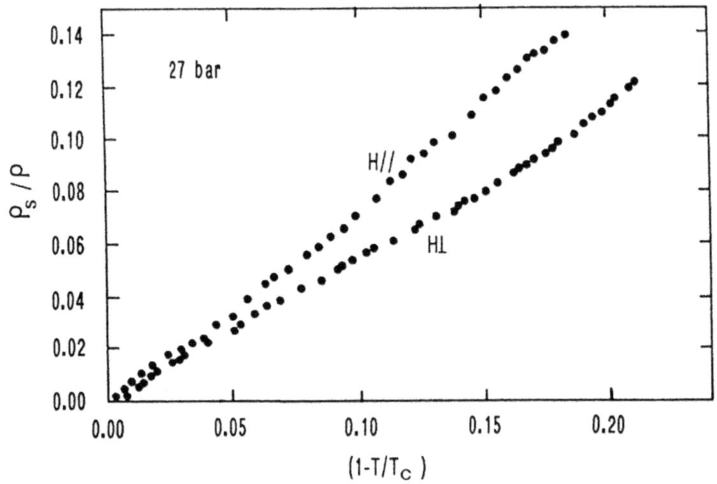

Figure 1.6 The superfluid mass of ^{3}He for two different orientations of the applied magnetic field, plotted as a function of $(1 - T/T_C)$. The ^{3}He is confined between parallel surfaces. (After J E Berthold et al. [6]).

other it is perpendicular to the plates. The moment of inertia of the system is measured as a function of temperature for both orientations of the magnetic field. It is found that the fluid exhibits greater superfluid properties with the magnetic field in one direction (parallel to the plates) than with it in the other direction. This demonstrates the intrinsic spatial anisotropy of the superfluid.

This concludes a few introductory comments about superfluid ^{3}He and I now wish to discuss cooling techniques. The idea of magnetic cooling involves demagnetizing a paramagnetic material. At high temperatures the magnetic moments are disordered (Figure 1.7). At lower temperatures there will be some order where the magnetic interaction between neighbours will produce some alignment of the magnetic moments. In some paramagnetic salts this ordering will occur at temperatures as low as 1 mK.

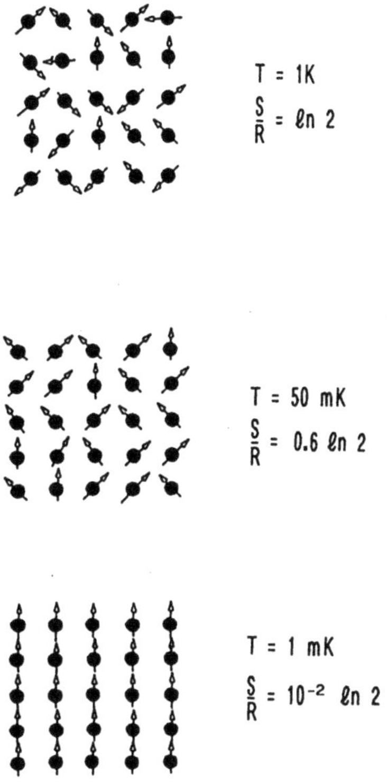

Figure 1.7 Disorder in paramagnets at different temperatures (B=0). The entropy decreases with decreasing temperature.

The cooling scheme which is used in magnetic cooling involves *tricking* the material by forcing a certain amount of order on the magnetic moments. Figure 1.8 (a) is a plot of entropy S versus temperature T for a paramagnetic material in several different magnetic fields. The upper curve shows the entropy in zero magnetic field. As a magnetic field is applied, the entropy is reduced. At low temperatures there is a temperature–dependent entropy down to fairly low temperatures which is proportional to $\ln 2$ for a spin of $\frac{1}{2}$. This continues until the strength of the magnetic interaction becomes comparable with the thermal energy, when some sort of order appears.

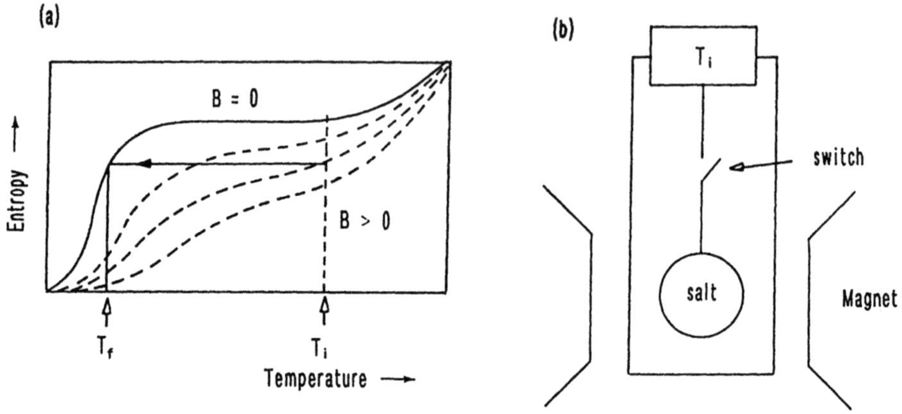

Figure 1.8 (a) Entropy—temperature diagram for a paramagnet in various applied fields. (b) Schematic representation of an adiabatic demagnetization stage.

Figure 1.8 (b) shows a schematic diagram of the system used to produce magnetic cooling. The working material can be a paramagnetic salt. This is isolated from the rest of the universe by placing it inside a vacuum can and the whole is placed in a variable magnetic field. A cold reservoir is used to absorb heat from the paramagnetic material during the cycle. A thermal switch permits contact to be made between the coolant and the thermal reservoir.

At some initial temperature the material is magnetized and the thermal switch is closed so that the heat of magnetization is removed. When the maximum amount of polarization has been achieved, the switch is opened and the magnetic field is gradually reduced with a minimum of thermal loss due to eddy currents, for example. If there is no entropy loss, on removing the magnetic field cooling occurs as the system moves to a point on the entropy curve corresponding to $B = 0$. In principle temperatures very close to the magnetic ordering temperature of the salt can be produced.

Experiments studying only the magnetic order of the spin system in a material have been performed in which a large polarization and temperatures well below the magnetic ordering temperature have been achieved. In the case of silver, in which the nuclear moments have a very low gyromagnetic ratio and therefore a very weak dipolar interaction, temperatures below 2 nK have been reached by Lounasmaa and co workers in Helsinki [7]. Serious problems arise in attempting to cool other materials to these temperatures using the silver nuclear spin system as the coolant. In silver, as with other metals, thermal equilibrium of the nuclei with the lattice occurs with a characteristic relaxation time which is very long at low temperatures. At 2 nK the thermal equilibrium time is estimated to be several hundreds of years. In virtually all metals the relaxation rate decreases as the temperature decreases.

A more sophisticated apparatus has recently been developed with more than one nuclear stage involving metal coolants (Figure 1.9). The primary cooling device is a dilution refrigerator, which depends on the properties of $^3He/^4He$ mixtures and which can obtain temperatures down to 2.5 mK. Two magnetic cooling stages are cascaded in series below the refrigerator mixing chamber with copper generally being used in the first stage and silver or copper in the second stage. For certain types of experiments a heat switch is not necessary in the second stage as the nuclear spin system has a very long relaxation time at the temperatures which are reached.

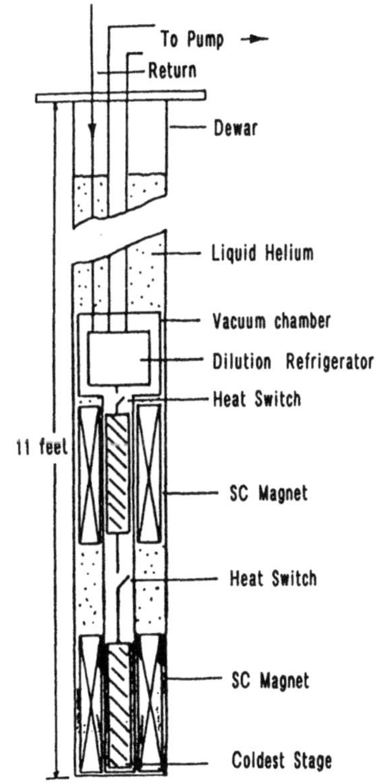

Figure 1.9 Sketch of a multistage nuclear demagnetization cryostat. High purity copper is typically used in nuclear stages.

Research today has been further aided by the availability of very high magnetic fields. In the 1930's a magnetic field of 1 T was considered adequate. Today magnets producing 8 or 9 T are relatively inexpensive, while those providing 6 or 7 T are easily constructed. Magnets larger than 7 or 8 T have not been used at the lowest temperatures.

Figure 1.10 shows a plot of the lowest temperature obtained as a function of the year and includes landmarks such as the liquefaction of hydrogen and helium and the discovery of superconductivity. While early workers claimed that the lowest temperature man could hope to achieve was 0.7 K, the development of magnetic cooling, nuclear demagnetization, and in the 1970's, two stage nuclear demagnetization cryostats has resulted in minimum temperatures of around 10 μK. There has been little progress in reaching still lower lattice temperatures than this in the last five years. Further research is necessary to determine whether the inability to achieve

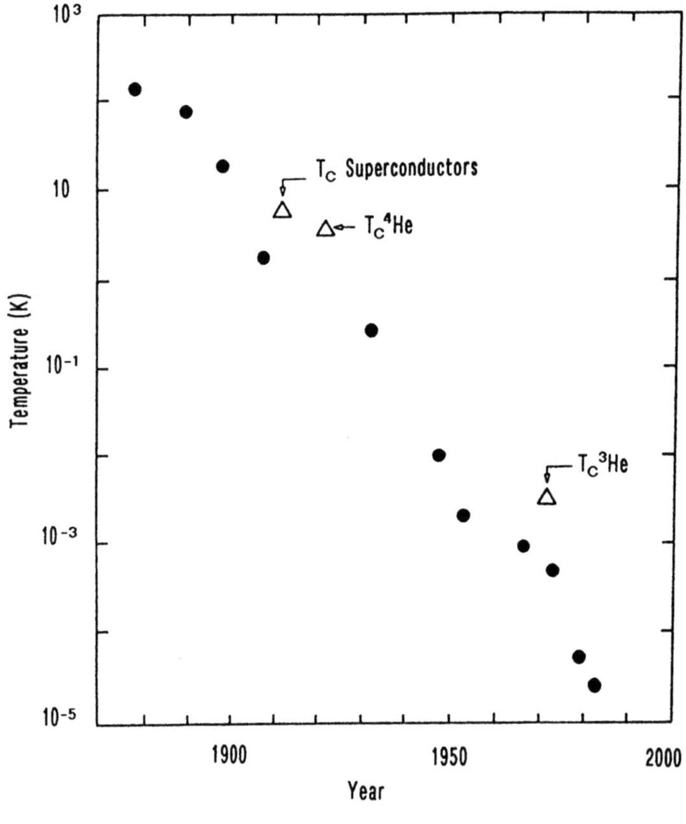

Figure 1.10 The lowest temperature achieved as a function of year showing the dates of discovery of superconductivity and superfluidity (^4He and ^3He).

temperatures below 10 μK is attributable to inadequate experimental techniques or to some underlying physics that we do not understand. I will return to this subject in my last lecture.

References 1

[1] J Bardeen, L W Cooper and J R Schrieffer, *Phys. Rev.* **108**, 1175 (1957).

[2] A J Leggett, *Phys. Rev.* **A140**, 1869 (1965).

[3] W P Halperin, C N Archie, F B Rasmussen, T A Alvesalo and R C Richardson, *Phys. Rev.* **B13**, 2124 (1976).

[4] D D Osheroff, *Phys. Rev. Lett.* **33**, 1009 (1974).

[5] D N Paulson, H Kojima and J C Wheatley, *Phys. Rev. Lett.* **32**, 1098 (1974).

[6] J E Berthold, R W Giannetta, E W Smith and J D Reppy, *Phys. Rev. Lett.* **37**, 1108 (1976).

[7] P J Hakonen, S Yin and O V Lounasmaa, *Phys. Rev. Lett.* **64**, 2707 (1990).

LECTURE 2 : ^{3}He IN CONFINED GEOMETRIES

Introduction

In this lecture I shall discuss investigations by Mark Freeman [1] at Cornell on superfluid ^{3}He in a narrow parallel plate geometry. The purpose of the experiments was to study the behaviour of ^{3}He for varying correlation lengths comparable to the spacing in the layers near the normal superfluid transition. E Thuneberg helped with the interpretation of the results. Both pure ^{3}He samples and dilute mixtures of ^{4}He in ^{3}He were examined. In the latter experiments the ^{4}He plates the surfaces confining the ^{3}He and leads to unexpected consequences. Good agreement with Ginzburg–Landau theory [2] has been obtained for the pure ^{3}He experiments. NMR has been used to examine the order parameter.

Review of some superfluid ^{3}He Theories

As described in lecture 1, ^{3}He has three phases: the A1 phase, the A phase, and the B phase. The A1 phase appears only in a large magnetic field. The present experiments are concerned with the A phase but I shall also mention some properties of the B phase.

In the A phase $s_z = \pm 1$. Two unit vectors, $\hat{d}$ and $\hat{\ell}$ respectively, determine the directions of the spin quantization axis and the orbital symmetry axis of the Cooper pairs. The Fermi surface of the A phase has a gap structure with a node at each end of the gap as depicted in Figure 2.1 (a). As one passes through the phase transition a shift in the frequency of the transverse nuclear magnetic resonance occurs. This shift may be attributed to the correlations occurring in paired spins. As a result of the strong boundary condition in the A phase, the preferred direction for the angular momentum $\hat{\ell}$ of the Cooper pairs is perpendicular to the wall (Figure 2.1 (b). The spin of the pairs along the spin vector $\hat{d}$ must equal zero, i.e.

$$\vec{s} \cdot \hat{d} = 0$$

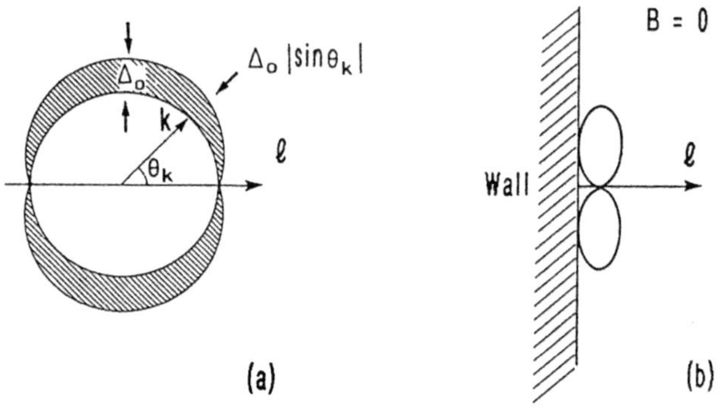

Figure 2.1 (a) The energy gap (shaded) for the axial (ABM) state. The gap vanishes at two points along $\underset{\sim}{\ell}$ as shown. **(b)** Schematic representation of the orientation of the orbital angular momentum for a Cooper pair in superfluid ^{3}He in the vicinity of a wall.

This leads to a preferred orientation of the spins relative to the angular momentum vector. The alignment property of the orbital part of the wavefunction gives rise to texture properties in ^{3}He. Quite a number of different textures can be created.

If the fluid is placed in a magnetic field it is energetically favorable to have the direction of the magnetization that is induced perpendicular to the orbital vector $\hat{\ell}$. Applying a pulse of radio frequency in an NMR experiment, the motion of the $\hat{d}$ vector takes the form of figures of eight around its equilibrium position along the $\hat{\ell}$ vector. This follows from the Leggett equations. A change in the size of tipping angle results in an effective change of the moment of inertia with a resultant change of the nuclear signals. The resonant frequency may be plotted as a function of the initial pulsing angle.

In the B phase $s_z = 1, 0, -1$. The Fermi surface of the B phase has a gap structure which is spherically symmetric. If the ^3He is placed in a container or in a magnetic field, distortions of this structure will occur.

Analogues of experiments in atomic physics can be reproduced with collective modes in ^3He. Experiments reproduced in superfluid ^3He include the Zeeman effect, the Stark effect and the Paschen–Back effect. The Zeeman effect occurs at a frequency such that a little of the J = 2 state is excited. People have had a lot of fun demonstrating atomic physics on a macroscopic scale.

Experiments have shown that the texture bends when a magnetic field is applied perpendicular to the wall of the container (Figure 2.2 (a) B). The characteristic bending length is ~10 μm. When the magnetic field is applied parallel to the wall, the texture is uniform (Figure 2.1 (a) A). The experiments I am going to describe involve films of thickness much less than 1 μm (~0.3 μm).

In low magnetic fields (< 50 Gauss) the dipole energy becomes dominant and the spin will not align parallel to the magnetic field. In fields larger than 50 G the magnetic moments can be rotated parallel to the $\boldsymbol{\ell}$ vector. This state is called the dipole–unlocked state. Most of the experiments I describe are in this regime.

Issues in Thin Slabs

Issues of concern in thin slabs include (a) the nature of the phase between the slabs and (b) the nature of the transition compared to bulk.

Figure 2.2 (b) shows two flat parallel plates, a distance d apart, with superfluid ^3He confined between them. Figure 2.2 (c) shows a plot of the correlation length ξ versus temperature T. As the transition temperature T_c is approached, the correlation length increases as follows:

$$\xi = \frac{\xi_o}{\sqrt{1 - T/T_c}}$$

where ξ_o, the zero temperature correlation length, is given by:

$$\xi_o = \frac{\hbar V_F}{k_B T_c} = \frac{V_F}{\omega_c}$$

with V_F the Fermi velocity.

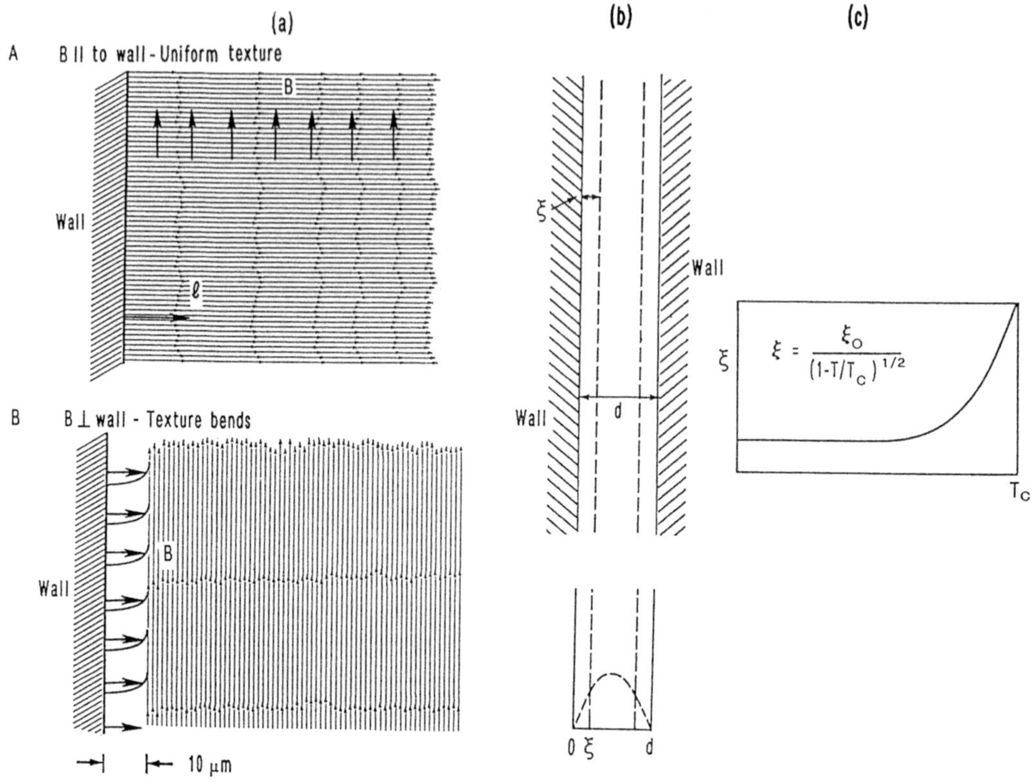

Figure 2.2 (a) Texture effects in flowing ^3He in the vicinity of a wall. In A a field is applied parallel to the wall while in B the field is perpendicular to the wall. **(b)** Flow between parallel plates of separation d. The correlation length ξ is indicated. **(c)** Behaviour of ξ as a function of temperature near T_C.

By adjusting the pressure of ^3He from 0 to 30 bar the size of ξ can be varied from 10 to 50 nm. We worked up to 25 bar before the delicate apparatus blew up.

Figure 2.3 shows the phase diagram for ^3He. It has been speculated that the B phase transition will be suppressed in a small magnetic field down to quite low temperatures when the ratio of the thickness of the film to correlation length is 10. Further speculation, which was rather widespread when we started this work, concerns the existence of another phase, the planar state, with a node line along the equator where the BCS gap is zero.

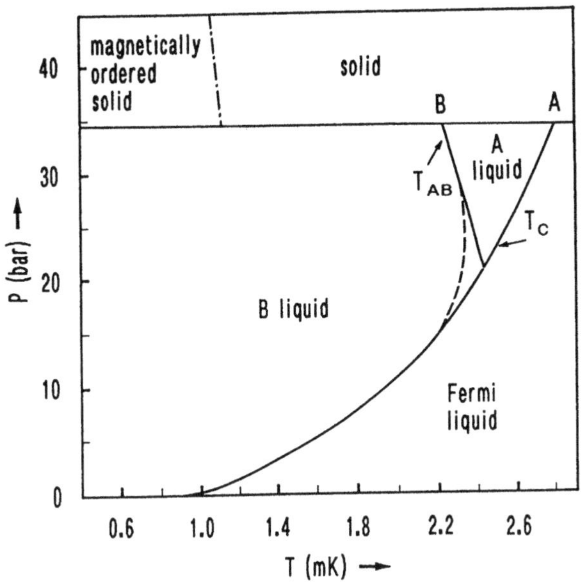

Figure 2.3 The phase diagram for ^3He in zero field.

Experiments in ^3He

Figure 2.4 shows the apparatus used in experiments at Cornell to identify the nature of the phase between the boundaries [1]. Both measurements of NMR frequency shifts, which identify the nature of the phase, and superfluid fraction measurements using a torsion oscillator have been made. The crucial central region of the apparatus consists of a stack of mylar sheets separated by 0.5 μm polystyrene microspheres to establish the spacing. These squash down slightly but give us a fairly uniform slab geometry.

In a study of the nature of the phase the boundary effects become very important when ^3He is placed in a container. The ^3He atoms that are first absorbed on the surface have a strong van der Waals interaction and the density on the surface is much higher than in the liquid. Several layers of atoms are localized. They have a very strong paramagnetic susceptibility which dominates in determining the magnetic properties of the layer. The inverse of the magnetization versus temperature is plotted to demonstrate the Curie–Weiss nature of the surface layer (Figure 2.5). The linear extrapolation of the high temperature data intersects the T axis at 0.5 mK, the Weiss temperature, and on most surfaces it appears that a two–dimensional ferromagnetic phase transition takes place in ^3He. Deviation of the measurements from this form, as indicated in the inset of the figure, occurs at low enough temperatures where the polarization becomes too large for the approximation to hold.

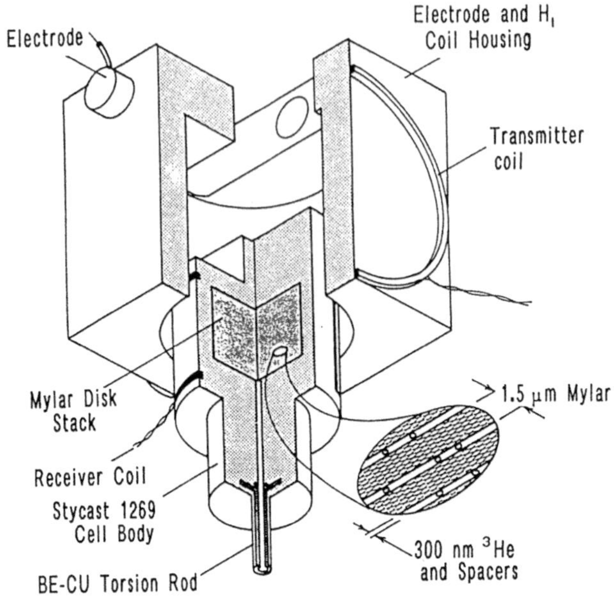

Figure 2.4 The combined torsion pendulum/NMR probe. The ^{3}He in the head of the oscillator fills and cools through the hole in the torsion rod, which terminates in a small heat exchanger linked to the nuclear refrigerant. The electrode structure is thermally sunk to the mixing chamber. (Drawing by Eric Smith.)

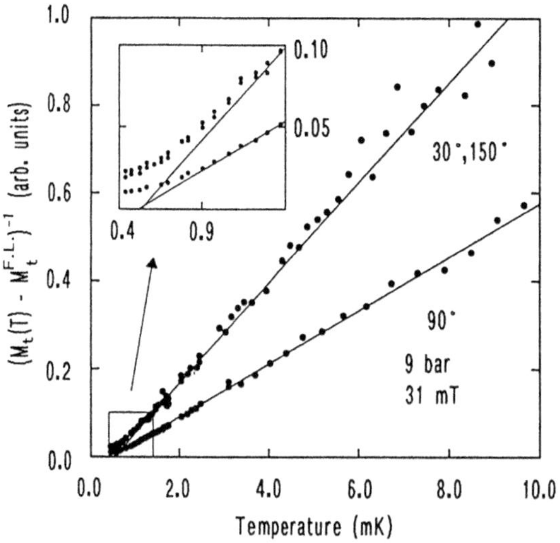

Figure 2.5 The inverse localized layer transverse magnetization following a tipping pulse. The localized layer magnetization is the temperature–dependent excess above the Fermi liquid component, $M_t^{F.L.}$. The solid lines are linear fits which reflect the Curie–Weiss nature of the surface layer susceptibility.

The temperature dependence of the frequency in the normal phase is shown in Figure 2.6. The pulse–independent behaviour and proportionality to inverse temperature can be attributed to magnetic effects associated with the protons in the mylar and allowance has to be made for this.

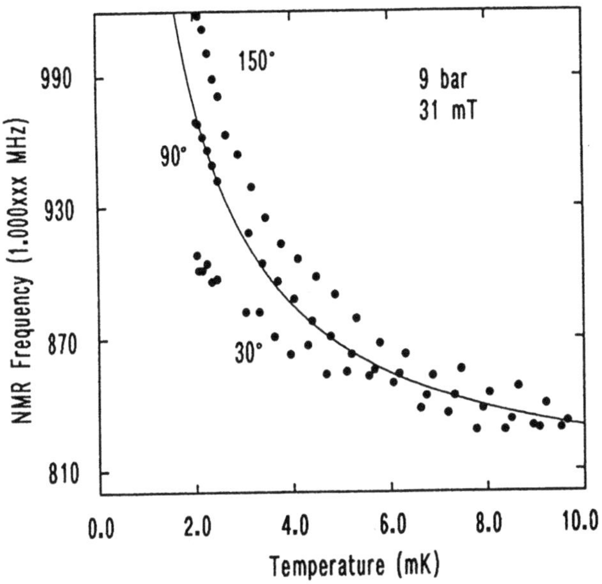

Figure 2.6 The tipping angle dependent NMR frequency in the normal phase. The solid line shows a shift proportional to inverse temperature, which we ascribe to increasing polarization in the mylar. The asymptotic high temperature limit is 1.000793 MHz.

Before we started the experiments we thought that one line would be associated with solid ^{3}He on the surface and another would split off in the superfluid phase. It has since been discovered that *motional averaging* of the two lines occurs. A single line is produced when the exchange frequency between ^{3}He atoms in bulk liquid and those on the surface is comparable to the Larmor frequency. In Figure 2.7 the NMR frequency versus inverse temperature is plotted for several tipping angles. The line through the 90° data is the fit to the data from Figure 2.6. The behaviour is typical of the A phase in the dipole–unlocked texture. The transverse resonance frequency

$$\nu^2 = \nu_L^2 + \frac{c}{4\pi^2}\,\Omega_L^2\left[\frac{T}{T_c}\right]$$

where Ω_L is the Leggett frequency [3] and ν_L is the unshifted Larmor frequency. The factor c is determined by the texture and the tipping angle, ϕ, in pulse NMR. In the dipole–unlocked case [4] $c = -\cos\phi$ and the frequency shift is symmetric about $\phi = 90°$. In the bulk or dipole–locked case [5] $c = \frac{1}{4} + \frac{3}{4}\cos\phi$ [4].

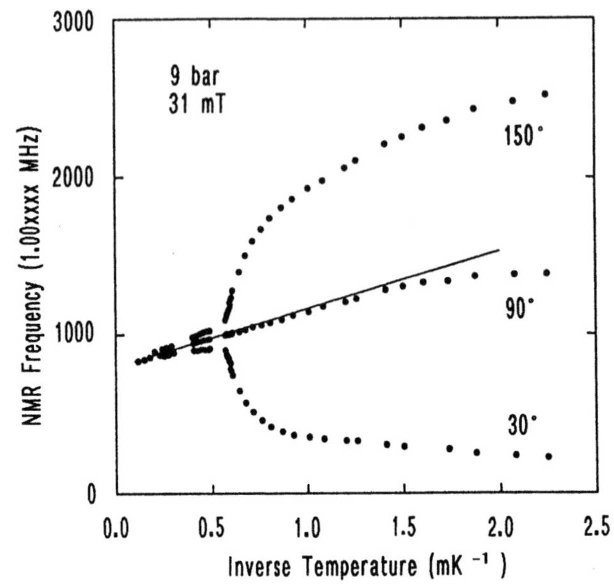

Figure 2.7 The NMR frequency for several tipping angles versus inverse temperature to emphasize the superfluid region. The line is the fit to the normal phase data from Figure 2.4. The bulk transition at this pressure occurs at $T^{-1} = 0.56$ mK^{-1}. The behaviour is characteristic of the A phase in the dipole—unlocked texture.

Figure 2.8 shows the spin precession frequency for several tip angles as a function of inverse temperature. The absolute value of the NMR frequency shift of the liquid in the superfluid phase is obtained by removing background effects caused by mylar proton and solid ^3He layer shifts and by correcting for motional averaging. This value establishes the symmetry about 90° of the 30° and 150° pulse angle results. The slope of the shift at the transition, conventionally referred to as $f^2(P)$, may be determined from the straight line. The Ginzburg–Landau theory predicts that this slope should be $\frac{2}{3}\, f^2_{bulk}(P)$ in the range $10 \leq \frac{d}{\xi} < \pi$, where $f^2_{bulk}(P)$ is the slope of the frequency shift for bulk, i.e.

$$f^2(P) = \left.\frac{d\Omega^2}{d(T/T_c)}\right|_{slab} = \frac{2}{3}\left.\frac{d\Omega^2}{d(T/T_c)}\right|_{bulk}$$

It may be concluded from the results that the properties of the phase present are those expected of the A phase. No transition to the B phase occurs down to the lowest temperatures. This may simply be due to the fact that in the confined geometry the A phase supercools and the B phase cannot nucleate. Further experiments are needed to establish whether this is so.

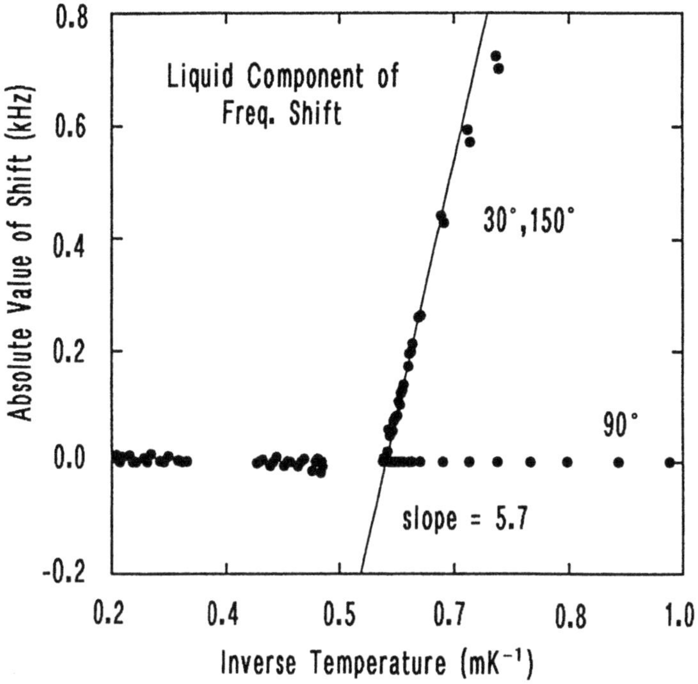

Figure 2.8 The NMR frequency shift of the ^3He liquid, liberated from the many burdens which encumber it in the raw data.

Measurements of ρ_n/ρ at various pressures using the torsional oscillator are shown in Figure 2.9, where ρ_n is the normal fluid density. A correction for bulk superfluid density, contributed by a gap of 10 μm at the edge of the cell, must be subtracted from the data. In the ^3He A phase in the transition region the Ginzburg–Landau prediction for the superfluid fraction of a ^3He film is [4]

$$\frac{\langle \rho_s \rangle_{slab}}{\rho_{s,bulk}} = 0.6209\ (1 - (\pi/\omega)^2)$$

Here $\omega = \frac{d}{\xi(T)}$ where d is the thickness of the film.

This prediction agrees well with our observations if we choose d = 290 nm, which is close to the value we expect for the packed disks. It has been observed that the correlation length decreases as pressure is increased and is also temperature– dependent. As the correlation length increases, T_c is suppressed below $T_{c,bulk}$. These results support the Ginzburg–Landau theory.

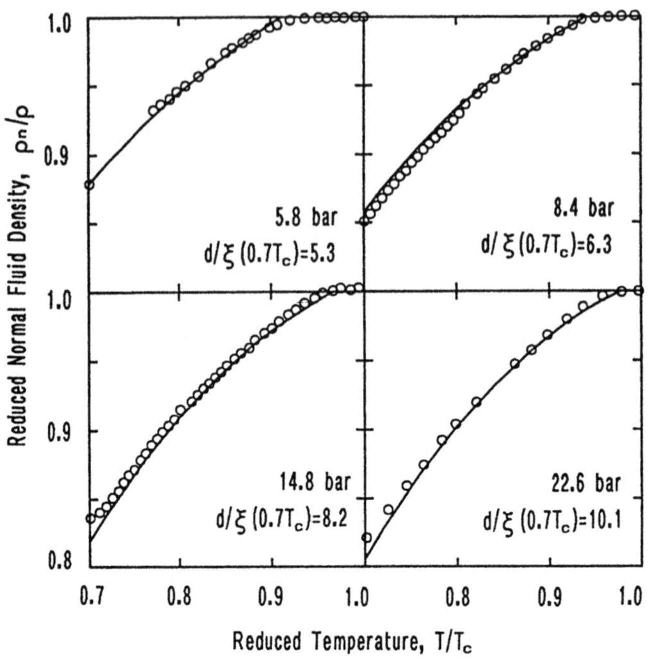

Figure 2.9 The normal fluid density of pure ^{3}He as measured by the torsional oscillator. The solid lines are the Ginzburg–Landau predictions based on Cornell data for the bulk B phase superfluid density, with no free parameters.

Measurements with ^{4}He Coated Walls

When a ^{3}He/^{4}He mixture is placed in a container, ^{4}He will coat the walls preferentially because of the slightly larger attraction between ^{4}He and the surface. This is due to the difference in zero point energy for the two isotopes. ^{4}He pre–plating results in the removal of the Curie–Weiss magnetization and the background effects of the polarized protons in the mylar and the ^{3}He in surface layers. The frequency remains constant for T slightly above T_c. With roughly two layers of ^{4}He the background effects are eliminated.

In Figure 2.10 NMR frequency shifts from the Larmor resonance at 1 MHz are plotted as a function of temperature. In ^{3}He the background effects are so large at low temperatures and low pressures that the superfluid transition in the NMR data cannot be identified. The addition of ^{4}He, indicated by the solid dots, resolves this problem. In studies of the effect of ^{3}He coverage on the Kosterlitz–Thouless transition in ^{4}He film on mylar, McQueeney et al. [6] found that 32 μmole m^{-2} coverage is the thickest layer that can be used without a superfluid ^{4}He film forming under the ^{3}He. The Kosterlitz–Thouless transition temperature can be used to calibrate the ^{4}He coverage.

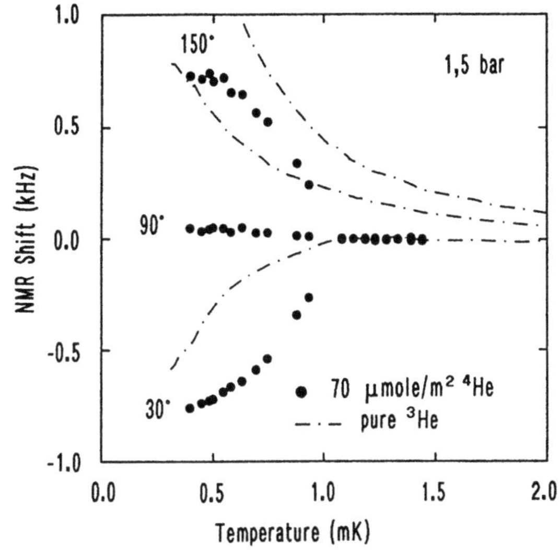

Figure 2.10 ^{4}He surface coverage dependence of the raw NMR shift for different tipping angles. The Larmor resonance as at 1 MHz.

Figure 2.11 shows a comparison of the dilute mixture NMR shifts with the liquid behaviour determined from the pure ^{3}He data at 8.5 bar. A surprising increase in the size of the order parameters occurs with an increase in the thickness of the ^{4}He layer. The 90 μmole m^{-2} surface is effectively specular. It behaves like a diffuse–walled film an order of magnitude thicker than it

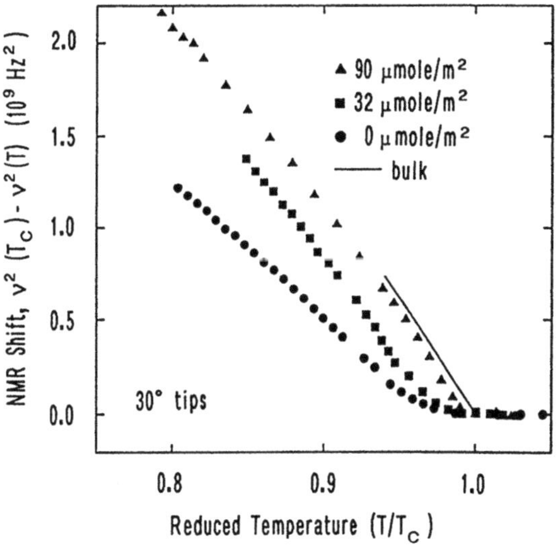

Figure 2.11 The NMR shift for three different boundary conditions, corresponding to 0.32 and 90 μmole m^{-2} ^{4}He coverages. The ^{3}He pressure is 8.5 bar. The ordinate represents the squared Leggett frequency, in the measurement spatially averaged across the width of the film.

actually is. Similar behaviour for the normal fluid density as a function of ^{4}He coverage is shown in Figure 2.12.

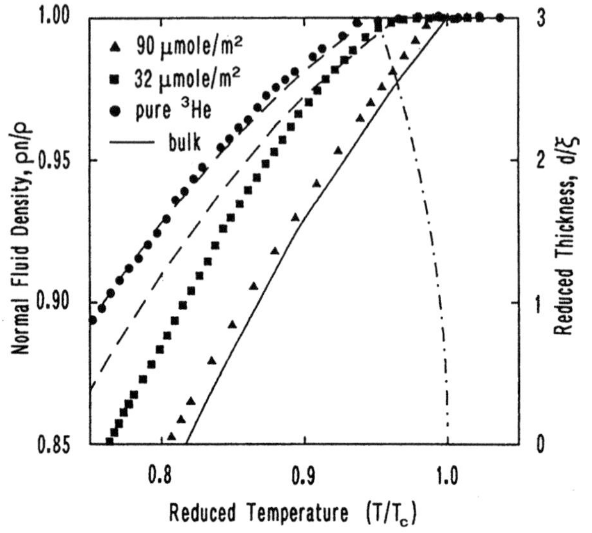

Figure 2.12 The ^{3}He normal fluid density at 8.5 bar, for three different surface coverages of ^{4}He. The two dashed lines are the Ginzburg–Landau results for d = 290 and 360 nm. The dot–dashed line indicates the temperature variation of the dimensionless thickness.

Conclusions

The behaviour of superfluid ^{3}He when confined to a film of thickness less than the correlation length of the order parameter has been examined. It has been concluded that the B phase, the stable bulk–low temperature phase, apparently cannot live in small spaces. The phase in the confined geometry has properties expected for the A phase. The character of the ^{3}He superfluid state in thin layers is profoundly affected by ^{4}He coating of surfaces and this leads to behaviour which approaches that of bulk ^{3}He. The reasons for this are not clear.

References 2

[1] M R Freeman, PhD dissertation, Cornell University, Ithaca, N.Y., 1988.

[2] V L Ginzburg and L D Landau, *ZETP* **20**, 1064, English translation in Collected Papers, 546 (1950).

[3] A J Leggett, *Ann. Phys.* (N.Y.) **85**, 11 (1974).

[4] E V Thuneberg, private communication.

[5] D D Osheroff and L R Corruccini, *Phys. Lett.* **51A**, 447 (1975); W F Brinkman and H Smith, ibid., 449.

[6] D F McQueeney, G Agnolet and J D Reppy, *Phys. Rev. Lett.* **52**, 1325 (1984).

LECTURES 3 : SURFACE PROBLEMS IN LIQUID ^{3}He

Introduction

Normal liquid ^{3}He exhibits a number of interesting magnetic properties at low temperatures. In this lecture we shall discuss the nuclear spin dynamics of ^{3}He atoms in contact with solid surfaces containing $I = {}^1/_2$ nuclei (^{19}F) at temperatures below 1 K. We shall examine the role of the surface in the enhancement of the thermal relaxation of liquid ^{3}He magnetization.

From the third law of thermodynamics we may expect that, as the absolute temperature T goes to zero, the thermal equilibrium time of a system tends to infinity. The nuclear spin–lattice relaxation time, T_1, in ^{3}He at low temperatures does not follow the curve predicted by simple theory and decreases as T decreases in the range 1 K – 1 mK. We have attempted to understand and explain this anomalous behaviour. A discussion of the conventional theory used to predict the behaviour of T_1 in ^{3}He is followed by a review of experiments performed at Cornell on thin films of ^{3}He on substrates of large surface area and of magnetic susceptibility experiments on these systems. Finally an explanation of the linear dependence of T_1 on temperature is provided by the Hammel–Richardson model.

Spin–lattice relaxation in liquid ^{3}He

The transition probability for a nucleus of spin I going from one spin state m to another state m′ via some interaction with the lattice, which surrounds the nucleus, may be written in terms of Fermi's Golden Rule as

$$W = \frac{2\pi}{\hbar} \left| \left\langle m' | V | m \right\rangle \right|^2 \rho(E) \tag{3.1}$$

where V is the interaction potential between the spin and the lattice and $\rho(E)$ is the density of lattice states at energy E.

For a magnetic dipolar interaction mechanism the relaxation rate is given by [1]

$$T_1^{-1} = (\gamma H_D)^2 f(\omega, \tau_c) \tag{3.2}$$

where $f(\omega, \tau_c)$ is a spectral function, γ is the nuclear magnetogyric ratio, H_D is the average dipolar field produced by neighbouring spins at each nuclear site, ω is the nuclear Larmor frequency in a magnetic field B, and τ_c is the characteristic time for fluctuations of the dipolar field.

In bulk ^3He the dominant interaction may be expected to be the dipole–dipole interaction between neighbouring nuclei which is modulated by the motion of the atoms involved. For Fermi particles the expression for T_1 may be written as [2]

$$\frac{1}{T_1} \simeq \left[\gamma H_D\right]^2 \left[\frac{a}{V_F}\right] \left[\frac{T}{T_F}\right]^2 \tag{3.3}$$

where a is the radius of the atom, V_F is the Fermi velocity and T_F is the Fermi temperature. In bulk liquid, which behaves like a Fermi liquid, T_1 should therefore increase as T^{-2}. Putting in numbers gives $T_1 = 10^5/T^2$ s mK^{-2}. At 1 mK we expect $T_1 \simeq 10^5$ s. When we did some early NMR experiments on ^3He at millikelvin temperatures in the 1970's we were surprised to find $T_1 \sim 1$ s.

Experimentally for normal ^3He it has been found that $T_1 \propto T$. The relaxation rate $1/T_1$ increases with the ratio of surface to volume, which suggests that all relaxation takes place at the surface. Figure 3.1 compares, in a schematic way, the theoretically predicted curve with the experimental results obtained for temperatures below 1 K. In order to preclude surface effects, ℓ, the linear dimension of the container, would have to be $\approx$ 3 m, where I put $\ell = \sqrt{DT_1}$, with D the spin diffusion constant. This quantity is probably more than all of the ^3He currently available in the world! A different approach has therefore to be used in the study of the surface relaxation in ^3He. One possible way is to introduce large internal surfaces containing another spin species into the sample and to examine the interaction between the ^3He nuclei and this other spin species.

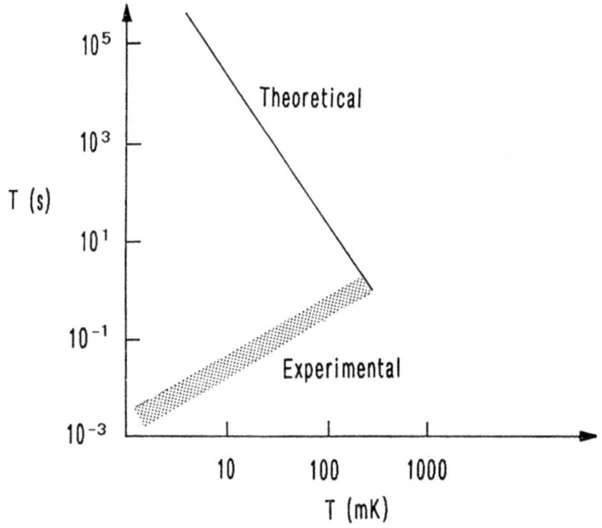

Figure 3.1 Nuclear spin–lattice relaxation time T_1 as a function of temperature in normal ^3He. Experimental results are compared with the results predicted by conventional theory.

Experiments with thin films

In these experiments a substrate of large surface area is covered with thin films of ^3He [3]. Pulsed NMR methods involving $\pi/2$ and π pulses are used to measure T_1. Further details on NMR methods can be found in the excellent book by Slichter [4]. We decided to use tiny teflon spheres as our substrate. These particles are available commercially (they are used in highly reflecting white paint) and the material is known as DLX–6000 (Du Pont). The chemical composition of DLX–6000 is shown in Figure 3.2. The substrate may be viewed as a frozen fluid of spin $^1/_2$ particles (^{19}F nuclei).

The spin diffusion rate of an excitation of ^{19}F spins to the surface via spin *flip–flop* interactions is given by

$$D \simeq \gamma H_D a^2 = \frac{\gamma^2 \hbar}{a^3} \cdot a^2 \qquad (3.4)$$

where a is the interspin spacing. For a particle of DLX–6000 of radius 250 nm the time taken for such an excitation to go from the centre of a sphere of radius r to the surface is given by

$$\tau \approx \frac{r^2}{D} = \frac{r^2 a}{\gamma^2 \hbar}$$

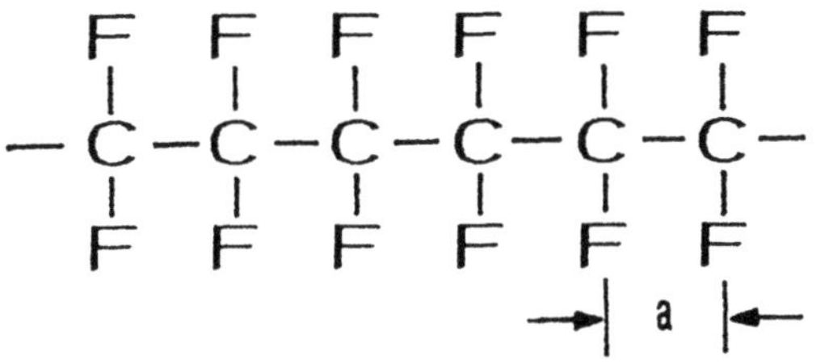

Figure 3.2 Chemical structure of the polymer $(CF_2)^n$ used in the experiments. The Du Pont Chemical company provided the material (DLX–6000) in the form of spheres of diameter 1 μm.

Putting in numbers gives $\tau \sim 5$ s. If the measured $T_1 > 5$ s, then all relaxation occurs at the surface. For a solid without surface relaxation T_1 should increase rapidly at low temperatures to values of minutes or hours. (In practice the value is dependent on paramagnetic impurities which are always present.) We can expect surface effects to be important for the DLX–6000 spheres in ^3He and this has been clearly demonstrated in a series of experiments performed at Cornell, at the Oersted Institute and at MIT..

Friedman [5] examined the dependence of the ^{19}F relaxation rate on the ^3He coverage at 1.2 K. T_1^{-1} is plotted as a function of the concentration of ^3He on the surface in Figure 3.3. As is shown, an increase in the relaxation rate occurs just below and just above monolayer completion. At one monolayer coverage a minimum in the relaxation rate is observed, presumably because of the inhibition of the motion of ^3He atoms, which are highly mobile in the surface layer on the substrate. This may be understood by analogy to a layer of marbles in the bottom of a box. When the bottom is completely covered, the mobility of the marbles is low.

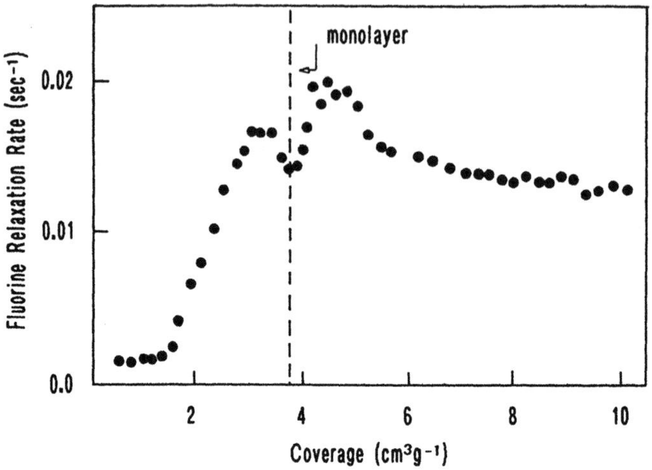

Figure 3.3 ^{19}F relaxation rate (T_1^{-1}) in DLX—6000 as a function of ^3He coverage. A monolayer coverage of ^3He is indicated at approximately 3.8 STP cm^3 per gram of DLX—6000.

The introduction of a small amount of ^4He into the chamber before the addition of the ^3He produced no change in the ^{19}F relaxation rate. One monolayer of ^3He was then introduced, followed by a further amount of ^4He. The results are shown in Figure 3.4. The addition of ^3He produces an increase in the relaxation rate, which decreases with the further addition of ^4He. This confirms previous results that ^4He plates preferentially due to the lower zero point motion [6]. Communication is cut off between the ^3He and ^{19}F by the layer of ^4He.

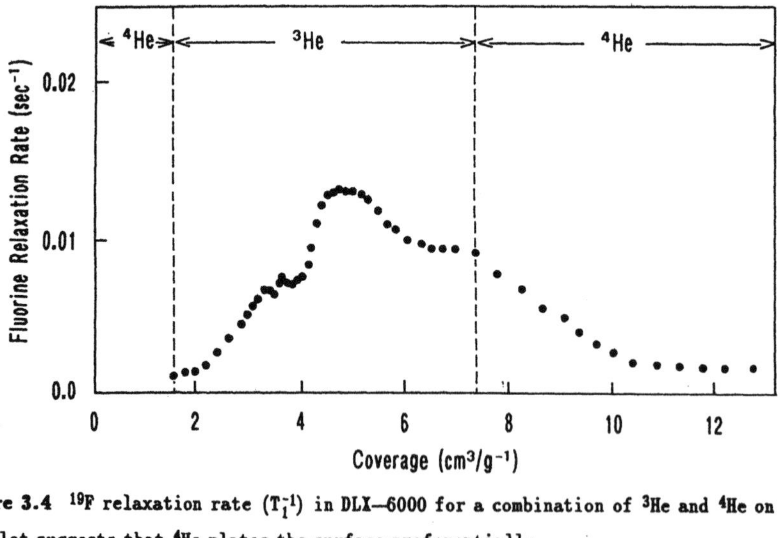

Figure 3.4 ^{19}F relaxation rate (T_1^{-1}) in DLX—6000 for a combination of ^3He and ^4He on the surface. The plot suggests that ^4He plates the surface preferentially.

Double resonance techniques of the type developed by Hahn [7] were then used by Friedman et al. in an attempt to measure the strength of coupling between the ^{19}F and ^{3}He spins. Sequences of pulses were applied to *mix* the spin magnetizations together and frequencies, phases and pulse widths were monitored. The results were not satisfactory. It turned out that the spins were doing the mixing themselves and our efforts were therefore superfluous.

In an experiment to demonstrate this mixing effect a π magnetization inverting pulse was applied at the ^{19}F resonance frequency and the magnetization recovered with $T_1 \approx 50$ s. The magnetization of the ^{3}He nuclei was found to follow the magnetization of the ^{19}F nuclei in the substrate. Magnetization is plotted as a function of time in Figure 3.5. A series of π pulses were applied to the ^{3}He nuclei and the recovery was monitored. Equilibration of the ^{3}He magnetization to the ^{19}F magnetization occurred rapidly with a time constant of 50 ms. The two spin systems have the same spin temperature [1]. Similar effects have been observed in organic materials containing protons. These systems are, however, homogeneous, with the spin transfer occurring between two protons, and the resonance frequencies are very close. The ratio of the resonance frequencies for ^{3}He and ^{19}F is given by

$$\frac{\gamma_{^3He}}{\gamma_{^{19}F}} \simeq \frac{3}{4}$$

The ^{19}F and ^{3}He resonance frequencies for the magnetic field used in the experiment are therefore quite different. The line widths are small ($\lesssim 10^{-3}$) and there is no obvious overlap of the resonance lines. This result is brought into perspective when one considers the possibility of two radio stations, broadcasting at different frequencies, communicating with each other.

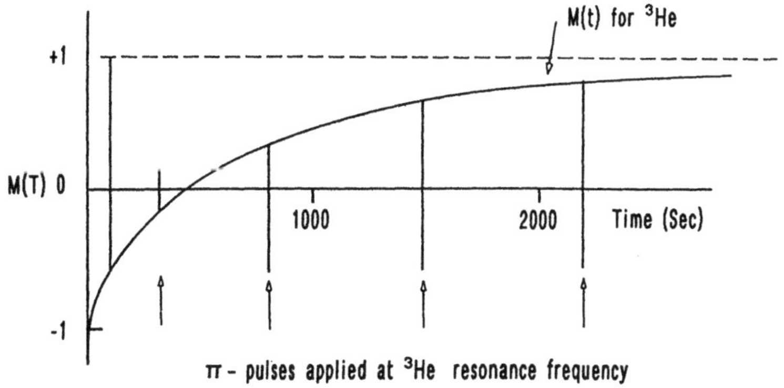

Figure 3.5 Schematic summary of the results obtained in the double resonance experiment discussed in the text. The magnetization of the ^{3}He nuclei follows the magnetization of the ^{19}F nuclei in the substrate (DLX—6000 particles). A series of π pulses applied at the ^{3}He resonance frequency results in the recovery of the ^{3}He magnetization to the ^{19}F magnetization with $T_1 = 50$ ms.

When Friedman applied π pulses to the ^{19}F nuclei and then monitored the ^{3}He signal, he found that the ^{3}He nuclei had the same relaxation time as the ^{19}F spins (50 s). From the results of this experiment it may be concluded that the polarized surface layer relaxes bulk ^{3}He liquid spins. The helium and fluorine spins rapidly reach the same spin temperature. However, the microscopic reasons for this process have not been clarified. Experiments done at different frequencies show that $1/T_1 \propto 1/\omega$ over a very wide range. This implies an unusual spectral distribution associated with the adsorbed ^{3}He atoms, reminiscent of 1/f noise. No detailed explanation of this observation has been given.

An analysis of the data of the thin film experiments results in the following conclusions:

(a) The motion of surface ^{3}He atoms relaxes spins on both sides of the boundary.

(b) Spectral contributions due to the ^{3}He dynamics cover an enormous range of frequencies.

Magnetic susceptibility experiments

Various experiments have been performed to probe surface properties of ^{3}He around 1 mK. To understand what occurs in these experiments it is necessary to discuss the related concept of surface magnetism. The magnetic susceptibility χ of normal liquid ^{3}He should theoretically imitate the Fermi gas behaviour. Above T_F χ should follow Curie's law ($\chi \propto 1/T$) but it should be temperature-independent for $T < T_F$ where Pauli susceptibility behaviour should apply.

A schematic plot of measured χ values as a function of temperature is shown in Figure 3.6 [10]. The Pauli susceptibility law applies for $T \lesssim T_F$ but χ increases rapidly at much lower temperatures, $T << T_F$. The effective Fermi temperature is $\approx$ 300 mK. The deviation of the susceptibility from the predicted behaviour may be ascribed to a surface effect – the ^{3}He atoms on the surface become polarized and dominate the magnetic susceptibility. This was shown in early

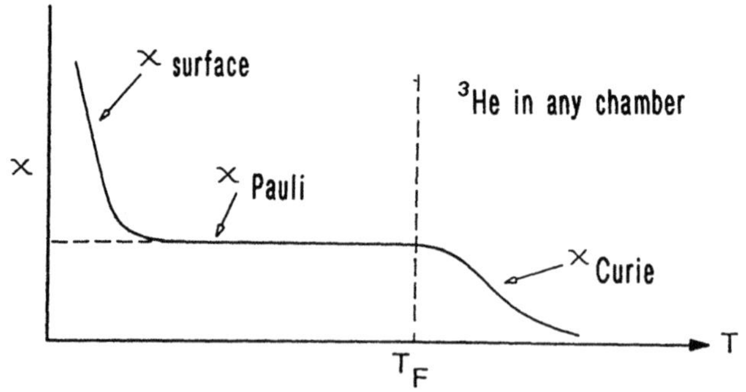

Figure 3.6 Schematic representation of the susceptibility of liquid ^{3}He as a function of temperature. The surface effect, $\chi_{surface}$, plays an important role in the ^{3}He NMR behaviour at low temperatures.

experiments by Keyston and Laheurte [8] and Wheatley et al. [9], who demonstrated that the behaviour of ^{3}He was modulated by a surface effect, which could be eliminated by the addition of ^{4}He in the experiments.

Ahonen et al. [11] studied magnetic susceptibility in superfluid ^{3}He–B with mylar as the substrate. They found that the temperature dependence of χ decreased with the addition of small quantities of ^{4}He. They concluded that the ^{3}He magnetic susceptibility is enhanced by the layers of ^{3}He adjacent to the substrate. Figure 3.7 shows a plot of χ as a function of T for ^{3}He. Besides mylar and silver, which show an enhanced susceptibility from five surface layers of ^{3}He, all substrates have an enhanced susceptibility corresponding to one or two surface layers.

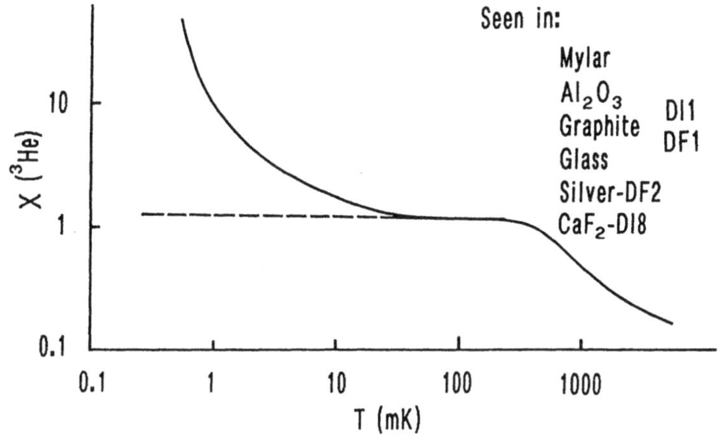

Figure 3.7 A plot of the susceptibility of liquid ^{3}He as a function of temperature. The increase in χ at low temperatures is observed for all of the substrates listed.

A possible model for the two–dimensional magnetic ordering in ^{3}He at low temperatures is illustrated in Figure 3.8. To account for the observed critical temperature of 0.5 mK an indirect exchange of two surface atoms is proposed. The probability of exchange between a surface and a liquid atom is high, while the frequency of a liquid atom – liquid atom exchange is very high. This *surface–liquid–liquid–surface* exchange may be compared to the RKKY coupling, which involves 'localized spin–conduction electron–conduction electron–localized spin' interactions in metals. This model has not yet been confirmed experimentally.

Surface magnetization χ_s has been studied as a function of temperature for layers of ^{3}He on very small CaF_2 crystals. Figure 3.9 shows the plot of $1/\chi_s$ versus T. As is shown, the surface magnetization follows the Curie–Weiss law, with a positive temperature intercept of

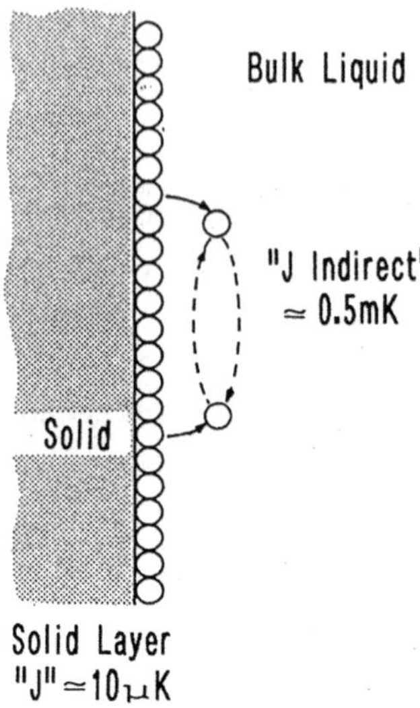

Figure 3.8 A possible model for the two–dimensional magnetic ordering in ^{3}He in the millikelvin regime. To account for the observed critical temperature an indirect exchange mechanism is proposed.

approximately 0.5 mK. This positive temperature intercept indicates that the surface layer is ordering as a two–dimensional ferromagnet. Experiments at the University of Southern California have provided evidence for low–frequency collective modes below 1 mK and a model based on 2–D ferromagnetism has been introduced to explain these measurements.

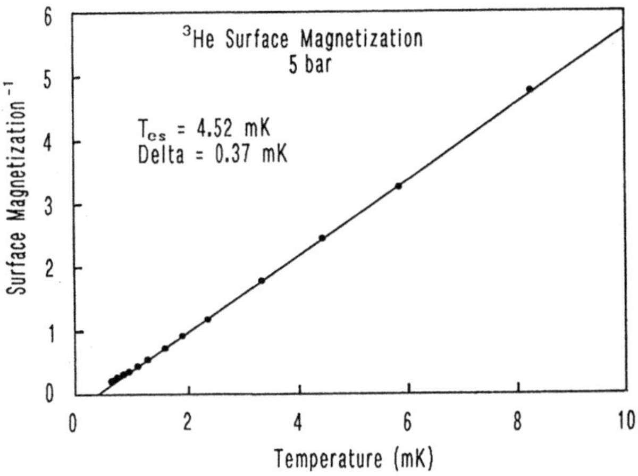

Figure 3.9 The inverse of the surface magnetization of ^{3}He at a pressure of 5 bar as a function of temperature. The positive temperature intercept Δ = 0.37 mK is evidence for a two–dimensional ferromagnetic ordering in the surface layer.

The existence of a surface susceptibility results in a simple explanation for the temperature dependence of the ^{3}He relaxation time. Hammel and I [12] proposed a model to explain this temperature dependence and a series of experiments have confirmed the validity of this model. The results of these experiments are illustrated graphically in Figure 3.10, which shows a plot of T_1 versus T for ^{3}He on CaF_2 crystals. For temperatures above 10 mK, T_1 is linear in temperature but becomes independent of temperature below 5 mK. The characteristic relaxation time of ^{3}He is approximately 10 ms. The result indicates that T_1 is dominated by the relaxation at the surface and is linked to the two–dimensional magnetic ordering of the ^{3}He film at low temperatures. The curve through the points is discussed below.

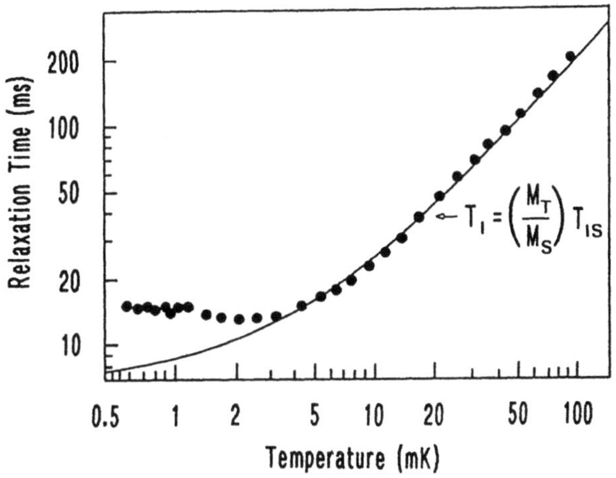

Figure 3.10 Spin–lattice relaxation time (T_1) for ^{3}He on a CaF_2 substrate at a pressure of 5 bar as a function of temperature. The solid line is a fit based on the Hammel–Richardson model using measured values of the magnetization. The curve is scaled to pass through the data point at 30 mK.

In Figure 3.11 the relaxation times for the ^{19}F substrate spins are plotted as a function of temperature. For temperatures above 2 mK there is little change in the relaxation rate, confirming the assumption of a temperature–independent relaxation rate above this temperature. The increase in T_1 at low temperatures can be attributed to heating effects. The data follow the curve

$$\frac{1}{T_{1_{19F}}} = \frac{1}{T_{1_{3He\ surf}}} \left[\frac{N_{surf}}{N_{19F}}\right] \tag{3.5}$$

where N_{surf} is the number of surface spins, N_{19F} is the number of fluorine spins, and $T_{1_{surf}}$ is the internal relaxation time of the surface ^{3}He spins.

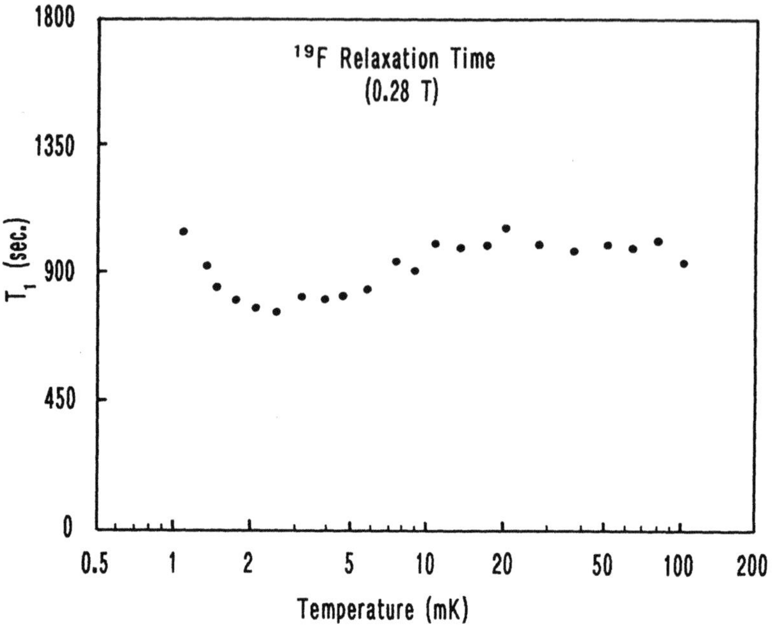

Figure 3.11 A plot of the ^{19}F relaxation time as a function of temperature in a magnetic field of 0.28 T and at a ^3He pressure of zero bar. This relaxation time is predicted to be temperature–independent by the Hammel –Richardson model. A heating effect accounts for the increase in T_1 at low temperatures.

A conservation of polarization takes place at the surface between the liquid ^3He above the surface and the surface atoms during the mixing of the two spin systems. The ratio of magnetization may be written as

$$\frac{M_s - M_s^o}{M_s^o} = \frac{M_{liq} - M_{liq}^o}{M_{liq}^o} \tag{3.6}$$

where M_s is the surface magnetization and M_s^o is the equilibrium magnetization.

At low temperatures bulk processes are negligible and one would expect

$$\frac{1}{T_1} = \frac{1}{T_{1_{surf}}} \left[\frac{M_s}{M_T}\right] \tag{3.7}$$

where M_T is the total magnetization, which is roughly temperature–independent. The surface magnetization M_s varies approximately as $1/T$ and this leads to a linear dependence of T_1 on T if it is assumed that $T_{1_{surf}}$ is constant. The relaxation rate is plotted as a function of temperature in Figure 3.10 and it can be seen that the predicted line fits the data.

It can be concluded that the Hammel–Richardson model is successful in explaining the temperature dependence of the ^3He relaxation time as the result of the localization of the relaxation to the interfacial layer of ^3He.

Conclusion

Measurements of nuclear spin relaxation and magnetic susceptibility in a system of liquid ^3He in contact with a substrate have been discussed. A strong coupling between ^3He nuclei in the surface layer and the ^{19}F nuclei in the substrate has been shown to exist. The relaxation of ^{19}F nuclei in the substrate has been shown to correlate with the coverage of ^3He on the surface. The magnetic susceptibility and magnetic ordering are affected by a surface contribution, which therefore contributes to the observed T_1 in ^3He at low temperatures. The temperature dependence of the ^3He relaxation time has been successfully explained by the Hammel–Richardson model. This temperature dependence can be attributed to the surface magnetization behaviour of ^3He.

References 3

[1] A Abragam, *The Principles of Nuclear Magnetism,* Clarendon Press, Oxford, 1961.

[2] D Vollhardt and P Wölfle, *Phys. Rev. Lett.* **47**, 190 (1981).

[3] L Friedman, P Millet and R C Richardson, *Phys. Rev. Lett.* **47**, 1078 (1981).

[4] C P Slichter, *Principles of Magnetic Resonance* (3rd Edn) Springer–Verlag, Berlin, Heidelberg 1990.

[5] L J Friedman, PhD dissertation, Cornell University, Ithaca, N.Y., 1982 (unpublished).

[6] D F Brewer, D J Cresswell and A L Thomson, *Low Temperature Physics,* **LT–13, Vol.1** 163 (1974).

[7] S Hartmann and E L Hahn, *Phys. Rev.* **128**, 2042 (1962).

[8] J R G Keyston and J P Laheurte, *Phys. Lett.* **24A**, 132 (1967).

[9] W C Black, A C Mota, J C Wheatley, J H Bishop and P M Brewster, *J. Low Temp. Phys.* **4**, 391 (1971).

[10] H M Bozler, T Bartolac, K Luey and A L Thomson, *Phys. REv. Lett.* **41**, 490 (1978).

[11] A I Ahonen, T Kodama, M Krusins, M A Paalanen, R C Richardson, W Schoepe, and Y Takano, *J. Phys.* **C9**, 1665 (1976).

[12] P C Hammel and R C Richardson, *Phys. Rev. Lett.* **52**, 1441 (1984).

LECTURE 4 : Heat Transfer at Boundaries

Introduction

Heat transfer at boundaries, the subject of my fourth lecture, is a complex process at low temperatures. A thermal boundary resistance exists between any two dissimilar materials. The most dramatic example of this occurs between liquid helium and a solid. There is a sharp region in space where a thermal gradient occurs. The thermal boundary resistance between a solid and liquid ^{3}He or ^{4}He or a mixture of ^{3}He and ^{4}He in all phases is called the Kapitza resistance and is defined as $R_k = (\Delta T/\dot{Q})$, where ΔT is the temperature difference across the liquid–solid interface and $\dot{Q}$ is the total heat flow. This simple linear form holds over a fairly wide range. The Kapitza resistance is therefore dependent on the area of the interface and varies from system to system. The Kapitza resistivity is independent of surface area and is defined as $\mathcal{R}_k = \Delta T/(\dot{Q}/A)$, where A is the area of the interface and $\dot{Q}/A$ is the heat flux.

Kapitza's discovery [1] of a thermal discontinuity across the boundary of a solid and superfluid ^{4}He in the presence of a heat flow initiated much theoretical and experimental work in the field of heat transport. A review of some of this work is presented below.

High Temperature Results

Khalatnikov [2,3] proposed the acoustic mismatch theory, which uses classical acoustic theory to calculate the heat transport across the boundary. It is very similar to the Planck radiation law for electromagnetic radiation but, of course, involves phonons.

This theory predicts a T^{-3} temperature dependence for the Kapitza resistance. In acoustic terms

$$\dot{Q} = W \frac{v_\ell}{2} A \frac{E}{V} \tag{4.1}$$

where W is the transmission probability, E is the energy, V is the volume and v_ℓ is the phonon velocity in the liquid. The energy per unit volume can be written as

$$\frac{E}{V} = \frac{(k_B T)^4}{(\hbar\, v_\ell)^3} \tag{4.2}$$

where k_B is Boltzmann's constant.

Phonon reflection and transmission at a boundary between a solid and a liquid involves Snell's law, $(\sin \alpha_\ell)/(\sin \alpha_s) = (v_\ell)/(v_s)$, where α is the angle between the normal to the interface and the direction of phonon propagation and v_ℓ and v_s are the phonon velocities in the liquid and solid respectively. The critical angle from liquid to solid is given by $\alpha_c = \sin^{-1}(v_\ell/v_s)$ The phonon velocity in liquid ^3He is approximately 200 ms^{-1}, while that in a typical solid is 5000 ms^{-1}. Only a small fraction f of phonons $\sim \frac{1}{2}(v_\ell/v_s)^2$ within a critical cone of angle $\sim 3^\circ$ can be transmitted through the interface. A second factor comes in for these phonons, i.e. the acoustic transmission coefficient t which is defined as

$$t = \frac{4 Z_\ell Z_s}{(Z_\ell + Z_s)^2} \approx \frac{4 \rho_\ell v_\ell}{\rho_s v_s} \tag{4.3}$$

where ρ is the density and $Z = \rho v$

In liquid ^3He $\rho_\ell = {}^1/_{10}\,\rho_s$ and $v_\ell = {}^1/_{10}\,v_s$. With these values we can estimate the probability that a phonon striking the interface will be transmitted into the solid as

$$W = ft = \frac{\rho_\ell v_\ell^3}{\rho_s v_s^3} < 10^{-5} \tag{4.4}$$

Therefore less than 10^{-5} phonons are transmitted through the interface. The Kapitza resistance takes the form

$$R_k = \frac{15 \hbar^3 \rho_s v_s^3}{16 \pi^5 k_B^4 T^3 \rho_\ell v_\ell}$$

$$= \frac{5 \times 10^{-2}}{T^3} KW^{-1} \tag{4.5}$$

The product of R_k and T^3 is therefore predicted to be a constant. However, for a typical metal a plot of RT^3 as a function of log T shows that RT^3 is less than the predicted constant value for T < 1 mK and that RT^3 decreases for T > 1 mK (Figure 4.1). The use of a clean surface, obtained by cleaving alkali halide crystals, resulted in experimental values close to the predicted value for R_k. It is found that the introduction of disorder on the surface produces a decrease in the boundary resistance of almost an order of magnitude. This decrease is possibly attributable to the disorder converting the longitudinal sound vibrations in the liquid to transverse sound vibrations in the solid.

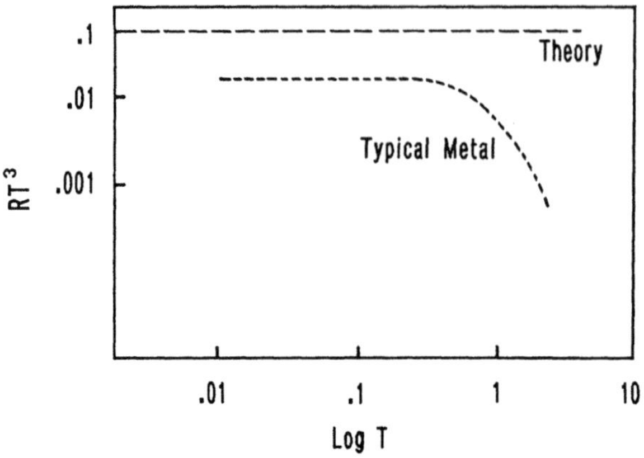

Figure 4.1 Kaptiza resistance multiplied by T^3 as a function of temperature T showing the difference found between theory and experiment.

If the Khalatnikov theory were valid, the millikelvin refrigerator would not be a reality today. Particle size is important in the measurement of the thermal equilibrium time and therefore in the design of heat exchangers for dilution refrigerators. Before experiments were performed, people concluded that particles of 1 micron were too small since the thermal wavelength of the ^{3}He phonons increases with decreasing temperature at low temperatures and it was felt that the particle size should not be smaller than the thermal phonon wavelength. However, it has subsequently been established that the smaller the heat exchange particle, the more efficient the cooling system. Today particles used are of the order of 500 Å and lower minimum temperatures have been obtained. Many experiments have been undertaken in the pursuit of a deeper understanding of the physics underlying heat transfer processes.

Ahonen et al. [4] measured the thermal resistance between sintered silver powder (approximately 70 nm) and liquid ^{3}He. At low temperatures density excitations exist in liquid ^{3}He but the dominant excitations are those of quasiparticles. When $\omega\tau << 1$, the excitations are ordinary sound waves, where ω is the excitation frequency and τ is the relaxation time. When $\omega\tau >> 1$, zero sound excitations occur. For thermally excited waves in ^{3}He the transition from ordinary to zero sound waves occurs at approximately 0.1 K. Resistance is plotted as a function of temperature in Figure 4.2. The measured results obtained within the sintered silver are indicated by the heavy line. The fine lines indicate the expected contributions to the total resistance from the electron–phonon resistance within the sinter, the boundary resistance and the zero–sound to quasiparticle resistance within the ^{3}He. The thermal boundary resistance and the electron–phonon resistance both show a T^{-3} temperature dependence. At 1 mK the measured resistance of approximately 10^5 KW^{-1} was found to be considerably less than the expected resistance of 3×10^6 KW^{-1} for the boundary. At 6 mK the expected and measured resistances had the same value. There are two schools of thought on the analysis of these results. Harrison [5] and his followers believe that the heat transfer process can be understood entirely in terms of non–magnetic effects. A second group, (which includes myself), supports the view that surface magnetic phenomena play an important role in the process.

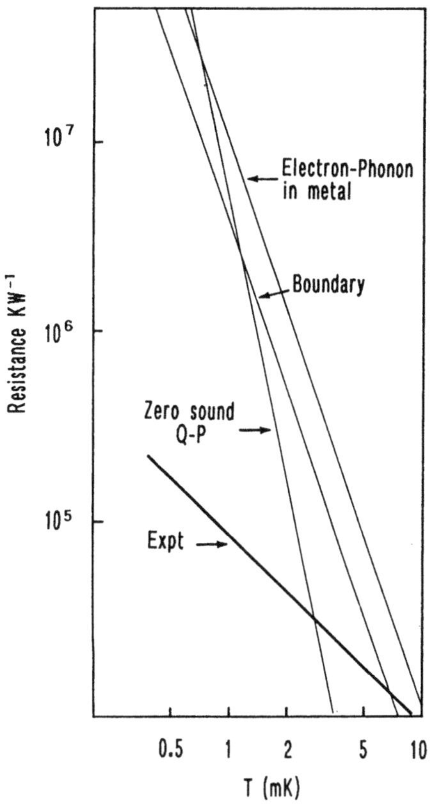

Figure 4.2 Boundary resistance between sintered silver powder (70 nm) and liquid ^{3}He. (After Ahonen et al. [4])

Magnetic effects were shown to be of importance in an early experiment by Wheatley et al. [6], who used adiabatic demagnetization of powdered cerium magnesium nitrate (CMN) to cool liquid ^{3}He at the millikelvin level. The magnetic effects were found to improve conductivity with decreasing temperature. Leggett and Vuorio [7], in a theoretical study of the problem, attributed this effect to interactions, involving magnetic fluctuations, between the paramagnetic moments of CMN and liquid ^{3}He quasiparticles.

Some years ago, when I spent a sabbatical at Bell Labs, Doug Osheroff and I [8] studied the temperature dependence of the Kapitza resistance in varying magnetic fields in a system of silver powder and liquid ^{3}He. The apparatus we used is sketched in Figure 4.3. Also shown in Figure 4.3 is the product of the boundary resistance and the temperature plotted as a function of

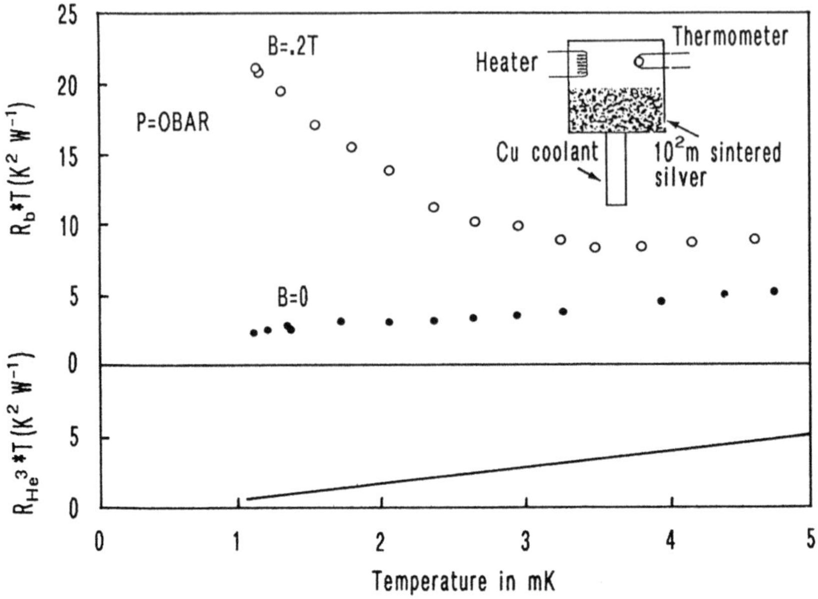

Figure 4.3 Plot of boundary resistance multiplied by temperature versus temperature at 0 bar in fields of 0 T and 0.2 T. The insert is a schematic representation of the apparatus used.

temperature at 0 bar in magnetic fields of 0 T and 0.2 T. At low temperatures it was found that $R_{^3He} \propto T^{-1}$. This result confirms earlier results. The Kapitza resistance was found to increase linearly with magnetic field up to 0.2 T but was insensitive to larger fields. Heat transfer is larger in this system in the absence of a magnetic field.

In a recent attempt to further our understanding of heat transport between ^{3}He and solids at low temperatures, Yue Hu [9] has examined the Kapitza resistance in a rather simple system consisting of liquid ^{3}He and vapour–deposited silver films of 5000 Å thickness deposited on a glass slide. The slide was used as a glass capacitor, schematically illustrated in Figure 4.4(a), to measure temperatures. The capacitor was immersed in liquid ^{3}He and thermal contact was made between liquid and capacitor through the silver electrodes. The fractional change of the glass capacitance, $(C - C_0)/(C_0)$, is plotted in Figure 4.4 (b) as a function of temperature, where C is the capacitance of the glass capacitor and C_0 is the reference capacitance at 1 K. The variation in C with temperature is linked to dielectric relaxation processes in the glass.

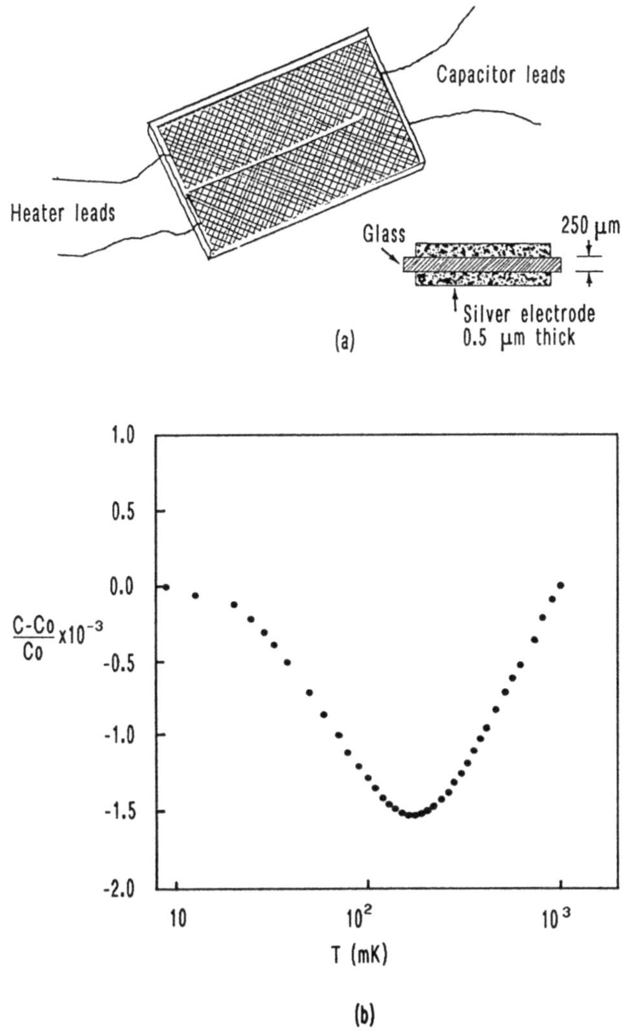

Figure 4.4 (a) Schematic representation of the glass capacitor with one of the electrodes partially split so that it can be used as a heater **(b)** Relative capacitance change as a function of temperature for a measuring frequency of 1 KHz. The minimum occurs for $\omega\tau \sim 1$ where τ is the relaxation time of the electric dipoles in the dielectric medium (Corning Cover Glass).

The silver film on one side of the glass was scratched, as shown, to produce a u-shaped conduction path for heater current. Figure 4.5 shows the response of the glass capacitance to heater power. The glass capacitance C was monitored with and without current through the heater and C was plotted as a function of time. The slight upward slopes of the two straight lines, fitted to the equilibrium values of C, are caused by the gradual upward temperature drift of the liquid ^{3}He. The ^{3}He was at an average temperature of 4.94 mK. The temperature change of the capacitor ΔT between the current on and current off states was 0.325 mK. The Kapitza resistivity was calculated from the equation

$$\mathscr{R}_k = \frac{\Delta T}{\dot{Q}/A} = \frac{\Delta T \cdot A}{I^2 r} \qquad (\dot{Q} = I^2 r) \qquad (4.6)$$

where I is the current and r the electrical resistance of the heater.

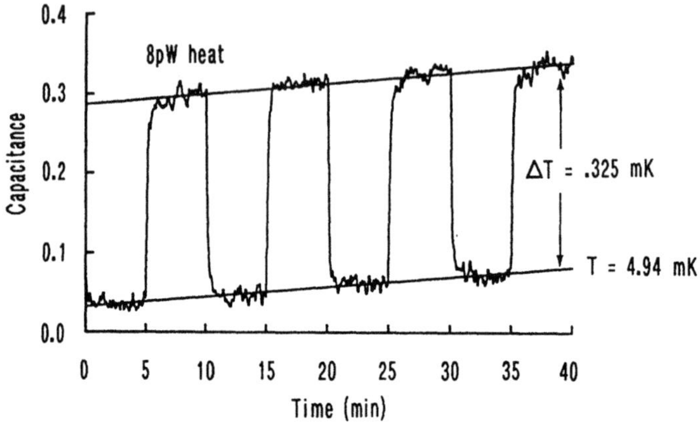

Figure 4.5 The response of the glass capacitance to periodically applied heater power. The temperature difference ($\Delta T = 0.325$ mK) between the power on and off conditions can be obtained from the capacitance change.

The Kapitza resistivity was measured at a ^{3}He pressure of 0 bar and 10 bars. The logarithm of the Kapitza resistivity is plotted as a function of temperature in Figure 4.6. As can be seen, the Kapitza resistivity at 10 bars is smaller by a constant factor than that at 0 bar at a

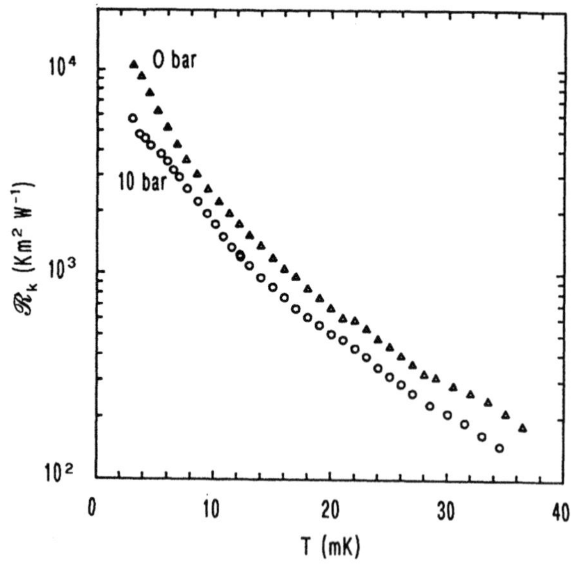

Figure 4.6 The Kapitza resistivity for the glass capacitor system at 0 bar and 10 bar as a function of temperature.

given temperature over a wide range of temperatures: $\mathscr{R}_k$ (10 bars) = α (10 bars) · $\mathscr{R}_k$ (0 bar, 7). The proportionality constant α appears to be independent of temperature. The value at 10 bars is numerically identical to the inverse of the ratio of the effective mass of ^3He at 10 bars to that at 0 bar, i.e.

$$\alpha\,(10\text{ bars}) = \frac{\mathscr{R}_k\ (10\text{ bars})}{\mathscr{R}_k\ (0\text{ bar})} = \left[\frac{m^*\ (10\text{ bars})}{m^*\ (0\text{ bar})}\right]^{-1} = 0.725 \tag{4.7}$$

In Figure 4.7 the product of the Kapitza resistivity and the square of the temperature is plotted as a function of temperature. It can be deduced that there are three parallel channels for heat conductivity, σ_1, σ_2, and σ_3 where $\sigma = \sigma_1 + \sigma_2 + \sigma_3$ and $R = 1/A\sigma$. For temperatures above 35 mK, $\mathscr{R}_k \propto T^{-3}$, $\sigma_3 \sim T^3$, and the heat transport is dominated by phonons. For temperatures between 15 and 35 mK, $\mathscr{R}_k \propto T^{-2}$, $\sigma_2 \sim T^2$, and no magnetic field dependence occurs.

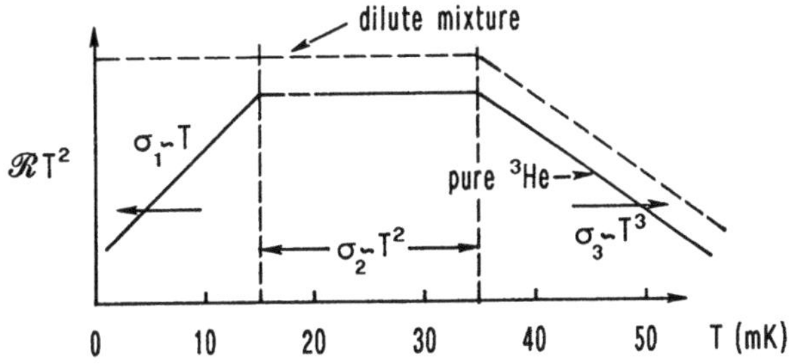

Figure 4.7 The ^3He Kapitza resistivity multiplied by T^2 as a function of temperature T. Three distinct regions can be indentified below 50 mK.

For temperatures below 35 mK, $\mathscr{R}_k \propto T^{-1}$, $\sigma_1 \sim T$, and a magnetic field dependence is seen. In the intermediate and low temperature regimes $\mathscr{R}_k$ is pressure dependent between 0 and 10 bar at a fixed temperature. $\mathscr{R}_k$ m* is a constant where m* is the effective mass (m* increases by ~ 30%). This is illustrated in Figure 4.8.

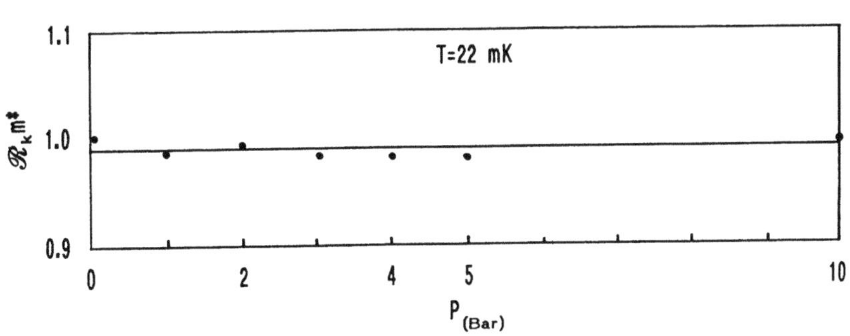

Figure 4.8 Plot of the Kapitza resistivity multiplied by the effective mass m* of liquid ^3He from 0 to 5 bar. $\mathscr{R}_k$ m* can be seen to be roughly constant over this range.

Similar results were obtained for dilute mixtures of ^3He and ^4He between 15 mK and 45 mK. In the high temperature regime $\sigma_3 \sim T^3$ and there is not much change if we apply a magnetic field or change the pressure. At lower temperatures $\sigma \sim T^2$ and does not change with magnetic field but does change with pressure. These results are consistent with an extended version of the theory of Leggett and Vuorio. In the inelastic scattering of a ^3He quasiparticle by a surface the quasiparticle changes momentum from a state $\hbar k$ to a state $\hbar k'$. This process involves two factors of $(\partial N/\partial E)_{FS}$ or $1/m^*$ where $(\partial N/\partial E)_{FS}$ is the density of states of liquid ^3He at the Fermi surface. It is expected that $\mathscr{R}_k$ should be proportional to $(m^*)^{-2}$ and an exchange enhancement factor. Yue Hu's results suggest that the exchange enhancement factor decreases with pressure, contrary to expectations. $\mathscr{R}_k$ was found to scale accurately as $1/T_F$, where T_F is the Fermi temperature. This is bad news for work on dilute mixtures and it has not been possible to cool such mixtures to less than 100 μK. Nature has been kinder in the case of pure ^3He.

Yue Hu studied the Kapitza resistivity at a constant temperature in magnetic fields from 0 to 80 gauss. Figure 4.9 shows the fractional change of the Kapitza resistivity, $\mathscr{R}(B) - \mathscr{R}(0)/\mathscr{R}(0)$ plotted as a function of magnetic field B at 7 mK, where R(0) is the Kapitza resistivity at zero field. To obviate any possible offset in the field, the direction of the field was reversed and the data represented as the open square points. No offset in the field was observed. As is shown, a

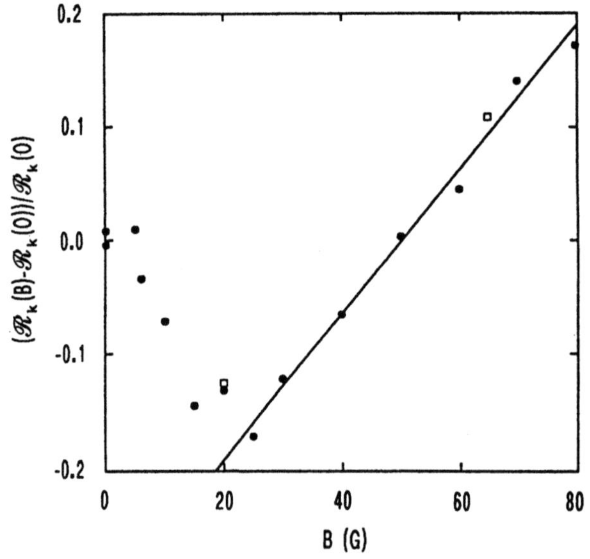

Figure 4.9 The fractional change in the Kapitza resistivity of ^3He in small magnetic fields. A minimum is seen near 25 G. The open squares are data taken with the magnetic field reversed. The straight line is a fit to the data for $B > B_{min}$.

minimum value in the Kapitza resistivity occurs at a field value of 25 gauss. The Kapitza resistivity decreases by about 20% relative to the zero field value and then increases linearly for larger fields (25–80 gauss). This minimum value, although predicted in many theories, has not been previously observed. Leggett and Vuorio predicted a minimum at $B \approx 3\ k_B\ T/2\ \mu_B$ or $\mu_B\ B/k_B\ T \approx 3/2$, where μ_B is the Bohr magneton. Taking B as 25 gauss(0.0025 T) at the minimum gives $\mu_B\ B/k_B\ T \approx 1/4$, which is of the same order of magnitude as the predicted result.

The Kapitza resistivity has been found to vary from system to system. This is possibly because of the different materials used. Bozler and Gould, using a system of $PrNi_5$ and ^3He, found that $\mathcal{R}_k$ was proportional to T^{-2} and decreased with increasing magnetic field down to the lowest temperatures. The use of Pt powder by Deconde resulted in a decrease in $\mathcal{R}_k$ with magnetic field. $\mathcal{R}_k$ was found to be proportional to T^{-1} in this study. Doug Osheroff and I, using Ag sinter, observed that $\mathcal{R}_k$ increased with magnetic field and was proportional to T^{-1}. We used fairly large field steps, however, and may have missed the minimum observed by Yue Hu.

Conclusions

At this stage it is not possible to draw any firm conclusions on the low temperature Kapitza resistance mechanisms. It is, however, evident that many different phenomena are involved in the problem of heat transfer at these temperatures. Surface effects involving magnetic spins may play a role but the details are far from clear. It is to be hoped that a coherent theoretical explanation can be found for the observed behaviour. The subject is of great importance in pursuing experimental goals such as searching for a BCS pairing transition in dilute mixtures of ^3He in ^4He.

References 4

[1] P L Kapitza, *Zh. Eksp. Teor. Fiz.* **11**, 1 *J. Phys.* (USSR) **4**, 181 (1941).

[2] I M Khalatnikov, *Zh. Eksp. Teor. Fiz.* **22**, 687 (1952).

[3] I M Khalatnikov, *An Introduction to the Theory of Superfluidity* (Benjamin, New York) 138.

[4] A I Ahonen, O V Lounasmaa and M C Veuro, in Proc. 15th Int. Conf. Low Temperature, *J. Phys.* (Paris) **39**, Suppl. 8, C6–264 (1978).

[5] J P Harrison, *J. Low Temp. Phys.* **37**, 540–558 (1979).

[6] W R Abel, A C Anderson, W C Black and J C Wheatley, *Phys. Rev. Lett.* **16**, 273 (1966).

[7] A J Leggett and M Vuorio, *J. Low Temp. Phys.* **3**, 359 (1970).

[8] D D Osheroff and R C Richardson, *Phys. Rev. Lett.* **54**, 1178 (1985).

[9] Yue Hu, PhD dissertation, Cornell University, Ithaca, N.Y., 1990.

LECTURE 5 : POSSIBLE LIMITS TO ULTRA–LOW TEMPERATURES IN MATTER. FUTURE DIRECTIONS IN LOW TEMPERATURE PHYSICS

Introduction

In the first part of this lecture the limits of cooling with copper nuclear demagnetization stages are examined This is followed by a discussion of possible topics in low temperature physics to be investigated in the future.

Temperatures of 10 μK have been achieved in adiabatic nuclear refrigeration [1] but it appears to be difficult to go below this. A nuclear refrigerator consists of a nuclear stage of 10 to 100 moles of copper in a field of 8 T, precooled by a ^{3}He – ^{4}He dilution refrigerator. The electronic temperature of the copper sample may be measured with Pt NMR thermometers.

The problem of cooling to below 10 μK has recently been investigated by Gloos, Smeibidl and Pobell [2] at Bayreuth in an experiment which determines the coupling between nuclei and electrons in copper at low temperatures. Contrary to expectation, it was found that this coupling weakened at temperatures below 1 mK and in fields between 2 mT and 440 mT. This has important implications for ultra–low temperature physics. The experiment is described below.

The limits to cooling with copper.

The two stage demagnetization cooling apparatus used by Gloos et al. [2] is schematically illustrated in Figure 5.1. A massive Aℓ superconducting heat switch is used between the two stages. The two magnets are far enough apart and have compensated windings to avoid field interactions. For this experiment the assumption has been made that the entropy S versus the temperature T can be calculated in a given magnetic field down to the ordering temperature of the copper nuclei. Figure 5.2 shows the entropy S as a function of T in varying magnetic fields. For the copper spins (I = $^3/_2$) there is a limiting high temperature entropy of R ℓn 4 per mole. The reduction of the magnetic field from 9 T to 0 T at various initial temperatures T_i produces a family of entropy curves. In the *high temperature* approximation the final temperature is given by

$$T_F = T_i \left[\frac{B_F}{B_i}\right]$$

where B_F is the final field and B_i is the initial field. The area indicated by Q is a measure of the total heat required to destroy the magnetic order, i.e. the magnetic enthalpy.

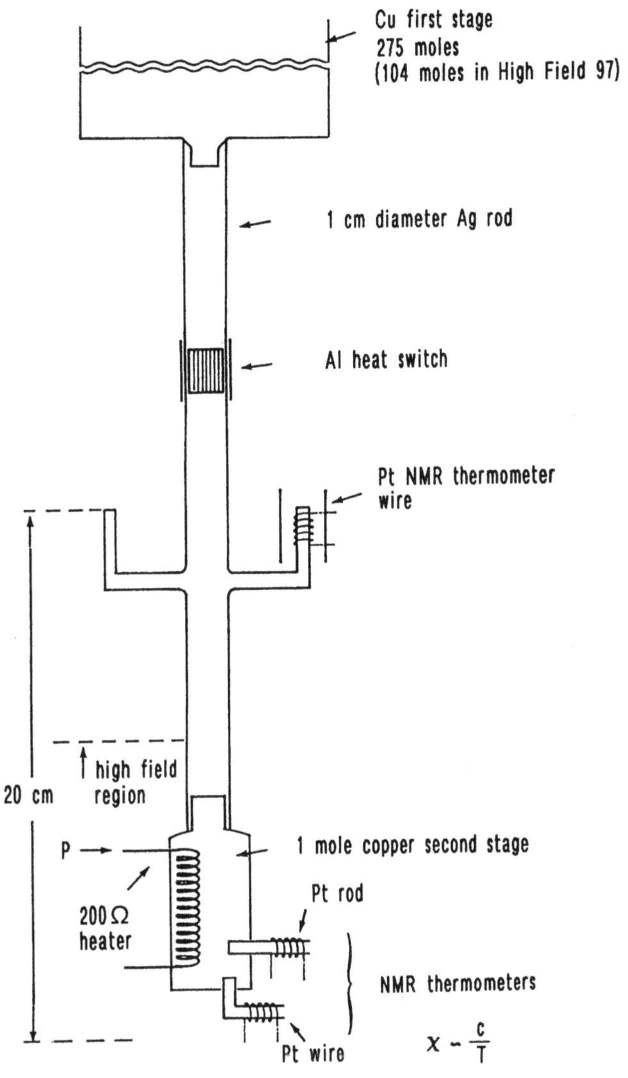

Figure 5.1 Lower end of two–stage nuclear demagnetization cryostat. The Cu sample consisted of a rectangular slab with slots milled in the sides to reduce eddy current heating. (After K Gloos et al. [2])

The experiment consisted of cooling the lower copper stage, which was effectively the sample, to temperatures between 2 mK and 6 mK in a field of 9 T by means of the first nuclear stage and, after opening the heat switch, demagnetizing the sample to the final field. Relaxation to minimum temperatures was observed overnight. The error in T_e, the electronic temperature, measured by the Pt thermometer, was estimated to be approximately 1 % and these thermometers proved to have a fairly short response time. The sample was then heated by applying power P at various levels (0.1 nW – 400 nW) to a 200 Ω heater mounted on the sample. The amount of energy needed to destroy the magnetic order could be measured in this way.

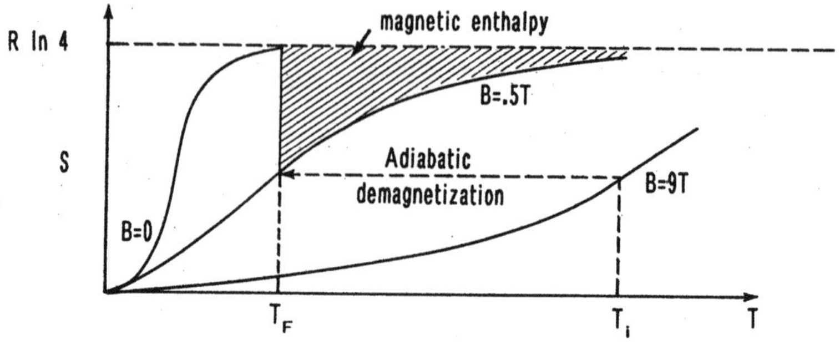

Figure 5.2 Nuclear entropy versus temperature curves for various applied magnetic fields. The shaded area is a measure of the total heat required to destroy the magnetic order following an adiabatic demagnetization process.

The total heat energy input Q is plotted as a function of the measured electronic temperature T_e in different applied fields in Figure 5.3. The heat leak L (≈ 50 pW) to the sample was measured from 1 mK to 10 mK and was taken into account in the analysis of the very low temperature experiments. The cosmic ray background heating was included in L.

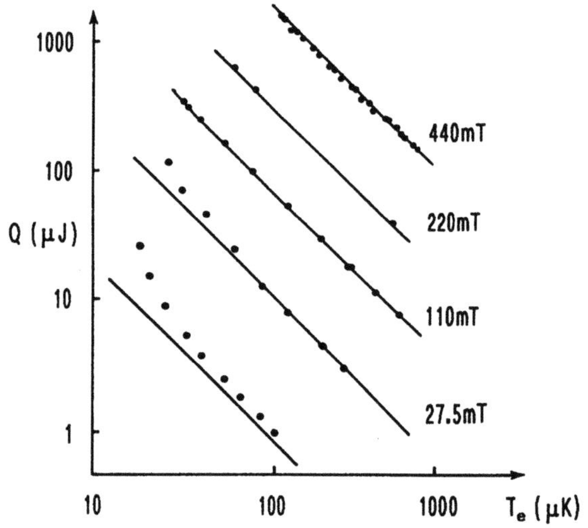

Figure 5.3 The heat content Q of the Cu sample as a function of the electronic temperature plotted for various applied fields. The electronic temperatures at the lowest fields are somewhat higher than predicted by the high temperature approximation (solid lines).
(After Gloos et al. [2]).

According to the high temperature approximation (HTA), the nuclear spin temperature $T_n = n\epsilon\, B^2/Q$, where n, the number of moles of copper, = 1.02 moles and ϵ, the nuclear magnetic Curie constant, = 3.19 μJK/(T^2 mole). B is the magnetic field. The error in T_e caused by the time lag of the thermometers was considered negligible.

The experimental data in Figure 5.3 show that at the largest fields applied $T_n = T_e$. At smaller fields the relative temperature difference between electrons and nuclei increased and the measured values for T_e were higher then expected from the calculation using Q and L. The solid lines depict the heat content of the Cu nuclei when no heat flow between electrons and nuclei is assumed.

The coupling between the electrons and the nuclear spin system is normally described by the Korringa relation. Thus the relaxation time $\tau = \kappa/T$, where κ is the Korringa constant. To determine this coupling an additional heating power P was applied to the sample over longer periods of time. For each field B, T_e was obtained as a function of Q and P.

The HTA predicts that the temperature difference between electrons and nuclear spins is given by

$$T_e - T_n = \frac{P\, T_n\, \kappa}{n\, \epsilon\, B^2}$$

Using $T_n = n\, \epsilon\, B^2/Q$ leads to $T_e\, Q = n\, \epsilon\, B^2 + \kappa P$. If κ is a constant, $T_e\, Q$ should not depend on T_e. Experimentally it was found that

$$T_e\, Q = n\, \epsilon\, B^2 + (\kappa_o + \Gamma/T_e^2)\, P$$

where Γ is a fit parameter and $\kappa_o = 1.2$ Ks [3].

The constant κ had to be replaced by $\kappa_o + \Gamma/T_e^2$ at low temperatures. Therefore, for a given magnetic field, the Korringa constant, while independent of T_n, shows a T_e dependency.

Figure 5.4 shows the Korringa constant κ as a function of the electronic temperature at different magnetic fields. As can be seen, $\kappa_o = 1.2$ Ks, which is the accepted literature value, at high electronic temperatures. κ increases with decreasing temperature for all fields. The solid lines represent $\kappa = \kappa_o + \Gamma/T_e^2$ with $\kappa_o = 1.2$ Ks and the parameter Γ proportional to the applied field, as is shown in Figure 5.5.

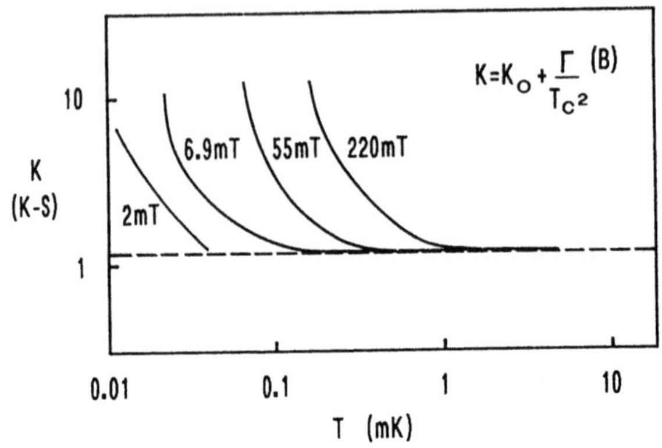

Figure 5.4 The measured Korringa constant for nuclear relaxation as a function of temperature for a number of applied magnetic fields. The solid lines are based on the formula $\kappa = \kappa_0 + \Gamma/T_e^2$ with $\kappa_0 = 1.2$ Ks.

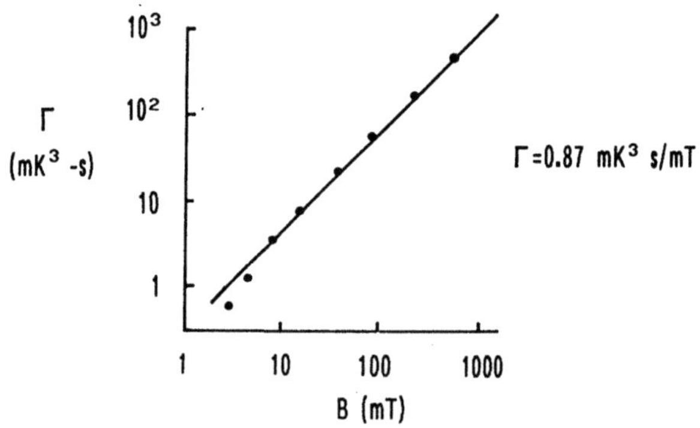

Figure 5.5 Plot of the parameter Γ used to account for the temperature variation of the Korringa constant of the Cu sample ($\kappa = \kappa_0 + \Gamma/T_e^2$) as a function of the applied magnetic field. The straight line fit is given by $\Gamma = 0.87$ B with T_e in mK and B in mT. (After Gloos et al [2]).

The measured Korringa constant is shown as a function of inverse electronic temperature for a field of 55 mT in Figure 5.6. The solid line corresponds to $\kappa = \kappa_0 e^{\Delta/T_e}$ where $\Delta = 0.15$ mK and $\kappa_0 = 1.0$ Ks for this field. The increase in κ with increasing electronic temperature may be attributable to the presence of a gap, Δ, in the density of electron states. In this picture κ increases because the number of conduction electrons free to transport energy at the Fermi surface decreases. The gap may be regarded as similar to the BCS gap. However Δ is dependent on magnetic field and the copper does not behave as a superconductor. It appears as if the copper electrons may have a new collective phenomenon associated with them.

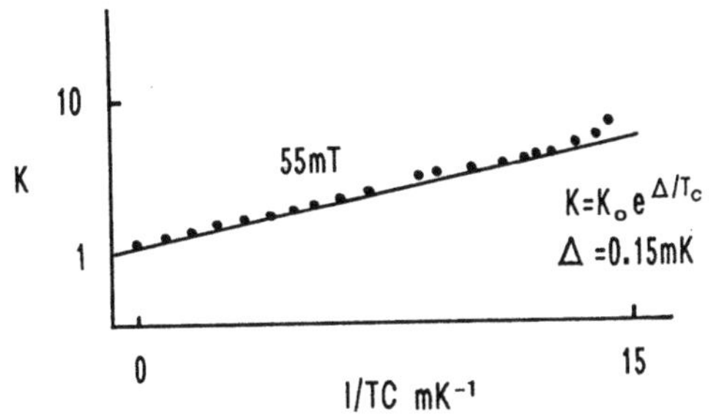

Figure 5.6 The measured Korringa constant κ for Cu as a function of $1/T_e$, in an applied field of 55 mT. The solid line fit is given by $\kappa = \kappa_0 \, e^{\Delta/T_e}$ with $\Delta = 0.15$ mK and $\kappa_0 = 1.0$ Ks. (After Gloos et al. [2]).

The authors have concluded that, if the electron density of states at the Fermi level does decrease with decreasing electronic temperatures, with a concomitant weakening of the nuclear spin–conduction electron coupling, copper may not be the best choice for refrigeration below 10 μK. Further work in this temperature region is clearly necessary in order to establish the details of the physics involved.

Related results have been reported for Tℓ [4] and Au In_2 [5] by Gloos et al.

It would of course be nice to be able to measure the nuclear spin temperature directly but this is difficult because of the long spin–lattice relaxation times ($10^5 - 10^6$ s). In addition one would like to see measurements of transport properties at the lowest temperatures reached but this is very difficult and has not been done yet.

For the present it appears that the 10 μK *brick wall* holds for Cu nuclear stages. Heat leaks will have to be substantially reduced if lower temperatures are to be reached with these systems.

Topics for the future

Topics which may reveal new phenomena at low temperatures may be categorized as follows:

I Quantum fluids

New superfluids may be observed.

(a) ^{4}He in a porous medium

If vycor glass, which is an open network with characteristic spacing of 50 Å, is covered with a thin film of ^{4}He, slightly thicker than the localized or solid layer, a superfluid transition occurs similar to that in bulk helium. As the layer decreases in thickness, the characteristic scaling parameter for the superfluid fraction changes with temperature. It appears that the value is tending towards the classical *ideal gas* Bose–Einstein value. The vycor – ^{4}He system is multiply–connected, however, and a much more dilute gas, with particles interacting still more weakly, cannot be obtained. The lowest transition temperature achieved has been 4 mK. If the temperature could be lowered to 200 to 300 μK, it is predicted that a clean Bose–Einstein gas condensation would be observed.

(b) ^{3}He diluted in liquid ^{4}He

Many predictions have been made concerning the temperature at which a superfluid transition will occur in this system. It is believed that a pairing transition will take place at a certain temperature. The interaction between the ^{3}He atoms in the mixture is much weaker than that between the ^{3}He atoms in pure ^{3}He. There is therefore uncertainty over the kind of pairing which may occur. It is difficult to cool the dilute mixture as a result of the thick film of ^{4}He on the wall of the heat exchanger, which increases the thermal boundary resistance. The lowest temperature yet achieved in these mixtures has been somewhat above 100 μK.

(c) Spin aligned atomic hydrogen (H↑)

If atomic hydrogen is placed in a sufficiently large field at low temperatures, the hydrogen atoms will not combine to form molecular hydrogen. Experiments to achieve Bose–Einstein condensation by decreasing the temperature of the gas at a constant density or by compressing the gas at a constant temperature have as yet been unsuccessful. Nature seems to be against us in both cases as three body interactions become important and cause recombination of the H–atoms.

(d) Low T_c superconductivity in materials such as heavy fermion systems. Measurements have been made similar to those in ^{3}He, where we have NMR frequency shifts or spikes in the ultrasound absorption due to excitation of special collective modes of the pairs. These experiments are analogous to optical spectroscopy of molecules. The characteristic frequency of the collective modes is $\omega \approx k_B T_c/\hbar$. At 1 mK this corresponds to a frequency of 20 MHz.

In conducting materials there are larger skin depths and fewer screening problems at lower frequencies. It is therefore simpler to do experiments at low temperatures. Heavy fermion materials have longer coherence lengths, ξ_o, at low temperatures ($\xi_o \sim 1/T_c$). Interesting low temperature transitions should occur in some of these materials.

II Normal metals

Many interesting problems occur in normal metals.

(a) Are metals really normal?

It appears that high temperature behaviour of metals often cannot be extrapolated to low temperatures. Observed behaviour does not correlate with the behaviour predicted in this way. An example of this is the breakdown of the Korringa relation described above.

(b) Localization

In the low temperature regime conductors are converted to insulators. The free flow of electrons may be inhibited by local variations in the electric potential. Interesting future experiments include the production of localization with micro–fabrication to study these phenomena. There is also the possibility of studying the discreteness of the electron energy levels in small geometries.

(c) Coherence phenomena in metals in 1, 2, and 3 dimensions.

Mesoscopic phenomena in small rings of a pure metal may be investigated. If the rings are small enough, flux quantization phenomena and the Aharonov–Bohm effect may be observed. Flux quantization occurs even in networks of small rings of copper. The characteristic flux quantum involved in the Aharonov–Bohm effect is h/e and not h/2e as in superconductors.

(d) Wigner crystal of electrons

2D quantum electron liquids can be formed in special systems such as at the interface at GaAs/GaAℓAs junctions or for electrons on ^{4}He surfaces. The characteristic magnetic length for an electron is $\ell_m = \sqrt{I_0/B}$, where I_0 is the flux quantum. When $\ell_m < a$, the inter–electron spacing, new phenomena may be observed. To achieve this, either the magnetic fields may be made very large with the conductor moderately dilute or the conductor may be considerably diluted and more modest magnetic fields used at very low temperatures.

(e) Noise

Noise in conductors – the so–called Nyquist or Johnson noise – obeys the Planck radiation law. White noise, with a uniform spectral density, will occur at frequencies below $k_BT/\hbar$. There is a thermal contribution below $k_BT/\hbar$ and a zero point motion noise or quantum noise of $1/2\ \hbar\omega$ above $k_BT/\hbar$. Observation of these types of noise in materials such as copper, cooled to very low temperatures, would be interesting and is necessary to confirm the theoretical prediction. As yet no transport measurements have been made below 1 or 2 mK. Nevertheless the frequencies become conveniently low and clever instrumentation may permit this to be done.

III Non conducting solids

The properties of glassy materials at low temperatures offer a challenging field of research. In some senses it is likely that a large number of materials will become glass–like at very low temperatures. While there are theories, such as the tunnelling two–level system theory, a number of workers, including Tony Leggett, believe that our understanding of these systems is incomplete.

(a) Mechanical properties at low temperatures.
In an experiment using silicon in oscillating motion, Q and the resonance frequency were measured as a function of T. It was found that below 0.5 K the motion became non linear ($Q \sim T^{-1}$). This effect, which becomes increasingly important with decreasing temperature, is not understood at present.

(b) Magnetic order
A large variety of metals with different spins exists for future studies of magnetic order. Nuclear ordering in non–cubic high spin systems is an example of the type of problem that can attract interest for many years.

(c) New lattice modes and order
It is postulated that types of lattice modes other than phonons may exist. One possible example is the supersolid. In the case of a solid with large zero point motion vacancies may exist at low temperatures. These vacancies become delocalized, have extended wave functions, and there could be a collective order associated with the *superfluidity* of the solid atoms. Andreev has argued that ^{4}He at the melting pressure has zero point vacancies and the superfluid solid should therefore exist at sufficiently low temperatures. These predictions have yet to be verified.

IV Technical Problems

Three technical problems have to be solved in very low temperature experiments.

- Cooling of the material. Each material is associated with certain problems, some of which have been dealt with in these lectures.
- Thermometry. It can take years to develop satisfactory methods for measuring low temperatures and generating a low temperature scale.
- Instrumentation. Dedication is required to develop the instrumentation necessary to achieve very low temperatures and to make measurements at these temperatures. As stated above, no transport measurements have yet been made below 100 μK.

Conclusion

There are many really interesting problems still to be studied in low temperature physics. The technology involved is difficult and new measuring techniques will have to be developed. The subject will continue to fascinate people for many years to come.

References 5

[1] K. Gloos, P Smeibidl, C Kennedy, A Singsaas, P Sekowski, R M Mueller, F Pobell, *J. Low Temp. Phys.* **73**, 101 (1988).

[2] K Gloos, P Smeibidl and F Pobell, *Phys. B Condensed Matter*, (1990).

[3] G C Carter, L H Bennett, D J Kahan, *Metallic Shifts in NMR* Oxford : Pergamon Press 1977.

[4] G Eska and E Schuberth; *Jpn. J. Appl. Phys.* [Suppl.3] **26**, 435 (1987).

[5] K Gloos, R König, P Smeibidl and F Pobell, *Europhys. Lett.* **12**, 661 (1990).

Acknowledgements

The author is very grateful to Mrs Renée Hoch for transcribing the recordings of the lectures into a manuscript. It is a difficult and time consuming task which she did with great skill. Much of the work done at Cornell which is described in these lectures was supported by the National Science Foundation through Grants DMR–89–21733 and DMR–88–18558.

INTRODUCTION TO STRONGLY CORRELATED SYSTEMS

Professor P Fulde
Max–Planck–Institut für Festkörperforschung, Stuttgart

1. INTRODUCTION

Electron correlations are strong when the on–site electron–electron repulsions U are much larger than the energies which are associated with the hybridization of atomic orbitals belonging to different atoms. The latter are characterized in a solid by the width W of the energy band under consideration. A large ratio U/W is expected in systems involving 4 f or 5 f electrons, i.e. rare earth or actinide atoms. But systems with d electrons can also be strongly correlated. A famous example is CoO. If one treated it within the independent electron approximation, one would find that this substance is metallic, with an odd number of electrons per unit cell and a partially filled d band. In reality, however, CoO is an insulator. The same holds true for La_2CuO_4, a prototype for a class of materials with high superconducting transition temperatures. CoO and La_2CuO_4 are not metallic because the strong electron correlations suppress the charge fluctuations which are required in order to have a nonvanishing conductivity.

Characteristic effects of strong correlation can be studied by looking at small systems, i.e. molecules. For example, the disappearance of a magnetic moment at low temperature, a feature which is usually associated with the Kondo effect, should be also found in molecules. When one is dealing with molecules containing for example one rare earth ion only, then the strong correlations can be treated by MC–SCF calculations. The configurations which are included in the SCF calculation must be those which are generated by the strong correlations. In this way high precision calculations for these strongly correlated systems can be performed.

For infinite systems strong correlations cannot be treated in the same way because infinitely many different configurations would have to be included. This aspect is discussed further in Section 4.

One class of materials which show a variety of effects based on strong correlations are the heavy fermion systems. They are discussed in Section 3. Another important class of materials with strong correlations are the high T_c superconducting materials discussed in Section 5. In this section we will concentrate on the Kondo effect and on model calculations based on the Hubbard Hamiltonian.

1.1 Molecules

The simplest example of a system of strongly correlated electrons consists of two electrons distributed over two orbitals. These orbitals are denoted by L (for ligand) and F (for example 4 f), respectively and it is assumed that the corresponding orbital energies are ϵ_1 and ϵ_f with $\epsilon_f < \epsilon_1$. When the two electrons are in the F orbital they are assumed to repel each other with an energy U >> $(\epsilon_1 - \epsilon_f)$. When they are both in the L orbital, or when one electron is in the L and the other in the F orbital, their Coulomb interaction is neglected. This is justified when the spatial extent of the L orbital is sufficiently large, for example when the ligand is a large molecule. The hybridization V between the two orbitals is assumed to be small, i.e. V << $(\epsilon_1 - \epsilon_f)$. This is indicated in Figure 1.1.1. The Hamiltonian of the system is

$$H = \epsilon_1 \sum_\sigma 1^+_\sigma 1_\sigma + \epsilon_f \sum_\sigma f^+_\sigma f_\sigma + V \sum_\sigma (1^+_\sigma f_\sigma + f^+_\sigma 1_\sigma) + U n^f_\uparrow n^f_\downarrow . \tag{1.1.1}$$

The $1^+_\sigma(1_\sigma)$, $f^+_\sigma(f_\sigma)$ create (annihilate) electrons with spin σ in the L and F orbitals, respectively. Furthermore, $n^f_\sigma = f^+_\sigma f_\sigma$. When V = 0, the ground state of the system has energy $E_o = \epsilon_1 + \epsilon_f$ and is four–fold degenerate. One electron is in the F orbital while the other one is in the L orbital. A double occupancy of the F orbital is prevented by the large Coulomb repulsion U. The four states are eigenstates of the total spin S and are of the form

$$\begin{aligned}
|\Phi_{S=0}\rangle &= \frac{1}{\sqrt{2}} (f^+_\uparrow 1^+_\downarrow - f^+_\downarrow 1^+_\uparrow) \, |0\rangle \qquad ; \\
|\Phi^1_{S=1}\rangle &= f^+_\uparrow 1^+_\uparrow \, |0\rangle \\
|\Phi^0_{S=1}\rangle &= \frac{1}{\sqrt{2}} (f^+_\uparrow 1^+_\downarrow + f^+_\downarrow 1^+_\uparrow) \, |0\rangle \\
|\Phi^{-1}_{S=1}\rangle &= f^+_\downarrow 1^+_\downarrow \, |0\rangle \qquad .
\end{aligned} \tag{1.1.2}$$

The system has one excited state, which is of the form

$$|\Phi_{ex}\rangle = 1^+_\uparrow 1^+_\downarrow \, |0\rangle \qquad . \tag{1.1.3}$$

The excitation energy is $E_{ex} = 2\,\epsilon_1$. The state $f^+_\uparrow f^+_\downarrow \, |0\rangle$ is left out from further considerations, because its energy is of order U and it is assumed that $U \rightarrow \infty$.

When the hybridization is turned on, the singlets $|\Phi_{S=0}\rangle$ and $|\Phi_{ex}\rangle$ are coupled with each other while the S = 1 states $|\Phi_{S=1}\rangle$ remain unchanged. The resulting 2 × 2 matrix

$$\begin{bmatrix} \epsilon_f + \epsilon_1 & \sqrt{2}\,V \\ \sqrt{2}\,V & 2\,\epsilon_1 \end{bmatrix} \tag{1.1.4}$$

is easily diagonalized. For small values of $(V/\Delta\epsilon)$ the eigenvectors are

$$|\psi_o> = \left[1 - (\frac{V}{\Delta\epsilon})^2\right] |\Phi_{S=0}> - \frac{\sqrt{2}V}{\Delta\epsilon} |\Phi_{ex}>$$

$$|\psi_{ex}> = \left[1 - (\frac{V}{\Delta\epsilon})^2\right] |\Phi_{ex}> + \frac{\sqrt{2}V}{\Delta\epsilon} |\Phi_{S=0}> \qquad (1.1.5)$$

with $\Delta\epsilon = (\epsilon_1 - \epsilon_f)$. The eigenvalue of $|\psi_o>$ is

$$\tilde{E}_o = E_o - \frac{2V^2}{\Delta\epsilon} \qquad (1.1.6)$$

while that of $|\psi_{ex}>$ is

$$\tilde{E}_{ex} = E_{ex} + \frac{2V^2}{\Delta\epsilon} \qquad (1.1.7)$$

The changes in the spectrum are depicted in Figure 1.1.2. For small values of V there is a low lying triplet excitation above the singlet ground state. One can attach a characteristic temperature T^* $(= -2V^2/(k_B\,\Delta\epsilon))$ with the energy gain associated with the singlet formation. The form of the ground state, as well as the low lying excitations, are both due to the strong correlations, which forbid a double occupancy of the F orbital. In the ground state the occupancy of the F orbital is

$$n^f = 1 - 2\,(\frac{V}{\Delta\epsilon})^2 < 1 \qquad (1.1.8)$$

For temperatures $T << T^*$ there exist two distinct types of excitations

(a) low lying spin excitations with a typical energy $k_B T^*$

(b) excitation of an f electron into a ligand orbital, with an excitation energy of order $\Delta\epsilon$. These excitations involve charge degrees of freedom.

A classification of excitations into those involving predominantly spin degrees of freedom and those involving charge degrees of freedom is a characteristic property of strongly correlated electronic systems.

For $T >> T^*$ the singlet and triplet states are equally populated and the singlet– triplet excitations become unimportant. The two electrons in the L and F orbital are then effectively decoupled. In this case only the high–energy excitation into the ligand orbital remains.

The above simple model contains key ingredients of the Kondo problem or of systems with heavy electrons (heavy fermions). The ground state is a singlet and the magnetic moment of the

Two electrons – two orbitals

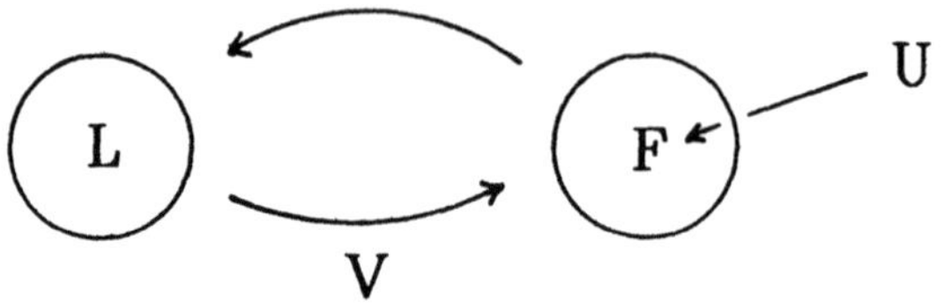

ℓ^{+}_{σ} F^{+}_{σ}

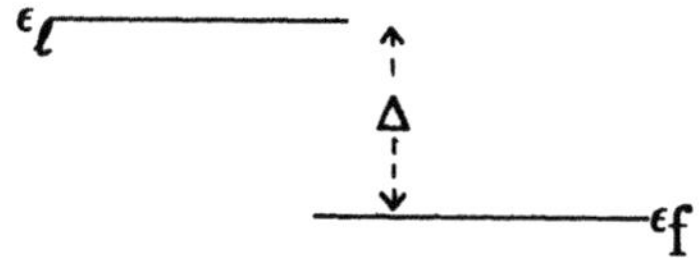

$V = 0$

ground state : quartet

$f\uparrow \ell\downarrow$, $f\downarrow \ell\uparrow$, $f\uparrow \ell\uparrow$, $f\downarrow \ell\downarrow$

excited state :

$\ell\uparrow \ell\downarrow$

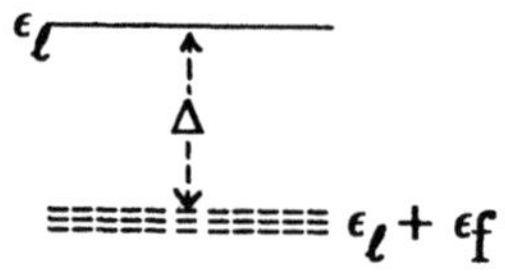

Figure 1.1.1 **Two electron system, V = 0.**

V ≠ 0

mixing of $f_\uparrow^\dagger \ell_\downarrow^\dagger$, $f_\downarrow^\dagger \ell_\uparrow^\dagger$, $\ell_\uparrow^\dagger \ell_\downarrow^\dagger$

formation of a singlet

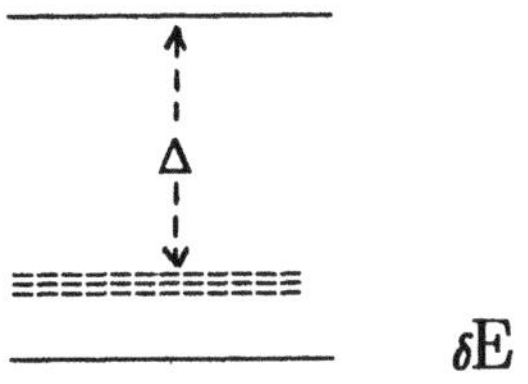

energy gain

$$\delta E = -\frac{2V^2}{\Delta}$$

ground state :

$$|\psi_0> = \frac{1}{\sqrt{2}}(f_\uparrow^+ \ell_\downarrow^+ - f_\downarrow^+ \ell_\uparrow^+)\cos\theta + \ell_\uparrow^+ \ell_\downarrow^+ \sin\theta$$

excited states :

low lying excitations :

⇒ spin degrees of freedom

high energy excitations :

⇒ charge degrees of freedom

Figure 1.1.2 Two–electron system, V ≠ 0.

partially filled F level is zero. As the temperature increases towards T*, the triplet states are thermally populated. Since they have a moment, the magnetic character of the f electron starts to appear. For T >> T* the magnetic moment is fully present. The singlet character of the ground state is noticeable only for T << T*. The low lying excitations are intimately connected with the degeneracy of the ground state in the absence of hybridizations.

It is instructive to solve the Hamiltonian (1.1.1) in the independent electron or SCF approximation. In this case the following replacement is made

$$U\, n_{\uparrow}^{f} n_{\downarrow}^{f} \rightarrow U\, (\langle n_{\downarrow}^{f}\rangle\, n_{\uparrow}^{f} + \langle n_{\uparrow}^{f}\rangle\, n_{\downarrow}^{f} - \langle n_{\uparrow}^{f}\rangle \langle n_{\downarrow}^{f}\rangle) \tag{1.1.9}$$

and the resulting SCF Hamiltonian is trivially diagonalized. The resulting eigenstates are bonding and antibonding states. The corresponding creation operators (B_{σ}^{+}) and (A_{σ}^{+}) can be expressed in terms of the operators 1_{σ}^{+} and f_{σ}^{+} as

$$\begin{aligned} B_{\sigma}^{+} &= 1_{\sigma}^{+} \cos\theta + f_{\sigma}^{+} \sin\theta \\ A_{\sigma}^{+} &= -1_{\sigma}^{+} \sin\theta + f_{\sigma}^{+} \cos\theta \end{aligned} \tag{1.1.10}$$

For U >> V one finds for the angle

$$\theta = -(V/U)^{1/3} \tag{1.1.11}$$

$$\begin{aligned} \epsilon_{\sigma}^{B} &= \epsilon_{1} - V\,(V/U)^{1/3} \\ \epsilon_{\sigma}^{A} &= \epsilon_{f} + (V^{2}U)^{1/3} \end{aligned} \tag{1.1.12}$$

One notices that in the SCF ground state

$$|\Phi_{SCF}\rangle = \prod_{\sigma} B_{\sigma}^{+}\, |0\rangle \tag{1.1.13}$$

the electrons are predominantly in the ligand orbital. Because there is a contribution of the F orbital to the molecular orbital $B_{\sigma}^{+}\,|0\rangle$, the probability is $\sin^{2}\theta$ that the F orbital is doubly occupied with a resulting large Coulomb repulsion. The ground state energy is

$$E_{SCF} = \sum_{\sigma} \epsilon_{\sigma}^{B} - U \langle n_{\uparrow}^{f}\rangle \langle n_{\downarrow}^{f}\rangle$$

$$= 2\epsilon_1 - 3V\,(V/U)^{1/3} \qquad (1.1.14)$$

and therefore considerably above the true ground state energy. Note that for $U \to \infty$ one finds $E_{SCF} = 2\epsilon_1$, while according to Equation (1.6) it is $E_o \simeq \epsilon_1 + \epsilon_f - 2V^2/\Delta\epsilon$. Due to the missing correlations, the SCF approximation gives unphysical results in the present case.

The situation is improved when an unrestricted SCF approximation is made. Thereby one allows for different orbitals for the two electrons of opposite spin. It is found that one electron is localized on the ligand while the other electron is predominantly in the F orbital but lowers its energy by hybridization. The ground state wave function is of the form

$$|\Phi_{USCF}\rangle = \left[(1 - \tfrac{1}{2}(\tfrac{V}{\Delta\epsilon})^2)\, f_\uparrow^+ + \tfrac{V}{\Delta\epsilon}\, 1_\uparrow^+\right] 1_\downarrow^+ \,|0\rangle \qquad (1.1.15)$$

and the corresponding energy is

$$\tilde{E}_{USCF} = E_o - \frac{V^2}{\Delta\epsilon} \qquad (1.1.16)$$

One notices that this energy is much better than the previous one $\tilde{E}_{USCF}$, but the wave function certainly does not describe correctly the real situation. It is instructive to reconfirm the conclusions that within the independent electron approximation, symmetry broken solutions strongly suppress charge fluctuations. They therefore simulate correlation effects as far as the energy is concerned, but the wave function remains grossly in error.

The above simple Hamiltonian describes the essence of the ground state of molecules like di–π–cyclo–octatetraene cerium or abbreviated cerocene. The structure of $(C_8H_8)_2Ce$ is shown in Figure 1.1.3. Because the total valence electron number is even, Ce is usually considered to be tetravalent. However this is grossly misleading. Instead the Ce ion is almost entirely in a $4f^1$ configuration corresponding to $Ce^{3+}(C_8H_8^{1.5-})_2$. The 4f electron forms a singlet with an electron of the highest occupied molecular orbital (HOMO), which has e_{2u} symmetry. The HO's are calculated by treating the f electron by a pseudopotential. The $4f^1e_{2u}^3$ singlet state resembles the state $|\psi_o\rangle$ in Equation (1.1.5), except that two more electrons are present because the HOMO is four–fold degenerate. This degeneracy is a consequence of the C_{80} symmetry. Large scale MC–SCF calculations with several hundreds basis functions confirm this picture. The multiconfigurations, on which the SCF calculation is based, must include $4f^0e_{2u}^4$, $4f^1e_{2u}^3$ and also $4f^2e_{2u}^2$. It is found that the ground state has an admixture of the f^0 and f^2 configurations with weights of 3.8% and 0.2%, respectively, to the f^1 configuration. The calculations provide also the low lying excitation energies to triplet states. For cerocene they are of the order of 0.3 eV. Because of this relatively large excitation energy the Van Vleck paramagnetic contribution is less than the diamagnetic one of the ring currents. If one replaced the C_8H_8 rings by more extended molecules, then the HOMO would couple much less to the 4f orbital and the excitation energies would correspondingly decrease. In this case the sandwich molecule would become paramagnetic.

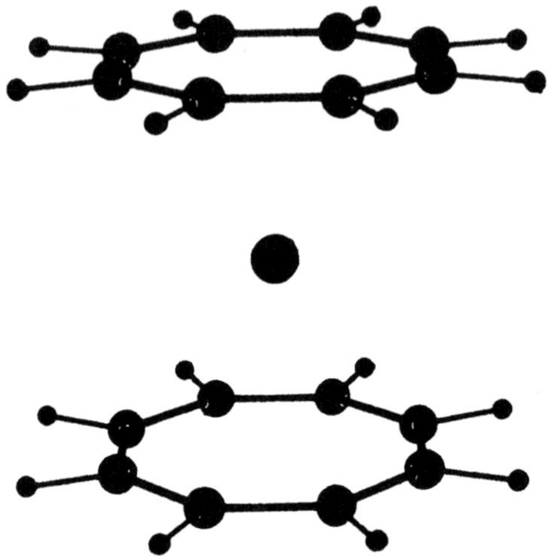

Figure 1.1.3 The structure of Cerocene, $Ce(C_8H_8)_2$

2. KONDO EFFECT

When transition metal ions or rare–earth impurities are dissolved in a metal, the metal often shows a number of anomalous transport properties. In particular it is found that the resistance, as function of decreasing temperature, does not show a monotonous decrease but rather has a minimum at a temperature T_K which is a characteristic property of the system. The latter may vary from a few [K] to a few hundred [K]. The explanation of this effect, as well as of a number of other anomalies (e.g. in the thermopower or heat conduction), was provided by Kondo. He showed that one is dealing here with an important correlation or many–body effect.

In order to link the Kondo problem to that of strong correlations discussed in the last section, it is instructive not to start out from the Kondo Hamiltonian but rather from the Anderson Hamiltonian

$$H = \sum_{\underline{k}\sigma} \epsilon(\underline{k})\, c^+_{\underline{k}\sigma} c_{\underline{k}\sigma} + \epsilon_f \sum_m n^f_m + \frac{U}{2} \sum_{m \neq m'} n^f_m n^f_{m'} + $$

$$+ \sum_{\underline{k}m} \left[V_{m\sigma}(\underline{k})\, f^+_m c_{\underline{k}\sigma} + V^*_{m\sigma}(\underline{k})\, c^+_{\underline{k}\sigma} f_m \right]. \qquad (2.1)$$

The magnetic impurity is assumed to be Ce. According to Hund's rule, the lowest j multiplet of an f electron has $j = 5/2$. The z component of j is denoted by m. The corresponding creation (annihilation) operators are f^+_m (f_m) and the number operator is $n_m = f^+_m f_m$. Due to crystal field splitting of the j manifold, the f electron degeneracy, ν_f can be less than (2j + 1). (The case of one and zero f electrons will be considered later.) The problem of the coupling between different f electrons can therefore be discarded by using (2.1). The total angular momentum j of an f electron is then the same as the total angular momentum J of the incomplete f shell. All energies are measured from the Fermi energy ϵ_F. The f orbitals hybridize with the conduction electrons. The resemblance of the Anderson Hamiltonian with the earlier one of Equation (1.1.1) is apparent. The differences are that here the f orbital is ν_f–fold degenerate and that the ligand orbital has been replaced by a partially filled band of conduction electrons. The interesting effects stem from the coupling of the impurity to the conduction electrons. Therefore only those degrees of freedom of the conduction electrons are kept, which couple to the impurity. The matrix elements $V_{m\sigma}(\underline{k})$ are varying rapidly with the direction of $\underline{k}$. Therefore the angular average is small except when m′ = m. One may therefore set

$$\sum_\sigma \int \frac{d\hat{k}}{4\pi} V^*_{m\sigma}(\underline{k})\, V_{m'\sigma}(\underline{k}) \simeq V^2(k)\, \delta_{mm'} \tag{2.2}$$

where $\hat{k} = \underline{k}/|\underline{k}|$. This suggests introducing the following orthogonal electronic basis

$$|k,m\rangle = \frac{1}{V(k)} \sum_\sigma \int \frac{d\hat{k}}{4\pi} V^*_{m\sigma}(\underline{k})\, |\underline{k}\sigma\rangle \quad . \tag{2.3}$$

When expressed in this basis, and provided that $\epsilon(\underline{k}) = \epsilon(|\underline{k}|) = \epsilon(k)$, the Anderson Hamiltonian is of the form

$$H = \sum_{km} \epsilon(k)\, c^+_{km} c_{km} + \epsilon_f \sum_m n^f_m + \frac{U}{2} \sum_{m \neq m'} n^f_m n^f_{m'} +$$
$$+ \sum_{km} V(k)\, (f^+_m c_{km} + c^+_{km} f_m) + \tilde{H}_0 \quad . \tag{2.4}$$

The Hamiltonian $\tilde{H}_0$ contains all those degrees of freedom of the conduction electrons which do not couple to the impurity. It is irrelevant for the following discussion.

In order to take advantage of the energy gain found in the earlier example when the ground state singlet is formed, one has to reduce the problem to a two–electron problem. Out of a given conduction electron state, one electron is moved to ϵ_F. The remaining electron and the f electron can form a singlet. In contrast to the earlier example, where only one ligand orbital was considered, the singlet formation can take place here with many different conduction electron states. The states close to ϵ_F are therefore the important ones, since it takes less energy to move one electron from these states to ϵ_F.

In order to put the above arguments onto a quantitative basis, one assumes that

$$|\epsilon_f| >> \nu_f \Gamma = \nu_f \pi N(0) V^2$$
$$V = V(k_F) \quad (2.5)$$

where N(0) is the conduction electron density of states per spin. The energy Γ is the width of the f level due to the coupling to the conduction electrons when Fermi's golden rule is applied. For the ground–state wave function the following ansatz due to Varma and Yafet is made:

$$|\psi_{S=0}> = A\left\{1 + \sum_{km} \alpha(k) f_m^+ c_{km}\right\} |\Phi_o> \quad . \quad (2.6)$$

The state $|\Phi_o>$ denotes the filled Fermi sea of the conduction electrons with the f orbital unoccupied. It is a singlet with total spin S = 0 and corresponds to $|\Phi_{ex}>$ in Section 1 (see Equations (1.1.5)). The linear combination of states $f_m^+ c_{km}$ must preserve the singlet character. The energy of $|\psi_{S=0}>$ has to be compared with that of the multiplet

$$|\psi_m> = f_m^+ |\Phi_o> \quad . \quad (2.7)$$

The electron number of the two states $|\psi_{S=0}>$ and $|\psi_m>$ differs by one, but this is unimportant because ϵ_F is the zero point of energy. As expected from Section 1, the energy of the singlet $|\psi_{S=0}>$ is found to be always lower than that of the multiplet $|\psi_m>$. The latter is

$$E_{magn} = E_o + \epsilon_f \quad (2.8)$$

where E_o is the energy of the filled Fermi sea $|\Phi_o>$. The energy of the singlet is

$$E_s = \frac{<\psi_{S=0}|H|\psi_{S=0}>}{<\psi_{S=0}|\psi_{S=0}>} = E_o - |\epsilon_f| + \epsilon \quad (2.9)$$

where ϵ has to be determined variationally. The stationarity conditions are

$$\frac{\delta<H>}{\delta A} = 0$$
$$\frac{\delta<H>}{\delta(A\alpha(k))} = 0 \quad . \quad (2.10)$$

They result in the coupled equations

$$\epsilon = |\epsilon_f| + \sqrt{\nu_f}\, V \sum_{k<k_F} \alpha(k)$$

$$\epsilon\, \alpha(k) = \sqrt{\nu_f}\, V - \epsilon(k)\, \alpha(k) \qquad . \qquad (2.11)$$

By replacing $\alpha(k)$ in the first equation one obtains

$$\begin{aligned} \epsilon &= |\epsilon_f| + \nu_f V^2 \sum_{k \le k_F} \frac{1}{\epsilon + \epsilon(k)} \\ &= |\epsilon_f| + \nu_F V^2 N(0) \int_{-D}^{0} \frac{d\epsilon(k)}{\epsilon + \epsilon(k)} \end{aligned} \qquad . \qquad (2.12)$$

where a constant density of states has been assumed. The lower cut–off D equals half the conduction–electron band width. For sufficiently small values of V three solutions are found. One of them has $\epsilon < 0$ and is the one which is sought. It is approximately determined from

$$|\epsilon_f| = \nu_f N(0) V^2 \ln (D/|\epsilon|) \qquad . \qquad (2.13)$$

The lowering of the energy due to the singlet formation is therefore

$$\epsilon = -D\, e^{-|\epsilon_f|/(\nu_f N(0) V^2)} \qquad . \qquad (2.14)$$

It is customary to associate with this energy gain a characteristic temperature T_K, i.e. the Kondo temperature

$$k_B T_K = D\, e^{-\pi|\epsilon_f|/(\nu_f \Gamma)} \qquad (2.15)$$

with Γ as defined in Equation (2.5). The condition $|\epsilon_f| >> \nu_f \Gamma$ ensures that T_K is sufficiently small (Kondo regime). In distinction to Equation (1.6) the energy gain is a nonanalytic function of V. Its origin is easily traced back to the fact that the singlet involves many different k states.

The normalization constant A is related to the f electron number by

$$|A|^2 = 1 - n_f \qquad (2.16)$$

where

$$n_f = \sum_m n_m^f \qquad . \qquad (2.17)$$

For $n_f \to 1$ the state $|\Phi_o>$ has therefore little weight in $|\psi_{S=0}>$. Again, this is very similar to the findings in Section 1. The requirement $<\psi_{S=0}|\psi_{S=0}> = 1$ yields a relationship between T_K and $(1-n_f)$. From Equation (2.6) it follows that

$$A = \frac{1}{\sqrt{1+C}} \tag{2.18}$$

where C is given by

$$C = \nu_f V^2 \sum_{k<k_F} \frac{1}{(\epsilon+\epsilon(k))^2}$$

$$\simeq \nu_f \frac{N(0)\ V^2}{|\epsilon|} \quad . \tag{2.19}$$

By combining Equations (2.16–2.19) together with Equations (2.15–2.16) it follows that

$$k_B T_K = \frac{1-n_f}{n_f} \nu_f \frac{\Gamma}{\pi} \quad . \tag{2.20}$$

Therefore, the closer n_f is to unity, the smaller is the Kondo temperature or the energy gain due to the singlet formation.

The ansatz (2.6) can be generalized to excited states. The hole state $|\psi^{ex}_{ph}>$ with quantum numbers p and n is of the form

$$|\psi^{ex}_{pn}> = A(1 + \sum_{km} \alpha(k)\, f^+_m c_{km})\, c_{pn}\, |\Phi_o> \quad . \tag{2.21}$$

The quasiparticle described by Equation (2.21) should be thought of as a dressed conduction–band state. The "dress" or "cloud" of the hole consists of the admixture of the f state at the impurity site. For energies $|\epsilon - \epsilon_F| << k_B T_K$ the weight of the "bare" state $c_{pn} |\Phi_o>$ in $|\psi^{ex}_{pn}>$ is $|A|^2 = 1 - n_f$ and therefore very small in the Kondo limit. It is important though for the interpretation of photoemission experiments where the f–like weight which is determined by the sum over the squared matrix elements $\Sigma_{pn} |<\psi^{ex}_{pn}|f_m|\psi_{S=0}>|^2 = |A|^2 n_f/\nu_f$, and therefore varies like $n_f(1-n_f)/\nu_f$. This is the weighting factor of the well known Abrikosov–Suhl or Kondo resonance which appears in the vicinity of the Fermi energy. The main contributions to the f spectral weight come from an energy regime near ϵ_f and are well separated from the Kondo resonance, provided T_K is small.

When the ground state singlet is formed, the magnetic susceptibility of the impurity χ^{imp} remains finite. It can become very large though, when the energy difference to the excited states is small (Van Vleck susceptibility). The susceptibility can be calculated by including an external field h in the Hamiltonian. It lifts the degeneracy by Zeeman splitting the f levels. Therefore ϵ_f in Equation (2.4) must be replaced by

$$\epsilon_f \longrightarrow \epsilon_f - g_J \mu_B mh \qquad ; \qquad -J \leq m \leq J.$$

The factor g_J denotes the Lande factor of the ground–state J multiplet, which is $J = {}^5/_2$ in the case of Ce. The energy $E_S(h)$ is written in the form

$$E_S(h) = E_o + E_f + \Delta E(h) \qquad . \tag{2.22}$$

By repeating the variational calculations one finds

$$\Delta E(h) = |E_f| + V^2 \sum_m \sum_{k \leq k_F} \frac{1}{\Delta E(h) + g_J \mu_B m h + \epsilon(k)} \quad . \tag{2.23}$$

This equation generalizes Equation (2.12) to finite magnetic fields. By taking the second derivative with respect to h one obtains

$$\begin{aligned} \chi^{imp} &= -\frac{\partial^2}{\partial h^2} \Delta E(h) \mid_{h \to 0} \\ &= (g_J \mu_B)^2 \frac{J(J+1)}{3} \frac{n_f}{\nu_f \Gamma} \frac{n_f}{1-n_f} \\ &= \frac{(g_J \mu_B)^2}{\pi} \frac{J(J+1)}{3} \frac{1}{k_B T_K} \qquad . \end{aligned} \tag{2.24}$$

This shows that the smaller T_K is, the larger is χ^{imp}. Experiments, which measure the magnetic susceptibility, find that the magnetic impurity loses its moment as the temperature decreases below T_K. This is a direct consequence of the singlet formation, which in turn results from the strong electron correlations.

3. HEAVY FERMION SYSTEMS

3.1 Introduction

Heavy–fermion systems or metals with heavy electrons have developed into a new branch of metal physics. In all cases these systems contain either Ce, Yb, U or Np as one of their constituents, i.e. there are always 4 f or 5 f electrons involved. Characteristic examples are $CeAl_3$, $CeCu_2Si_2$, $CeRu_2Si_2$, $CeCu_6$, CeB_6, $YbAl_3$, $YbCu_2Si_2$, UBe_{13}, UPt_3, UCd_{11}, U_2Zn_{17} and $NpBe_{13}$. At low temperatures, i.e. below a characteristic temperature T^*, heavy–fermion systems show Fermi liquid behaviour with huge effective masses m^* of the quasiparticles. In fact, m^* may equal several hundred times the free electron mass. Consequently, the low temperature specific heat $C = \gamma T$ has an enormously large γ coefficient (i.e. γ is of the order of 1 J/(mol.K^2) rather than 1 mJ/(mol.K^2) as found for example in metallic sodium). The Pauli paramagnetic susceptibility is similarly enhanced in that temperature range. As the temperature increases to values above T^* the quasiparticles lose their heavy masses. The specific heat levels off as indicated in Figure 3.1.1 and the susceptibility changes from a Pauli–like to a Curie–like behaviour. With increasing temperature the Ce and U ions behave more and more like ions with well localized f electrons. Another interesting phenomenon is the following: with respect to the thermodynamics at low temperatures the f electrons, of Ce for example, seem to be placed right at the Fermi energy giving rise to the large density of states (see Figure 3.1.2(a). However, in a photoemission experiment it takes approximately 2 eV in order to excite a 4 f electron of Ce into an unoccupied conduction electron state above the Fermi energy (see Figure 3.1.2(b). What seems to be a contradiction at first sight is in fact not so. The two plots in Figure 3.1 merely demonstrate that there are two types of electronic excitations, namely low energy excitations involving predominantly spin degrees of freedom and high energy excitations involving charge degrees of freedom of the 4 f electrons. By means of a simple model consisting of two electrons and two orbitals it was shown in Section 1 that such a separation of excitations is a characteristic feature of strongly correlated electrons.

The anomalous low temperature properties are related to the weak hybridization of the f electrons with the electrons of the neighbouring atoms. Because the overlap between the corresponding wave functions is so small, the strong Coulomb repulsions between the f electrons suppress charge fluctuations to a large extent. This results in low energy excitations. The simplest way of seeing this is by means of the molecular model described in Section 1. It is shown that the weak hybridization results in a singlet ground state with an energy gain given by the Kondo temperature T_K. The heavy–fermion systems can be considered as lattices of Kondo ions. Singlets are formed at the different sites which "lock" together. The way in which this occurs on a microscopic level is not well understood. This does not imply that T_K and T^* are the same temperatures. Instead, one expects $T^* < T_K$ in most cases. This is due to the magnetic interactions between different Ce or actinide ions. This interaction energy is lost when nonmagnetic singlets are formed and therefore the energy gain is expected to be less than that for a single ion. This argument suggests that singlet formation should not take place when the magnetic interaction energy per site exceeds the singlet formation energy. In systems like $CeAl_2$, $CePb_3$ and $NpBe_{13}$ this seems to be the case. At low temperatures they are antiferromagnets.

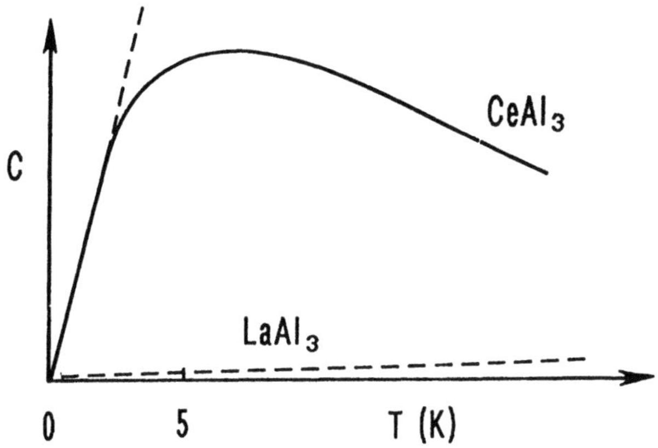

Figure 3.1.1 Qualitative plot of the specific heat C(T) of a heavy fermion system like $CeA\ell_3$. Also shown is the corresponding plot when Ce is replaced by La.

Separation of spin and charge excitations

(a) low T thermodynamics

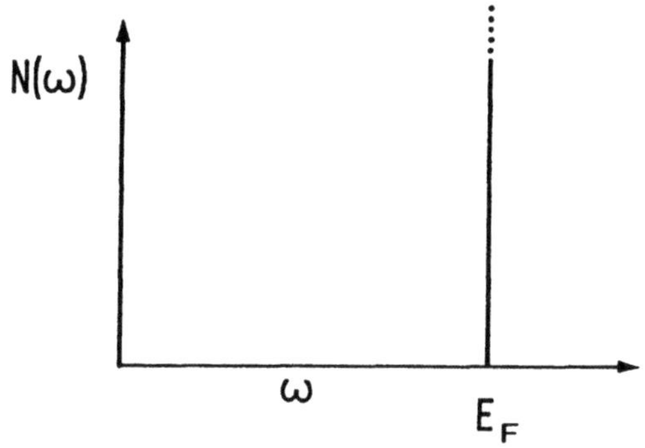

Figure 3.1.2 (a) Quasiparticle density of states as obtained from the low temperature thermodynamics. The high values near ϵ_f are due to the f electrons and are absent when e.g. Ce is replaced by La.

(b) photoemission

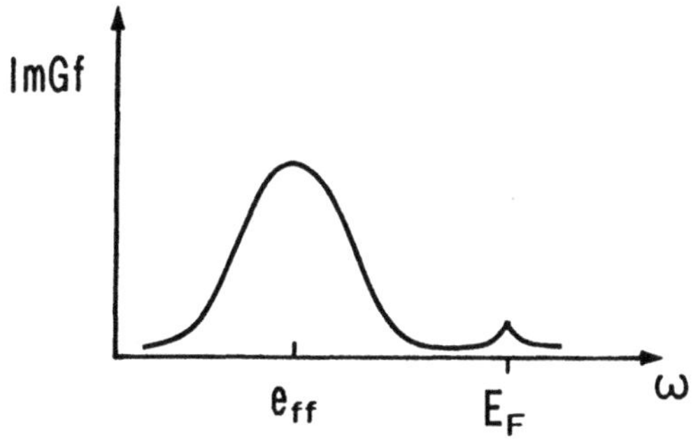

Figure 3.1.2 (b) Imaginary part of the f electron Green's function which is measured by photoemission experiments. It takes an energy of order eV in order to remove a 4 f electron from $CeAl_3$ or $CeCu_2Si_2$. The structure at ϵ_F has small weight (Kondo peak).

Singlets, which have formed, unlock and break up as the temperature increases. When this is the case the weak hybridization does not play an important role anymore and the Ce or U ions behave as ions with a well localized moment.

The delocalized nature of the f electrons due to the small hybridization has been convincingly demonstrated by de Haas–van Alphen measurements. They show the existence of strong mass anisotropies at the Fermi surface. This raises the question of how one can calculate the Fermi surface and the anisotropic masses for a system of strongly correlated electrons. Renormalized band structure calculations have proven able to provide a successful computational scheme for doing this. Fermi surfaces as well as the strongly anisotropic effective masses can be calculated by using one adjustable parameter only.

In order to obtain a better microscopic understanding of heavy–fermion systems, model Hamiltonians have to be used. They contain the strong interactions of the f electrons as ingredients but make simplifying assumptions, e.g. about the size of the basis set which is used. Various approximation schemes have been developed for solving these model Hamiltonians. This includes the introduction of auxiliary slave boson fields which are treated in mean–field approximations as well as a variational ansatz for their ground state wave function.

Of further interest are electron–phonon interactions based on the strong correlations. They result from the fact that the energy gain due to the formation of singlet states is a sensitive function of the hybridization V between the f electron and atomic orbitals of neighbouring atoms (see Equation 2.14). But V is strongly pressure–dependent, because the overlap between wave functions changes with pressure. This results in a coupling between volume changes (phonons) and electronic degrees of freedom, and therefore in an electron–phonon interaction which is solely due to electron correlations. Other phenomena of interest are quasiparticle interactions. They influence normal–state properties but may also result in superconductivity.

3.2 Fermi Surface and Quasiparticle Excitations

There is experimental evidence that at low temperatures heavy–fermion systems behave like Fermi liquids. This implies that there is a one–to–one correspondence between the excitations of these strongly correlated electron systems and those of free electrons in an external periodic potential. This information can be used to develop a semi–phenomenological theory for the computation of the Fermi surface and the low energy excitations. The quasiparticles are moving in the periodic potential set up by the lattice ions and electrons. This potential can be completely described by a set of energy dependent phase shifts $\left\{\eta_\ell^A(\epsilon)\right\}$. Here A denotes the different atoms in the unit cell and ℓ is the orbital angular momentum quantum number. The phase shifts contain all information about the periodic potential. Consider, for example,

$CeRu_2Si_2$. The phase shifts at the Fermi energy ϵ_F, i.e.

$$\left\{\eta_\ell^A(\epsilon_F)\right\} = \left\{\eta_\ell^{Ce}(\epsilon_F),\ \eta_{\ell'}{}^{(Ru)}\nu\,(\epsilon_F),\ \eta_{\ell''}{}^{(Si)}\mu\,(\epsilon_F)\right\};\ \nu, \mu = 1{,}2 \tag{3.2.1}$$

determine the Fermi surface of that material. The partial electronic densities n_ℓ^A are given by

$$n_\ell^A = \frac{2}{\pi}(21 + 1)\ \eta_\ell^A(\epsilon_F) \tag{3.2.2}$$

and knowing all the n_ℓ^A is the same as knowing the Fermi surface. The discussions in Section 2 have shown that the strong correlations are associated with a strong reduction of the bare f electron fluctuations as compared with uncorrelated electrons. Conduction electrons which couple to these fluctuations must have $\ell = 3$ when an expansion in terms of spherical harmonics around a given Ce or U site is made. This particular scattering channel is therefore linked to the strong many–body effects and one expects that the associated phase shift $\eta_{\ell=3}^{Ce}(\epsilon)$ differs from the one given by the LDA. All the other phase shifts should remain essentially uneffected by the strong correlations and therefore are taken from a LDA calculation. This neglects the coupling of conduction electrons to different configurations of the 4 f and 5 f shell with <u>fixed</u> f electron number. For example, conduction electrons with different values of ℓ may cause transitions between different j multiplets or crystalline electric field split levels within a j multiplet. The conduction electron mass enhancement in Pr metal falls into that category. When all phase shifts except $\eta_{\ell=3}^{Ce}(\epsilon)$ are determined from a LDA calculation by treating the f electron of Ce as part of the core, the Fermi surface is already fixed to a considerable extent. What remains to be determined is the f phase shift. According to Hund's rules the ground state multiplet of the 4 f^1 configuration is $j = {}^5/_2$. The $j = {}^7/_2$ multiplet is usually higher in energy by an amount of the order of 0.3 eV. Therefore one may set $\eta_{j={}^7/_2}^{Ce}(\epsilon_F) = 0$, because there is no f electron density of $j = {}^7/_2$ character inside the Fermi surface. The $j = {}^5/_2$ multiplet is split by the presence of the crystalline electric field (CEF). For Ce this splitting is usually of the order of a few hundred K or a few tens of meV. Assume that $k_B T^*$ is much smaller than this CEF splitting. At temperatures $T < T^*$ the f electron density must then have the symmetry of the CEF ground state. Except in special cases the latter consists of a doublet which can be characterized by a pseudospin $\tau = \pm 1$. Therefore only the phase shift $\eta_\tau^{Ce}(\epsilon_F)$ among the different $\ell = 3$ channels differs from zero. An example of the 4 f charge density is shown in Figure 3.2.1 where it is seen that it reflects the cubic symmetry. Close to the Fermi energy the phase shift $\eta_\tau^{Ce}(\epsilon)$ can be expanded as

$$\eta_\tau^{Ce}(\epsilon) = \eta_\tau^{Ce}(\epsilon_F) + \frac{1}{k_B T^*} (\epsilon - \epsilon_F) + \sum_{\tau' i, \epsilon'} \Phi^i_{\tau\tau'} \delta n^i_{\tau'}(\epsilon'). \tag{3.2.3}$$

The slope of the phase shift at ϵ_F is used to define a characteristic temperature T*. Its size fixes the width of the resonance at ϵ_F and therefore the effective mass of the quasiparticles of f character. The last term in Equation (3.2.3) describes the effect of other quasiparticles which may be present either on neighbouring sites i of the Ce site or on the site itself (i = 0). Pauli's principle prevents two f electrons from occupying a Ce site with the same quantum number τ. Since the quasiparticles have predominantly 4 f character this implies that to good approximation $\Phi^0_{\tau\tau} = 0$. Here we are interested only in the energy dispersion $E(\underline{k})$ of a single quasiparticle, i.e. when no other quasiparticles are present. Therefore the last term in Equation (3.2.3) is zero. The theory contains then only two parameters, i.e. $\eta_\tau^{Ce}(\epsilon_F)$ and T*. Moreover, one parameter, namely $\eta_\tau^{Ce}(\epsilon_F)$, is fixed by the requirement that the volume enclosed by the Fermi surface does not

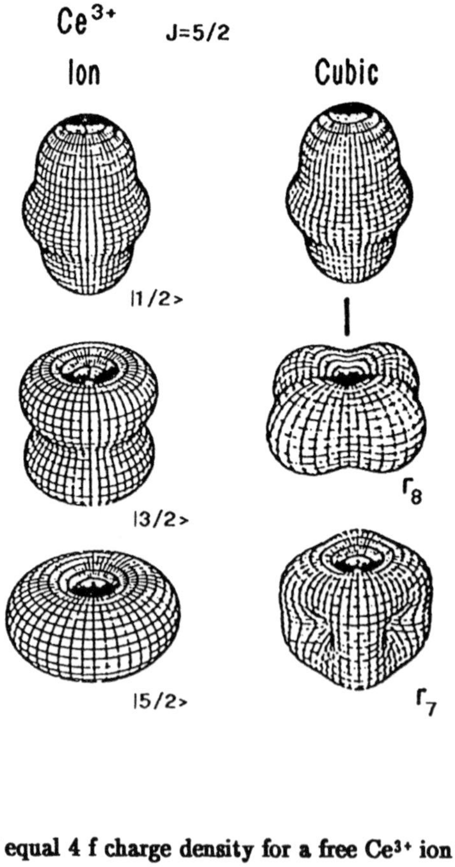

Figure 3.2.1 Contours of equal 4 f charge density for a free Ce^{3+} ion and for a Ce^{3+} ion in cubic symmetry (by courtesy of U Walter).

change when the strong correlations are included by renormalization of the f phase shifts. The volume of the Fermi sea depends on the electron number only and is independent of the interactions (Luttinger's theorem). As the 4 f electrons are part of the Fermi liquid, they participate in the counting of this volume. With one 4 f electron per Ce site $\eta_{\Gamma}^{Ce}(\epsilon_F) \simeq \pi/2$. Having fixed one of the two parameters the only remaining one is T*. It can be adjusted by requiring for example that the calculated density of states is in agreement with the linear specific heat coefficient γ.

After the phase shifts have been determined one can calculate the Fermi–surface cross sections, which are measured in a de Haas–van Alphen experiment, as well as the effective masses. The technical execution of such calculations is not simple. One reason is the high energy resolution which is required, and which is parts of 1 meV. But the results are very encouraging. Consider $CeRu_2Si_2$ as an example.

Experimental findings show that, as far as the Fermi surface topology is concerned, the deviations between LDA calculations and renormalized band calculations are relatively small. However, in distinction to the former, the renormalized band calculations reproduce well also the large measured mass anisotropies. The extremal cross sections δ, α and ϵ belong all three to the same portion of the Fermi surface (see Figure 3.2.2). The corresponding experimental values for the ratio of the effective mass m* to the free electron mass m_0 are $m^*/m_0 = 4$, 12 and 20 respectively. They have to be compared with the values obtained from a renormalized band calculation which are $m^*/m_0 = 2$, 10 and 20. They cannot be obtained by simply scaling the LDA results. These portions of the Fermi surface with the extremely large masses ($m^* > 200\ m_0$) have not been seen yet in the experiments. Their contributions dominate the specific heat coefficient γ, to which T* has been fitted. Therefore one may take the point of view that the anisotropic masses are obtained without a fit parameter.

A special case of interest is when the CEF splitting is much smaller than $k_B T^*$. Then there is again only one phase shift for the various $\ell = 3$ channels which has to be known, i.e. $\eta_{j=5/2}^{Ce}(\epsilon)$.

In a LDA calculation the hybridization is always found to be larger than the CEF splitting. Therefore because of Luttinger's theorem, the phase $\eta_{j=5/2}^{Ce}(\epsilon_F)$ at the Fermi energy is the same when a renormalized band calculation is done and when the LDA is applied. This presumes that the $j = 1/2$ multiplet is sufficiently high in energy, so that the corresponding phase shift at ϵ_f is also zero within the LDA. In this case all phase shifts at ϵ_F are the same in a renormalized band calculation and in one based on the LDA. Therefore the Fermi surface comes out identically the same in both cases. The effective masses are different though, because the derivative $(d\eta_{j=5/2}^{Ce}(\epsilon)/d\epsilon)$ at ϵ_F differs in the two theories. This seems to be the case in UPt_3.

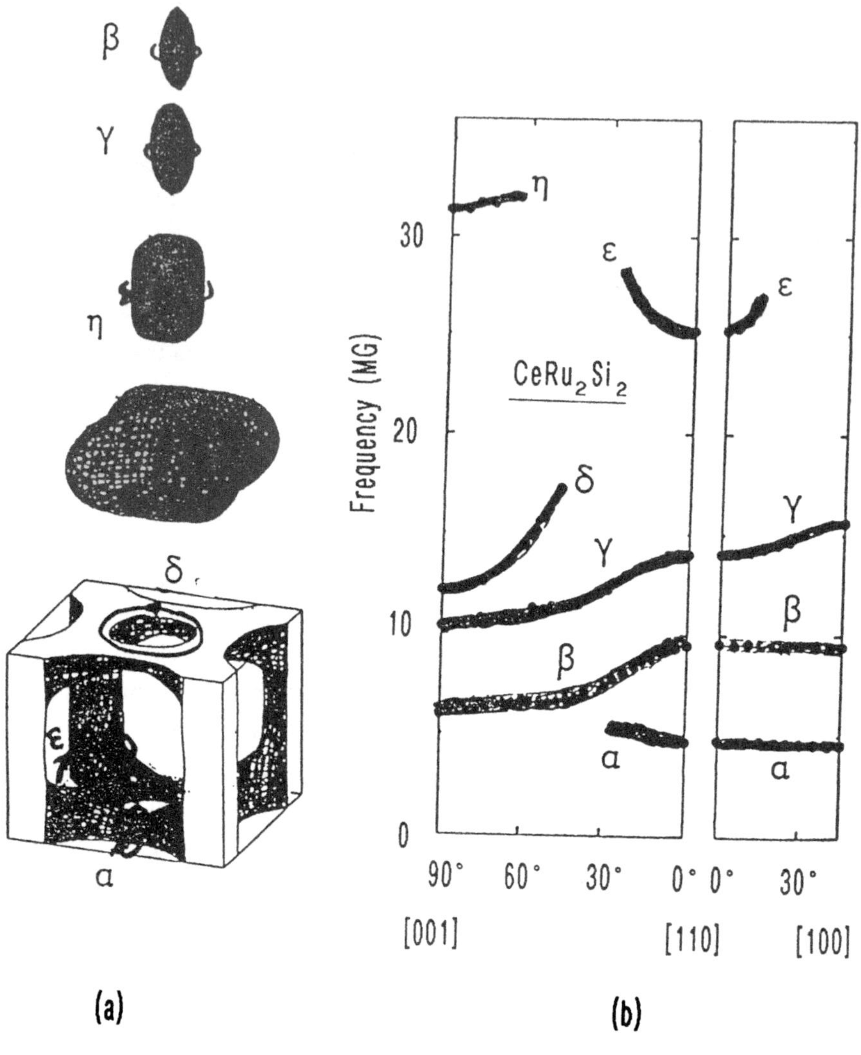

Figure 3.2.2 (a) Calculated Fermi surface of $CeRu_2Si_2$ (Runge, Zwicknagl and Christensen)

(b) Results of de Haas–van Alphen measurements.

It is instructive to leave the quasiparticle picture for a moment and to go over to the microscopic picture of bare f electrons. Their effective hybridization with the conduction electrons is strongly renormalized. The reason is very simple. A conduction electron can hop onto a 4 f orbital of a Ce ion only when the latter is empty. Otherwise the large Coulomb repulsion between the two 4 f electrons would come into play. But when the f electron number n_f is close to one i.e. $n_f \lesssim 1$, the f orbital is unoccupied only with probability $(1 - n_f)$. Therefore the effective hybridization is strongly reduced. This results in the large density of low lying excitations, or, alternatively, in a self–energy $\Sigma(\omega) = - A\omega$ with $A >> 1$. In fact, it turns out that $A \simeq n_f/(1 - n_f)$. The large coefficient A is closely related with the large slope $(k_B T^*)^{-1}$ of the phase shift $\eta_T^{Ce}(\epsilon)$ of the quasiparticles at ϵ_F.

We conclude by briefly discussing the form of the f electron Green's function $G_f(\underline{k}, \omega)$ close to the Fermi surface. By assuming only one band of heavy quasiparticles with dispersion $\epsilon_{qp}(\underline{k})$ and in accordance with the self–energy discussed above, it reduces to

$$G_f(\underline{k},\omega) = \frac{1 - n_f}{\omega - \epsilon_{qp}(\underline{k}) + i\eta \text{ sgn } \omega} + G_{inc}(\underline{k},\omega) . \qquad (3.2.4)$$

The first part contains the quasiparticle pole while the second part $G_{inc}(\underline{k},\omega)$ describes a structureless incoherent background. In accordance with the discussion in Section 2, the f–spectral weight near ϵ_F vanishes as $(1 - n_f)$ for small values of T^*. The form of Equation (3.2.4) should be set in contrast to the Green's function of the quasiparticles. According to Landau's Fermi liquid theory there is a one–to–one correspondence between the low energy excitations of a heavy fermion system and those of a non–interacting electron gas, provided that certain parameters are renormalized. Therefore the quasiparticle Green's function is of the form

$$G_{qp}(\underline{k},\omega) = \frac{1}{\omega - \epsilon_{qp}(\underline{k}) + i\eta \text{ sgn } \omega} \qquad (3.2.5)$$

as for free electrons. The renormalized parameters are contained in the dispersion $\epsilon_{qp}(\underline{k})$.

3.3 Model Hamiltonian and Slave Bosons

A proper model Hamiltonian for the description of a system of heavy electrons is the Anderson lattice Hamiltonian. By generalizing Equation (2.1) it is written in the form

$$H = \sum_{\underline{k}n\sigma} \epsilon_n(\underline{k})\, a^+_{\underline{k}n\sigma} a_{\underline{k}n\sigma} + \sum_{mi} \epsilon_{fm} f^+_m(i) f_m(i)$$

$$+ \frac{1}{\sqrt{N_o}} \sum_{im\underline{k}n\sigma} V_{m\sigma}(\underline{k},n) \left(a^+_{\underline{k}n\sigma} f_m(i)\, e^{-i\underline{k}\underline{R}_i} + h.c.\right)$$

$$+ \frac{U}{2} \sum_{i,m\neq m'} n^f_m(i)\, n^f_{m'}(i) \qquad . \qquad (3.3.1)$$

The index i labels the N_o f sites at positions $\underline{R}_i$. Furthermore, the conduction electron creation operators are denoted here by $a^+_{\underline{k}n\sigma}$ where n is a band index. Otherwise the notation is the same as in Equation (2.1). In most treatments the limit of large Coulomb repulsion U of the f electrons is considered. In this case the f electron number at a site is assumed to be either 1 or 0. Double occupancies of f orbitals are strictly excluded. One way of taking this exclusion directly into account is by the introduction of standard basis operators $X_{om} = |0\rangle\langle m|$, $X_{mo} = |m\rangle\langle 0|$ and $X_{mm} = |m\rangle\langle m|$ for each site. Here $|m\rangle = f^+_m|0\rangle$ where $|0\rangle$ is the vacuum i.e. the empty f orbital. For example, the operator $X_{om}(i)$ annihilates at site i the singly occupied f state with quantum number m and generates an empty f state. When expressed in terms of these operators, H is of the form

$$H = \sum_{\underline{k}n\sigma} \epsilon_n(\underline{k})\, a^+_{\underline{k}n\sigma} a_{\underline{k}n\sigma} + \sum_{mi} \epsilon_{fm} X_{mm}(i) +$$

$$+ \frac{1}{\sqrt{N_o}} \sum_{im\underline{k}n\sigma} V_{m\sigma}(\underline{k},n) \left[a^+_{\underline{k}n\sigma} X_{om}(i)\, e^{-i\underline{k}\underline{R}_i} + h.c.\right]. \qquad (3.3.2)$$

The $X_{om}(i)$ are no longer fermion operators because they do not obey simple fermionic anti–commutation relations. For example, $[X_{mo}(i), X_{om}(i)]_+ = X_{mm}(i) + X_{oo}(i) \neq 1$. In addition they must fulfill the subsidiary condition:

$$\sum_{m=1}^{\nu_f} X_{mm}(i) + X_{oo}(i) = 1 \qquad . \qquad (3.3.3)$$

The subsidiary condition ensures that one remains in the Hilbert space of singly occupied or empty f sites, respectively. Because the hybridization term is small one would like to treat it by perturbation expansion. For that purpose H is decomposed into $H = H_o + H_1$ with

$$H_o = \sum_{\underline{k} n \sigma} \epsilon_n(\underline{k})\, a^+_{\underline{k} n \sigma} a_{\underline{k} n \sigma} + \sum_{mi} \epsilon_{fm} X_{mm}(i)$$

$$H_1 = \frac{1}{\sqrt{N_o}} \sum_{im} \sum_{\underline{k} n \sigma} V_{m\sigma}(\underline{k},n) \left[a^+_{\underline{k} n \sigma} X_{om}(i)\, e^{-i\underline{k}\underline{R}_i} + h.c. \right].$$

(3.3.4)

But now one is facing a problem. Because the $X_{mm}(i)$ do not satisfy fermionic anticommutation relations, expectation values of the form $<T_\tau\ H_1(\tau_1)\ \ldots\ H_1(\tau_n)>_o$ which enter the free energy or the ground state energy computation cannot be easily evaluated by standard Green's functions techniques. Wick's theorem does not hold in this case. Corresponding calculations therefore become rather complex. This suggests looking for alternative ways of handling the problem.

The popular method of treating the Hamiltonian (3.3.1) is by a mean field approximation. Thereby it is assumed that the strong Coulomb repulsion between f electrons can be taken into account by a renormalization of the hybridization matrix element

$$V_{m\sigma}(\underline{k},n) \longrightarrow r V_{m\sigma}(\underline{k},n) = \tilde{V}_{m\sigma}(\underline{k},n) \quad . \tag{3.3.5}$$

The renormalization factor takes into account the feature that a conduction electron can hop onto a f orbital only when that f site is empty. With $n_f \leq 1$, the chance of finding a site unoccupied is $(1 - n_f)$. The rate of hops (and back) of conduction electrons onto f sites is, according to Fermi's golden rule,

$$\begin{aligned} \tilde{\Gamma} &= (1 - n_f)\, \pi\, V^2\, N(0) \\ &= \pi\, \tilde{V}^2\, N(0) \end{aligned} \tag{3.3.6}$$

where N(0) is the density of states at the Fermi level. This should be compared with Equation (2.5). The mean field approximation replaces the operator of f occupancy of a site $n_f^{op}(i)$ by its average value n_f. Whether or not a hop of a conduction electron onto an f site takes place depends in reality on whether or not the site is <u>actually</u> empty, i.e. the probability varies from configuration to configuration. By averaging over all configurations as the mean field approximation does, this special feature is lost. From Equation (3.3.6) one obtains the relation

$$r^2 = 1 - n_f \quad . \tag{3.3.7}$$

The parameter r has to be determined self–consistently, i.e. by minimizing the free energy or the ground state energy, depending on whether one is considering finite temperatures or T = 0, respectively. The subsidiary condition (3.3.7) is included by multiplication with a Lagrange parameter Λ. The mean field Hamiltonian H_{MF} is then of the form

$$H_{MF} = \sum_{\underline{k}n\sigma} \epsilon_n(\underline{k})\, a^+_{\underline{k}n\sigma} a_{\underline{k}n\sigma} + \sum_{m\underline{k}} \tilde{\epsilon}_{fm} f^+_{\underline{k}m} f_{\underline{k}m}$$

$$+ \sum_{nm\underline{k}\sigma} V_{m\sigma}(\underline{k},n)\, r(a^+_{\underline{k}n\sigma} f_{\underline{k}m} + \text{h.c.}) + \Lambda N_0(r^2 - 1) \qquad (3.3.8)$$

with $\tilde{\epsilon}_{fm} = \epsilon_{fm} + \Lambda$. The Fourier transform $f^+_{\underline{k}m}$ of the operators $f^+_m(i)$ has been introduced and the condition (3.3.7) has been written in the form $1 - n_f^{op}(i) = r^2$. The Hamiltonian H_{MF} is a one–particle Hamiltonian and therefore can be easily diagonalized as a function of the two unknowns Λ and r. It is written as

$$H_{MF} = \sum_{\underline{k}1\tau} E_1(\underline{k})\, c^+_{1\tau}(\underline{k})\, c_{1\tau}(\underline{k}) + \Lambda N_0(r^2 - 1) \qquad (3.3.9)$$

where the $c^+_{1\tau}(\underline{k})$ denote the creation operators of quasiparticles in branch 1 with pseudospin τ. We may speak of quasiparticles, because the complex many–body problem has been mapped onto a one–particle problem, as before, when phase shifts were used. The $E_1(\underline{k})$ are the quasiparticle energies. They depend on Λ and r. In terms of the $c^+_{1\tau}(\underline{k})$ the ground state $|\Phi_0\rangle$ of H_{MF} is written as

$$|\Phi_0\rangle = \prod_{\substack{1,\tau \\ |\underline{k}| < k_F}} c^+_{1\tau}(\underline{k})\, |0\rangle \quad . \qquad (3.3.10)$$

In order to relate the operators $a^+_{\underline{k}n\sigma}$ and $f^+_{\underline{k}m}$ to the $c^+_{1\tau}(\underline{k})$ and vice versa one expands

$$f^+_{\underline{k}m} = \sum_{1\tau} y_m(\underline{k};1\tau)\, c^+_{1\tau}(\underline{k})$$

$$a^+_{\underline{k}n\sigma} = \sum_{1\tau} x_{n\sigma}(\underline{k};1\tau)\, c^+_{1\tau}(\underline{k}). \qquad (3.3.11)$$

The coefficients $x_{n\sigma}(\underline{k};1\tau)$ and $y_m(\underline{k};1\tau)$ must satisfy the normalization condition

$$\sum_{m} |y_m(\underline{k};1\tau)|^2 + \sum_{n\sigma} x_{n\sigma}(\underline{k};1\tau)|^2 = 1 \qquad (3.3.12)$$

They are obtained from the set of equations

$$(\tilde{\epsilon}_{fm} - E_1(\underline{k}))\, y_m(\underline{k};1\tau) + \sum_{n\sigma} \tilde{V}_{m\sigma}(\underline{k},n)\, x_{n\sigma}(\underline{k};1\tau) = 0$$

$$\sum_{m} \tilde{V}^*_{m\sigma}(\underline{k},n)\, y_m(\underline{k};1\tau) + (\epsilon_n(\underline{k}) - E_1(\underline{k}))\, x_{n\sigma}(\underline{k};1\tau) = 0$$

(3.3.13)

The latter follow when the expansions (3.3.11) are set into Equation (3.3.8) and it is required that the off–diagonal terms vanish. By elimination of $x_{n\sigma}(\underline{k};1\tau)$ one obtains

$$[\tilde{\epsilon}_{fm} - E_1(\underline{k})]\, y_m(\underline{k};1\tau) + \sum_{n\sigma m'} \frac{\tilde{V}_{m\sigma}(\underline{k},n)\ \tilde{V}^*_{m'\sigma'}(\underline{k},n)}{E_1(\underline{k}) - \epsilon_n(\underline{k})} \cdot y_{m'}(\underline{k};1\tau) = 0$$

(3.3.14)

The $y_m(\underline{k};1\tau)$ are the probability amplitudes of finding a quasiparticle positioned in band 1 with pseudospin τ in a f state with quantum number m. Because of the weak hybridization $\tilde{V}_{m\sigma}(\underline{k},n)$ one finds that the solutions of Equation (3.3.14) yield one quasiparticle branch with an energy close to $\tilde{\epsilon}_{fm}$, while the other branches have energies close to that of free conduction electrons. For the special case of one conduction electron band and a f orbital degeneracy of $\nu_f = 2$, one finds two quasiparticle bands with energies

$$E_1(\underline{k}) = \frac{1}{2}\left\{(\epsilon(\underline{k}) + \tilde{\epsilon}_f) \mp W(\epsilon(\underline{k}))\right\}$$

$$W(\epsilon(\underline{k})) = \sqrt{(\epsilon(\underline{k}) - \tilde{\epsilon}_f)^2 + 4\,\tilde{V}^2} \qquad . \qquad (3.3.15)$$

The –(+) signs refer to the bands 1 = 1(2). By requiring that $E_1(k_F) = \mu$, where μ is the chemical potential, one obtains for $\tilde{\epsilon}_f$

$$\tilde{\epsilon}_f = \mu + \frac{\tilde{V}^2}{\epsilon(k_F) - \mu} \qquad . \qquad (3.3.16)$$

Because $\tilde{V}$ is very small, the renormalized f electron energy $\tilde{\epsilon}_f$ lies slightly above the Fermi energy. In order to determine the two unknowns r and Λ, consider the case of T = 0. One equation is obtained by requiring that the ground state energy $E_{MF} = <\Phi_o(r)\ |H_{MF}|\ \Phi_o(r)>$ is minimized with respect to r. Because of $<\Phi_o(r)\ |\ \Phi_o(r)> = 1$ this condition reduces to

$$<\Phi_o\ |\frac{\partial}{\partial r} H_{MF}|\ \Phi_o> = 0 \tag{3.3.17}$$

or

$$\sum_{nm\underline{k}\sigma} \tilde{V}_{m\sigma}(\underline{k},n)\ <a^+_{\underline{k}n\sigma}\ f_{\underline{k}m}> + N_o \Lambda r^2 = 0 \tag{3.3.18}$$

with the obvious notation that $<.....> = <\Phi_o|....|\Phi_o>$. A second equation follows from the condition (3.3.7) which is rewritten in the form

$$\sum_{\underline{k}m} <f^+_{\underline{k}m}\ f_{\underline{k}m}> + N_o(r^2 - 1) = 0 \tag{3.3.19}$$

By making use of Equation (3.3.10) for $|\Phi_o>$ and of Equations (3.3.11), as well as the solutions of Equations (3.3.14), one obtains for the expectation values in Equations (3.3.18–19) the following results

$$\frac{1}{N_o} \sum_{\underline{k}m} <f^+_{\underline{k}m}\ f_{\underline{k}m}> = \frac{\nu_f}{2} N(0)[\epsilon(k_F) + W(\epsilon(k_F)) - W(0)] = n_f$$

$$\frac{1}{N_o} \sum_{\underline{k}m} \tilde{V}_{m\sigma}(\underline{k},n)\ <a^+_{\underline{k}n\sigma}\ f_{\underline{k}m}> = -\nu_f\ N(0)\ \tilde{V}^2\ \ln\left[\frac{\epsilon(k_F - \tilde{\epsilon}_f + W(\epsilon(k_F))}{-\tilde{\epsilon}_f + W(0)}\right] . \tag{3.3.20}$$

To leading logarithmic approximation the two Equations (3.3.18 – 19) determine r and Λ as

$$r^2 = 1 - \nu_f\ N(0)\ ((\epsilon(k_F) - \mu) - \frac{\tilde{V}^2}{\mu}) \tag{3.3.21a}$$

$$\Lambda = \nu_f\ N(0)\ V^2 \ln\left[\frac{(\epsilon(k_F) - \mu)\ \mu}{\tilde{V}^2}\right] \tag{3.3.21b}$$

It is useful to define a characteristic temperature T* through

$$k_B\ T^* = \mu \exp\left(-\frac{\Lambda}{\nu_f\ N(0)\ V^2}\right) \tag{3.3.22}$$

From Equations (3.3.16) and (3.3.21b) one obtains

$$\tilde{\epsilon}_f = \mu + k_B T^* \quad . \tag{3.3.23}$$

Similarly, by using $r^2 = (\tilde{V}/V)$ and Equation (3.3.21b), one can eliminate $\epsilon(k_F)$ from Equation (3.3.21a) and obtain

$$n_f = 1 - \frac{\pi k_B T^*}{\nu_f \Gamma} \tag{3.3.24}$$

where Γ is defined according to Equation (2.5). The characteristic temperature T^* is related to the energy gain of the system due to the hybridization V. In order to see this, one computes the ground state energy per site

$$\begin{aligned} E &= \frac{1}{N_o} <H_{MF}> \\ &= \frac{\nu_f}{N_o} \sum_{|\underline{k}|<k_F} E_1(\underline{k}) - \Lambda n_f \end{aligned} \tag{3.3.25}$$

and subtracts from it the energy per site $E^{(0)}$ in the absence of the hybridization. After some algebra one finds that

$$E - E^{(0)} = - k_B T^* \quad . \tag{3.3.26}$$

The temperature T^* plays the role of a Kondo temperature for a lattice. Its dependence on microscopic parameters is very similar to that of Equation (2.15) since to leading order, Λ is given by $\Lambda = |\epsilon_f|$ when the energy is counted from the Fermi energy ϵ_F. This follows from Equation (3.3.23) and the relation $\tilde{\epsilon}_f = \epsilon_f + \Lambda$. The only difference between T_K as defined by Equation (2.15) and T^* defined by Equation (3.3.22) is that half the conduction electron bandwidth D has been replaced by the chemical potential μ. These findings show that the mean field approximation to the Anderson lattice Hamiltonian reproduces results which resemble those in Section 2 for the single impurity problem. It turns out that the mean field theory also contains the strong one–site repulsion which prevents two quasiparticles from occupying the same site. However, other features of quasiparticle interactions, e.g. those which enhance the magnetic susceptibility, are not contained in the theory. The above considerations were limited to T = 0, but the theory is easily extended to finite temperatures. The averages <....> in Equations (3.3.18–19) are then thermodynamic averages with respect to H_{MF}. Only for temperatures T less than a critical temperature T_c does one find a solution of those equations with $r \neq 0$. For $T > T_c$ one finds that $r = 0$ and the conduction electrons decouple completely from the f electrons.

The above derivation of the mean field theory was based on Equations (3.3.5) and (3.3.7). It can be put onto firmer ground by the introduction of an auxiliary– or slave boson field b(i) at each site i. The boson operator $b^+(i)$ is assumed to create an empty f state at site i. Accordingly, $b^+(i)\, b(i)$ is the operator for the probability that the f orbitals at site i are empty. In the limit of large U the operator

$$Q(i) = \sum_m f_m^+(i)\, f_m(i) + b^+(i)\, b(i) \qquad (3.3.27)$$

is conserved and the physically relevant eigenvalue is 1. Either the f orbitals at site i are occupied by one electrons and $\sum_m f_m^+(i) f_m(i) = 1$ or they are empty and $b^+(i)b(i) = 1$. In terms of these boson operators and taking the subsidiary condition (3.3.27) into account by a Lagrange multiplier, the Hamiltonian (3.3.1) reads

$$H = H_{band} + \sum_{mi} (\epsilon_{fm} + \Lambda_i)\, f_m^+(i)\, f_m(i) + \sum_i \Lambda_i (b^+(i)\, b(i) - 1)$$

$$+ \frac{1}{\sqrt{N_o}} \sum_{i m \underline{k} n \sigma} \left[V_{m\sigma}(\underline{k},n)\, b^+(i)\, a^+_{\underline{k}n}\, f_m(i)\, e^{-i\underline{k}\underline{R}_i} + h.c. \right]$$

(3.3.28)

where H_{band} denotes the conduction electron part of H. It is worth noticing that, by the introduction of the auxiliary boson field, one can re–establish Wick's theorem even when H_o includes the strong correlations. The Hamiltonian $H_o = (H - H_{hyb})$, where H_{hyb} is the hybridization term in the last equation, is diagonalized by operators with simple fermionic and bosonic commutation relations. Therefore the perturbation expansion, which will be described in Section 5.2, can be used and Feynman diagrams can be attached to each order of the perturbing Hamiltonian H_{hyb}. The price one has to pay is that one must deal with an additional boson field which is interacting with the conduction and f electron fields.

In mean field approximation the condition $Q(i) = 1$ is replaced by the weaker one

$$\langle Q(i) \rangle = 1 \qquad . \qquad (3.3.29)$$

This relation is satisfied then by replacing $b^+(i)$ by the site–independent mean value of the field operator, i.e.

$$b^+(i) \Rightarrow \langle b^+(i)\rangle = r \qquad (3.3.30)$$

In this way one ends up again with the mean field Hamiltonian (3.3.8).

A mean field theory usually breaks a symmetry of the Hamiltonian. In the present case the Hamilton operator (3.3.28) is invariant with respect to phase changes $b \longrightarrow be^{i\phi}$ and $f_m \longrightarrow f_m e^{i\phi}$. This symmetry is broken when b(i) is replaced by a c–number r. The fluctuations $\delta b(i) = \langle b(i)\rangle - b(i)$, which are neglected in a mean field theory, prevent a decoupling of the conduction and f electrons for $T > T_c$. With increasing orbital degeneracy ν_f the influence of the fluctuations decreases and in the limit $\nu_f \longrightarrow \infty$ the mean field theory becomes exact.

The method of introducing auxiliary fields is an important and powerful concept. It enables one to work with mean field approximations without eliminating the strong correlations as in a conventional SCF approximation. Instead, their most important effects are still contained in the theory. Finally, we want to establish a connection between the mean field approach and the renormalized band structure calculations presented in Section 3.2. The resemblance between the two approaches is apparent. The reduction of the bare hybridization V to $\tilde{V}$ has its equivalent in the large slope $(k_B T^*)^{-1} (\epsilon - \epsilon_F)$ of the f phase shift $\eta^i_\tau(\epsilon)$ at ϵ_F (see Equation 3.2.3)). Similarly, the positioning of $\tilde{\epsilon}_f$ just above ϵ_F corresponds to a value of $\eta^i_\tau(\epsilon_F)$ of slightly less than $\pi/2$. In fact, it can be shown quite rigorously by starting from the secular equation for the renormalized bands that by a "downfolding" or reduction procedure an effective Hamiltonian of the form of Equation (3.3.8) is obtained, with the only difference that the f electron energy is also $\underline{k}$–dependent, i.e. $\tilde{\epsilon}_{fm} \longrightarrow \tilde{\epsilon}_{fm}(\underline{k})$. When the CEF ground state is only two–fold degenerate and when spin–orbit effects for the conduction electrons can be neglected, it reduces to

$$H_{eff} = \sum_{\underline{k}\tau} \tilde{\epsilon}_f(\underline{k})\, f^+_{\underline{k}\tau} f_{\underline{k}\tau} + \sum_{\underline{k}n\sigma} \left[\tilde{V}_{n\tau}(\underline{k})\, f^+_{\underline{k}\tau} a_{\underline{k}n\tau} + \text{h.c.}\right] + H^{eff}_{band} \qquad (3.3.31)$$

where H^{eff}_{band} is an effective conduction electron Hamiltonian.

3.4 Quasiparticle Interactions

The interactions between quasiparticles play an important role in heavy fermion systems. They influence strongly not only static thermodynamic quantities such as the compressibility or spin susceptibility but also dynamic properties like low–energy plasmon excitations or hydrodynamic fluctuations. The basic ideas of Landau's phenomenological Fermi liquid approach make the crucial assumption that there is a one–to–one correspondence between the elementary excitations of the strongly correlated electron system and those of an ideal Fermi gas. These excitations are the quasiparticles. Their entropy is of the same form as for an ideal gas of fermions. This implies that the quasiparticles obey a Fermi distribution law. Let $\delta n_{\underline{p}\sigma}$ denote the deviation of the distribution $n_{\underline{p}\sigma}$ from a step function $\theta(|\underline{p}| - p_F)$. Let δE be the change of the electronic system due to the change $\delta n_{\underline{p}\sigma}$ in the distribution function. The function $f_{\sigma\sigma'}(\underline{p},\underline{p}')$ describes the interactions between the quasiparticles. In the phase shift formulation of the excitation spectrum (see Equation (3.2.3)) the quasiparticle interactions enter in the form of an additional contribution to the f phase shift of, for example, a Ce site

$$\delta\eta_{\tau}^{Ce}(\epsilon) = \sum_{\tau' i, \epsilon'} \Phi^{i}_{\tau\tau'} \, \delta n^{i}_{\tau'}(\epsilon'). \tag{3.4.1}$$

The $\delta n^{i}_{\tau'}(\epsilon')$ specify the quasiparticle distribution at sites i which are in the neighbourhood of the Ce site under consideration. The parameters $\Phi^{i}_{\tau\tau'}$ characterize the quasiparticle interactions. An important relation between these parameters follows from the requirement that the narrow quasiparticle band with the large density of states is pinned to the Fermi surface. One way of understanding this is by returning to the simple molecular model described in Section 1. The low lying singlet–triplet excitations which involve predominantly spin degrees of freedom, as in heavy fermion systems, depend only very weakly on the difference of the orbital energies $(\epsilon_1 - \epsilon_f)$. In a metal the role of the latter is taken by the energy difference $(\epsilon_F - \epsilon_f)$ and therefore the occurrence of a high density of states at the Fermi energy should move with ϵ_F when the latter changes. A different way of stating the same point is to notice that the f electron number per Ce site n_f should remain almost unchanged when electrons are added to the system. A narrow quasiparticle band, which is not pinned to the Fermi energy ϵ_F, fills up when electrons are added to the system. But when the narrow band moves with ϵ_F when the Fermi energy changes, then the additional electrons occupy conduction band states. A relation for the parameter $\Phi^{i}_{\tau\tau'}$ is obtained by requiring that the contributions from the last two terms in Equation (3.2.3) cancel when the Fermi energy is shifted. This leads to

$$\frac{1}{k_B T^*} + N^*(0) \sum_{\tau' i} \Phi^{i}_{\tau\tau'} = 0 \tag{3.4.2}$$

where $N^*(0) = m^* p_F / 2\pi^2$ is the large quasiparticle density of states. Pinning the narrow quasiparticle band to the Fermi energy implies that the electronic compressibility must be of a size similar to that for an ordinary metal. It is not enhanced by the large density of states $N^*(0)$ or quasiparticle mass m^*. For <u>homogeneous</u> systems one would conclude that the Landau parameter F_0^s must be large, i.e.

$$F_o^s \simeq \frac{m^*}{m}. \qquad (3.4.3)$$

Here we are dealing with strongly inhomogeneous systems and therefore the subscript 0 should not be taken to refer to an s–wave quasiparticle scattering parameter. Instead, it only aids the comparison. The large value of F_0^s is the most important effect of quasiparticle interactions in heavy fermion systems.

The effect of quasiparticle interactions on the spin susceptibility χ_s reflects itself in the Sommerfeld–Wilson ratio

$$R = \frac{\pi^2 k_B^2}{3\mu_{eff}^2} \frac{\chi_s}{\gamma} \qquad (3.4.4)$$

where μ_{eff} is the effective magnetic moment of the quasiparticles and γ is the linear specific heat coefficient. The ratio χ_s/γ no longer contains the large quasiparticle density of states because it is cancelled out but it depends on the quasiparticle interactions which enter χ_s. For the single Kondo ion case, $R = 2$, provided the orbital degeneracy is $\nu_f = 2$. For higher orbital degeneracies $R = \nu_f/(\nu_f - 1)$. Therefore in the limit of large orbital degeneracy one finds $R \rightarrow 1$ and the quasiparticle interactions do not affect the susceptibility. In the Kondo lattice case it is no longer possible to derive a definite value for R. Instead one finds

$$\frac{1}{R} = 1 - \sum_j (\Phi^j_{\uparrow\downarrow} - \Phi^j_{\uparrow\uparrow}) \quad . \qquad (3.4.5)$$

Measuring the susceptibility fixes therefore only a particular combination of the $\Phi^j_{\tau\tau'}$ parameters. In passing we note that there is an additional complication when the electron–phonon interaction is strong. It is known that this interaction contributes a factor of $(1 + \lambda)$ to the specific heat coefficient γ but not to χ_s. The dimensionless parameter λ is hereby a measure of the electron–phonon interaction strength. Therefore one has to redefine $R_{eff} = R(1 + \lambda)$.

3.5 Quasiparticle–Phonon Interactions Based on Strong Correlations

An unexpected finding is that in strongly correlated electron systems the dominant interaction of quasiparticles with phonons is purely based on correlations. Therefore electron–phonon interactions in these systems are of special interest. Experimentally they show up in the elastic properties of heavy fermion systems. For example, the elastic constants show a temperature dependence at low T which is much stronger than usual. An anomalous ultrasonic attenuation has also been observed in some cases. The experimental situation is reviewed.

In order to discuss the coupling of f electrons to phonons consider first temperatures $T >> T^*$. In this case the f electrons can be assumed to be localized. They couple to phonons because the crystal field is changed when the neighbourhood of an ion, e.g. Ce, is deformed by lattice vibrations (phonons). Changes in the crystal field imply changes in the CEF eigenstates and eigenvalues of the f electron shell. Formally the coupling of f electrons to phonons is described by virtual excitations of CEF levels by phonons. The physical implications of these interactions have been studied in great detail and can be found in recent reviews. When $T << T^*$ the situation is quite different. The f electrons form singlets with the conduction electrons and the corresponding energy gain is $k_B T^*$ per f site. It may be noticed, for example from the mean field expression (3.3.22), that T^* depends exponentially on the hybridization V which itself is a sensitive function of pressure. The latter changes the overlap between wave functions of neighbouring atoms and therefore affects V. This implies that phonons have a strong influence on electronic energies. The result is a strong coupling of phonons to electronic degrees of freedom and therefore to quasiparticle excitations. The phonons can be thought of as generating quasiparticle–quasihole excitations out of the Fermi sea. This process has the same effect as a deformation potential has in an ordinary metal. A measure of the electron–phonon coupling is the electronic Grüneisen parameter

$$\eta = -\frac{d(\ln T^*)}{d(\ln \Omega)} \quad . \tag{3.5.1}$$

It is dimensionless and quantifies the changes in T^* due to volume Ω changes. In heavy fermion systems η is found to be of order $10 - 100$, i.e. it is extremely large.

It is not difficult to derive a Hamiltonian for the quasiparticle–phonon interaction by starting from Equation (3.3.8) for example. However, it is more convenient to use $\underline{r}$–space instead of $\underline{k}$–space representation. For simplicity, an orbital degeneracy of $\nu_f = 2$ is assumed. The starting Hamiltonian is therefore

$$H_{eff} = H^{eff}_{cond} + \sum_{ij\tau} t_{ij}\, f^+_\tau(i)\, f_\tau(j) + \sum_{i\tau} r_i\, V(f^+_\tau(i)\, a_\tau(i) + a^+_\tau(i)\, f_\tau(i)) + \Lambda \sum_i (r_i^2 - 1) \tag{3.5.2}$$

In accordance with Equation (3.3.31) a finite dispersion of the f electrons has been assumed. The Hamiltonian H_{eff} changes with changing volume Ω. In order to compute these changes one should notice that $r^2 \sim k_B T^*/\Gamma$, (this follows from Equations (3.3.7) and (3.3.24)), and that the f electron bandwidth is of order $k_B T^*$. The changes of T^* with Ω are much larger than those of V and therefore the latter are neglected.

In the presence of a lattice deformation $\epsilon_\Omega(i)$ one expands

$$t_{ij}(\epsilon_\Omega) = t_{ij} + \frac{1}{2}\left[\frac{\partial t_{ij}}{\partial \epsilon_\Omega(i)}\epsilon_\Omega(i) + \frac{\partial t_{ij}}{\partial \epsilon_\Omega(j)}\epsilon_\Omega(j)\right]$$

$$r_i = r(\epsilon_\Omega = 0)\left(1 + \frac{1}{2T^*}\frac{\partial T^*}{\partial \epsilon_\Omega}\epsilon_\Omega(i)\right). \tag{3.5.3}$$

The volume strain $\epsilon_\Omega(i)$ can be expressed in terms of longitudinal phonon creation and annihilation operators $b^+_{\underline{q}}$, $b_{\underline{q}}$ as

$$\epsilon_\Omega(i) = \sum_{\underline{q}} \frac{|\underline{q}|}{\sqrt{2N_o M\omega_{\underline{q}}}}(b_{\underline{q}} + b^+_{-\underline{q}})\, e^{i\underline{q}\underline{R}_i} \quad . \tag{3.5.4}$$

The ions of mass M are positioned at $\underline{R}_i$. The dispersion of the phonons is given by $\omega_{\underline{q}}$. When Equations (3.5.3 – 4) are inserted into Equation (3.5.2), the following quasiparticle–phonon interaction Hamiltonian is obtained to first order in $\epsilon_\Omega(i)$

$$H^{(1)}_{qp-ph} = -\eta \sum_{\underline{q}} \frac{|\underline{q}|}{\sqrt{2N_o M\omega_{\underline{q}}}}(b^+_{-\underline{q}} + b_{\underline{q}})\, A_{\underline{q}} + \frac{\partial}{\partial \epsilon_\Omega}[\Lambda(r^2 - 1)] \cdot N_o\, \epsilon_\Omega(\underline{q} = 0)\,. \tag{3.5.5}$$

The operator $A_{\underline{q}}$ is hereby given by

$$A_{\underline{q}} = \sum_{\underline{k}\tau} \left\{ \frac{\tilde{V}}{2}(a^+_{\underline{k}+\underline{q}\tau} f_{\underline{k}\tau} + f^+_{\underline{k}+\underline{q}\tau} a_{\underline{k}\tau}) + \frac{1}{2}(\tilde{\epsilon}_f(\underline{k}) + \tilde{\epsilon}_f(\underline{k}+\underline{q}))\, f^+_{\underline{k}+\underline{q}\tau} f_{\underline{k}\tau} \right\} . \tag{3.5.6}$$

In deriving $H^{(1)}_{qp-ph}$ the derivative $\partial T^*/\partial \epsilon_\Omega(i)$ has been replaced by its value in the presence of a homogeneous strain, i.e. by $\partial T^*/\partial \epsilon_\Omega = -\eta T$. As before $\tilde{V} = rV$ while $\tilde{\epsilon}_f(\underline{k})$ is the Fourier transform of t_{ij}. The last term in Equation (3.5.5) leads to changes in the equilibrium position of the ions and disappears when the new ion positions are introduced, although it is needed when the next–higher order term in the phonon operators is calculated. Interactions which are bilinear in the phonon operators have to be considered, for example, when the electronic contribution to the bulk modulus, Δc_B, is calculated.

The electronic operator $A_{\underline{q}}$ can be re–expressed in terms of the quasiparticle operators $c^+_{1\tau}(\underline{k})$, $c_{1\tau}(\underline{k})$ (see Equation (3.3.11). One finds that

$$A_{\underline{q}} = \sum_{\underline{k}\tau} \tilde{v}(\underline{k}, \underline{q})\, c^+_{1\tau}(\underline{k}+\underline{q})\, c_{1\tau}(\underline{k}) + \ldots \tag{3.5.7}$$

with a coupling function

$$\tilde{v}(\underline{k}, \underline{q}) = E_1(\underline{k}) - \mu + \frac{1}{2}\left(\tilde{\epsilon}_f(\underline{k}+\underline{q}) - \tilde{\epsilon}_f(\underline{k})\right) \tag{3.5.8}$$

and $E_1(\underline{k})$ given by Equation (3.3.15). For $\underline{q} = 0$ and $|\underline{k}| = k_F$ the coupling $\tilde{v}(\underline{k}, 0)$ vanishes. For finite values of $|\underline{q}|$ it is of order $k_B T^*$. The coupling constant $(\eta\, k_B T^*)$ in $H^{(1)}_{qp-ph}$ has to be compared with the deformation potential coupling constant $-(2\epsilon_F/3)$ in ordinary metals. In heavy fermion systems $k_B T^*$ corresponds to the Fermi energy because it characterizes the width of the narrow quasiparticle band. Since $k_B T^*$ is multiplied by the large Grüneisen parameter η, the coupling is much stronger than a conventional deformation potential when the small Fermi energy is used. One should be aware of the fact that in many applications of the electron–phonon interaction the density of states of the quasiparticles enters as a multiplicative factor. Because the latter is large, the interactions derived here can be important, as has been experimentally confirmed. We recall that η is solely due to correlations because the energy scale $k_B T^*$ is a correlation effect and therefore also changes with pressure or volume.

4. UNIFYING DESCRIPTION OF STRONGLY AND WEAKLY CORRELATED SYSTEMS

4.1 Introduction

It was pointed out in Chapter 1 that strong correlations in infinite systems cannot be treated using the methods described there, because infinitely many different configurations would have to be included. In order to introduce the possible difficulties consider the problem outlined below.

Let H denote the Hamiltonian of a system. One would like to separate H into $H = H_0 + H_1$ and include in H_0 the large electron repulsion e.g. on a rare earth ion. The Hamiltonian H_1 is supposed to include the comparatively weak hybridizations of, say, the 4 f electrons of a rare earth ion with the electrons of the neighbouring atoms. Now consider the operator $e^{-\lambda H}$ where λ is either $\lambda = it$ (time evolution operator) or $\lambda = (k_B T)^{-1}$ (statistical operator). The conventional way of separating off the H_0 term is by going over to the interaction picture and writing

$$e^{-\lambda(H_o+H_1)} = e^{-\lambda H_o} T_\tau e^{-\int_o^\lambda d\tau H_1(\tau)} \quad . \qquad (4.1.1)$$

T_τ is a "time" ordering operator and $H_1(\tau) = \exp(iH_o\tau)\ H_1\ \exp(-iH_o\tau)$. When this equation is expanded in powers of $H_1(\tau)$ and when the expectation value is taken, e.g. with respect to the ground state $|\Phi_o>$ of H_o, one must deal with expressions of the form

$$\int_o^\lambda d\tau_1 \ldots\ldots \int_o^\lambda d\tau_n <\Phi_o|T_\tau H_1(\tau_1) \ldots H_1(\tau_n)|\Phi_o> \quad . \qquad (4.1.2)$$

They can be evaluated by breaking them up into products of simpler expectation values provided Wick's theorem holds. This theorem holds when the operators which diagonalize H_o satisfy simple fermionic or bosonic commutation relations. This is not the case when H_o contains the strong electron interactions. The operators, which diagonalize H_o, have then complex commutation relations with the operators causing transitions between the eigenstates. For this reason, the usual diagrammatic rules are not applicable and alternative techniques are required. This is one of the difficulties to be faced when dealing with strongly correlated electrons. A way out is to give up the decomposition (4.1.1), and to start instead from the decomposition given below in Equation (4.2.3). We develop this approach in detail in the next section.

4.2 Projection Technique

Here we formulate a conceptually simple, mathematical framework which allows one to compute the correlated ground state of an electron system. The theory presented is quite general and can be applied to weakly–correlated as well as strongly–correlated electron systems. It will be used later to derive the correlation energy in terms of local operators (Local Ansatz).

Our starting point is again a Hamiltonian of the form

$$H = H_o + H_1 \quad . \qquad (4.2.1)$$

We assume that the eigenstates and eigenvalues of H_o are known and that the effect of H_1 on the ground–state energy is relatively small compared with that of H_o. No further assumptions with respect to the above decomposition of H will be made. The ground state $|\Phi_o>$ of H_o, i.e.

$$H_o|\Phi_o> = E_o|\Phi_o> \qquad (4.2.2)$$

is assumed, for convenience, to be nondegenerate. The theory, however, can be extended also to degenerate ground states.

In our discussion we will use the decomposition

$$e^{-\lambda H} = e^{-\lambda(H_1+L_0)}\, e^{-\lambda H_0} \quad , \tag{4.2.3}$$

with the Liouville operator L_0 belonging to H_0 and defined through

$$L_0 A = [H_0, A]_- \quad , \tag{4.2.4}$$

where A is an arbitrary operator. Equation (4.2.3) is obtained by setting

$$R(\lambda) = e^{-\lambda H}\, e^{\lambda H_0} \quad ; \quad R(0) = 1 \tag{4.2.5}$$

and by integrating the equation of motion for $R(\lambda)$,

$$\begin{aligned} \frac{d}{d\lambda} R(\lambda) &= - HR(\lambda) + R(\lambda)\, H_0 \\ &= - (H_1 + L_0)\, R(\lambda) \quad . \end{aligned} \tag{4.2.6}$$

Consider the function

$$\begin{aligned} f(\lambda) &= \ln \langle\Phi_0|\, e^{-\lambda(H-E_0)}\, |\Phi_0\rangle \\ &= \ln \langle\Phi_0|\, e^{-\lambda(H_1+L_0)}\, |\Phi_0\rangle \end{aligned} \tag{4.2.7}$$

and its Laplace transform

$$f(z) = -\int_0^\infty d\lambda\, e^{\lambda z}\, f(\lambda) \quad ; \quad \mathrm{Re}\, z < 0. \tag{4.2.8}$$

We want to show that the shift δE_0 in the ground-state energy due to H_1 is given by

$$\delta E_0 = \lim_{z\to 0} z^2 f(z) \quad . \tag{4.2.9}$$

For that purpose Equation (4.2.7) is written in the form

$$f(\lambda) = \ln \sum_m |\langle\Phi_o|\psi_m\rangle|^2 e^{-\lambda(E_m - E_o)} \tag{4.2.10}$$

where the $|\psi_m\rangle$ are eigenstates of H with eigenvalues E_m, i.e.

$$H|\psi_m\rangle = E_m|\psi_m\rangle \quad . \tag{4.2.11}$$

By using the notation $\Delta E_m = E_m - E_o$ with $E_o = E_o + \delta E_o$, one can also write

$$f(\lambda) = -\lambda\delta E_o + \ln \sum_m |\langle\Phi_o|\psi_m\rangle|^2 e^{-\lambda\Delta E_m} \tag{4.2.12}$$

Taking the Laplace transform and integrating by parts leads to

$$f(z) = \frac{1}{z^2}\left[\delta E_o - z\int_0^\infty d\lambda\, e^{\lambda z} \frac{\sum_m |\langle\Phi_o|\psi_m\rangle|^2 \Delta E_m e^{-\lambda\Delta E_m}}{\sum_m |\langle\Phi_o|\psi_m\rangle^2 e^{-\lambda\Delta E_m}}\right] \tag{4.2.13}$$

Provided that $\langle\Phi_o|\psi_o\rangle \neq 0$, Equation (4.2.9) immediately follows from this relation.

Next we show that f(z) can be written as

$$f(z) = \frac{1}{z^2}\langle\Phi_o|\, H_1 \left[1 + \frac{1}{z - (H_1 + L_o)} H_1\right] |\Phi_o\rangle^c . \tag{4.2.14}$$

The index c implies taking the cumulant of this expectation value. The fraction $(z - (H_1 + L_o))^{-1}$ stands for the corresponding power expansion.

Cumulants are common in solid state theory. Kubo has been pivotal in demonstrating and emphasizing their usefulness. The cumulant $\langle\Phi_o|A^n|\Phi_o\rangle^c$ of the n'th power of an operator A is defined by

$$\langle\Phi_o|A|\Phi_o\rangle = \frac{\partial^n}{\partial\lambda^n} \ln \langle\Phi_o|e^{\lambda A}|\Phi_o\rangle\Big|_{\lambda=0} \quad . \tag{4.2.15}$$

From this definition, it can be easily verified that

$$
\begin{aligned}
<A>^c &= <A> \\
<A^2>^c &= <A^2> - <A>^2 \\
<A^3>^c &= <A^3> - 3<A^2><A> + 2<A>^3 \qquad .
\end{aligned}
\tag{4.2.16}
$$

For convenience the abbreviation $<...> = <\Phi_o|...|\Phi_o>$ has been used.

By expanding Equation (4.2.7) and using Equation (4.2.15), one obtains

$$
\begin{aligned}
f(\lambda) &= -\sum_{m=1}^{\infty} \frac{(-\lambda)^m}{m!} \frac{m}{(-\lambda)^m} \left[\ln <\Phi_o| e^{-\lambda(H_1+L_o)} |\Phi_o> \right]_{\lambda=0} \\
&= \sum_{m=1}^{\infty} \frac{(-\lambda)^m}{m!} <\Phi_o|(H_1 + L_o)^m|\Phi_o>^c \\
&= <\Phi_o|(e^{-\lambda(H_1+L_o)} - 1)|\Phi_o>^c \qquad .
\end{aligned}
\tag{4.2.17}
$$

When the Laplace transform is taken, one finds

$$
f(z) = <\Phi_o| \left[\frac{1}{z - H_1 - L_o} - \frac{1}{z} \right] |\Phi_o>^c
\tag{4.2.18}
$$

and by rearrangement Equation (4.2.14) is obtained. The energy shift (4.2.9) can therefore be written as

$$
\delta E_o = \lim_{z \to o} <\Phi_o|H_1 \left[1 + \frac{1}{z - (H_1 + L_o)} H_1 \right] |\Phi_o>^c .
\tag{4.2.19}
$$

For more details see Appendix B. The last equation is a general and very useful result. Since the subdivision of H into H_0 and H_1 has remained unspecified, it can be applied to weakly–correlated as well as strongly–correlated electronic systems. The result (4.2.19) can be condensed by introducing a bilinear form in Liouville space.

$$
(A|B) = <\Phi_o|A^+B|\Phi_o>^c \qquad .
\tag{4.2.20}
$$

Note that because of the properties of cumulants, expressions like (AB|AB) need not be positive. Thus Equation (4.2.20) is not a scalar product in the usual sense.

With the definition (4.2.20), one may write

$$\delta E_o = \lim_{z \to o} \left[H_1 | 1 + \frac{1}{z - (H_1 + L_o)} H_1\right] \tag{4.2.21}$$

or

$$\delta E_o = (H_1 | \Omega) \tag{4.2.22}$$

with

$$|\Omega) = \lim_{z \to o} |1 + \frac{1}{z - (H_1 + L_o)} H_1 \quad . \tag{4.2.23}$$

The operator Ω transforms the ground state of H_o into the exact ground state. The operator Ω considered here has to be seen in conjunction with the particular bilinear form (4.2.20). Knowing $|\Omega)$ is equivalent to knowing the exact ground state. Specifying the ground state by $|\Omega)$, i.e. an element of Liouville space, and not by $|\psi>$, i.e. an element of Hilbert space, is an advantage which will become clear later.

In general, Equation (4.2.21) cannot be evaluated without simplifying approximations. In cases where the effect of H_1 on δE_o is sufficiently small, we can expand powers of H_1. From Equation (4.2.21) one obtains by expansion

$$\delta E_o = \lim_{z \to o} \sum_{n=0}^{\infty} (H_1 | \left[\frac{1}{z - L_o} H_1\right]^n) \quad . \tag{4.2.24}$$

Incorporating the special properties of L_o shown in Appendix B, this equation is equivalent to

$$\delta E_o = \lim_{z \to o} \sum_{n=0}^{\infty} (H_1 | \left[\frac{1}{z - H_o} H_1\right]^n) \quad . \tag{4.2.25}$$

This is the Rayleigh–Schrödinger perturbation expansion written in terms of cumulants; the equivalence can be checked term by term. In order to cast δE_o into a form familiar from textbooks, one replaces Equation (4.2.25) by the equivalent form (see Appendix A)

$$\delta E_o = \lim_{z \to o} \sum_{n=0}^{\infty} (H_1 | \left[\frac{1}{z + E_o 1^c - H_o} H_1\right]^n) \quad . \tag{4.2.26}$$

Here 1^c is a unit operator which is subject to cumulant formation, i.e. $(A|1^c) = 0$. When H_0 is a single particle Hamiltonian and H_1 contains the two–particle interactions, the last equation is known as Goldstone's linked cluster expansion. In the literature, Equation (4.2.26) is usually derived for a homogeneous electron gas, in which case H_0 describes noninteracting electrons and H_1 their interactions.

When we analyze the corresponding Goldstone diagrams we find that only **linked** diagrams, i.e. those which do not separate into several disconnected parts, contribute to δE_0. Computing the cumulant of an expectation value is the same as computing only the associated linked diagrams. Equation (4.2.24) can be considered a generalization of Goldstone's theorem to arbitrary Hamiltonians H_0 and H_1; it does not require the use of any diagrams.

Now we will discuss the approximate evaluation of Equation (4.2.21). The approximation applied consists in dividing the Liouville space R into a **relevant** part R_0 and an irrelevant part R_1 and neglecting the effect of R_1 on δE_0. By successively increasing the size or dimension of R_0, one can improve the quality of the approximation. In the limit of R_0 covering the full Liouville space, we obtain the exact energy increment δE_0. For a given tolerance in the calculated energy δE_0 we aim at choosing the restricted space R_0 as small as possible. Let R_0 be of dimension R_0 and spanned by a set $\{A_\nu\}$ of elements $|A_\nu)$. Without loss of generality, the $|A_\nu)$ may be assumed to be orthogonal to each other, i.e.

$$(A_\nu|A_\mu) = \chi_\nu \delta_{\nu\mu} \quad . \tag{4.2.27}$$

The projector P, which projects onto R_0, thus takes on the form

$$P = \sum_\mu |A_\mu) \frac{1}{\chi_\mu} (A_\mu| \quad . \tag{4.2.28}$$

It can be easily verified that $P^2 = P$. When the effect of R_1 on δE_0 is neglected, Equation (4.2.21) is replaced by

$$\delta E_0 = (H_1|1) + \lim_{z \to 0} \left[H_1 | P \frac{1}{z - P(L_0 + H_1)} P H_1 \right] \quad . \tag{4.2.29}$$

The approximation may be improved in two ways: either by enlarging R_0 sufficiently or by successively taking parts of R_1 into account. This can be done by applying the Mori–Zwanzig projection method. It is described in Appendix B, and with its help δE_0 can be expressed in the form of Padé approximates. The following important point should be noticed. The conventional projection method of Mori and Zwanzig applies to **dynamical** phenomena and processes. The method described above, on the contrary, expresses a **static** quantity, namely δE_o in a form which is suitable for the application of projection techniques, thus extending considerably their usefulness. If we assume that $H_o = H_{SCF}$ and $H_1 = H_{res}$ ($= H - H_{SCF}$), then $(H_1|1) = 0$ and δE_o is simply the correlation energy $E_{corr}(R_o)$ of the ground state, which results from the reduced Liouville space R_o.

A particularly simple form of $E_{corr}(R_o)$ is obtained when all $\{A_\nu\}$ couple directly to H_1, i.e. when

$$(A_\nu|H_1) \neq 0 \text{ for all } \nu \quad . \tag{4.2.30}$$

This occurs when A_ν describes two–particle excitations out of the ground state. Then Equation (4.2.29) goes over into

$$E_{corr}(R_o) = -\sum_{\nu\mu} (H_1|A_\nu)\, L^{-1}_{\nu\mu}\, (A_\mu|H_1) \tag{4.2.31}$$

with

$$L^{-1}_{\nu\mu} = \lim_{z\to 0} \frac{(-1)}{\chi_\nu} \left[A_\nu| \frac{1}{z - P(L_o + H_1)} P\, A_\mu\right] \frac{1}{\chi_\mu} \quad , \tag{4.2.32}$$

which is a matrix of dimension R_o. It can be inverted resulting in

$$L_{\rho\tau} = (A_\rho|H\, A_\tau) \quad . \tag{4.2.33}$$

One can verify this relation through Equation (B.17) (note that the memory matrix vanishes as Q projects onto R_1, which is neglected here). Because of the properties of cumulants, we have $(A_\rho|L_o\, A_\tau) = (A_\rho|H_o\, A_\tau)$.

The computation of the correlation energy $E_{corr}(R_o)$ reduces, therefore, to a matrix inversion and matrix multiplication. Within the same approximation, the wave operator is given by

$$|\Omega) = |1 - \sum_\nu \eta_\nu A_\nu) \quad . \tag{4.2.34}$$

The coefficients η_ν follow from a comparison of Equations (4.2.22) and (4.2.34) with Equation (4.2.31), i.e.

$$\eta_\nu = \sum_\mu L^{-1}_{\nu\mu} (A_\mu | H_1) \quad . \tag{4.2.35}$$

Despite the simple form of $|\Omega)$, size consistency is ensured by the measure (4.2.20). In Hilbert space, such a simple form for Ω is not possible. Instead, Ω has to be of an exponential form in order to obtain size–consistent results.

Using relation (4.2.35), Equation (4.2.31) can be also written in the form

$$\begin{aligned} E_{corr}(R_o) &= -\sum_\nu \eta_\nu (H_1 | A_\nu) \\ &= \sum_\nu \epsilon_{corr}(\nu) \quad . \end{aligned} \tag{4.2.36}$$

The correlation energy separates into a sum of contributions $\epsilon_{corr}(\nu)$ from the different elements $|A_\nu)$ which span R_o. The Equations (4.2.31) – (4.2.36) are based on the assumption that all $\{A_\nu\}$ which span R_o couple to H_1 (see Equation (4.2.30)). This limitation of the $\{A_\nu\}$ is not always justified if R_o is supposed to give quantitatively good results for $E_{corr}(R_o)$. When R_o contains a subset $\{A_\rho\}$ with $(A_\rho | H_1) = 0$, the previously derived expression (4.2.31) for the correlation energy can be generalized by employing the projection technique described in Appendix B.

A key result of the previous considerations was Equation (4.2.22), which describes the shift in the ground state energy brought about by H_1. As shown in Appendix A, this expression is equivalent to

$$\begin{aligned} E &= E_o + (H_1 | \Omega) \\ &= (\Omega | H \Omega) \quad . \end{aligned} \tag{4.2.37}$$

The last equation is interesting because $|\Omega)$ specifies the exact ground state. The formal proof of the equivalence of the two expressions is not obvious.

4.3 Projection Technique for Correlation Functions

In the following it is shown how the Green's function, or, more precisely, a closely related correlation function, can be evaluated with the help of projection techniques. This has the advantage that Wick's theorem is not required, because within this method there is no need to introduce a time dependence of the perturbing Hamiltonian. The projection method is also applicable to strongly correlated electronic systems where Wick's theorem does not hold. First it is shown that $G_{ij}(t)$ can be written in a form which allows for the application of projection techniques. For this purpose the Hamiltonian H is split into $H_0 + H_1$ and it is assumed that the ground state $|\Phi>$ of H_0 is known. A correlation function

$$R_{ij}(\tau) = <\psi_0^N | c_i(\tau)\, c_j^+(0) | \psi_0^N> \tag{4.3.1}$$

is introduced for imaginary times τ, i.e.

$$c_i(\tau) = e^{\tau H}\, c_i\, e^{-\tau H} \quad . \tag{4.3.2}$$

One notices that $R_{ij}(\tau)$ is closely related to the Green's function $G_{ij}(t)$ defined in real time.

By generalizing the formalism presented in Section 4.2 it can be shown that $R_{ij}(\tau)$ can be written in the form

$$R_{ij}(\tau) = \left[\Omega | c_i\, e^{-\tau(L_0 + H_1)}\, c_j^+ \Omega\right] \quad . \tag{4.3.3}$$

The round brackets are previously defined by Equation (4.2.20). The operator $|\Omega)$ is given by Equation (4.2.23) and characterizes the exact ground state $|\psi_0^N>$ of the system. Equation (4.3.3) expresses the function $R_{ij}(\tau)$ in the form of an expectation value with respect to the ground state of H_0.

Taking the Laplace transform of $R_{ij}(\tau)$, i.e.

$$R_{ij}(z) = -\int_0^\infty d\tau\, e^{z\tau}\, R_{ij}(\tau) \quad ; \quad \text{Re } z < 0 \tag{4.3.4}$$

results in

$$R_{ij}(z) = \left[\Omega |\, c_i \frac{1}{z - (L_0 + H_1)} c_j^+\, \Omega\right] \quad . \tag{4.3.5}$$

The poles of $R_{ij}(z)$ give the excitation energies of the system like the Green's function does. One notices that $R_{ij}(z)$ is expressed in terms of cumulants from the very beginning and therefore any approximation which is made in evaluating it preserves size consistency. Equation (4.3.5) is very suitable for applying projection or partitioning methods.

Assume that the Liouville space R is again divided into a relevant part R_o and an irrelevant part R_1, as in Section 4.2. The subspace R_o is supposed to be spanned by a set of operators $\{A_\nu\}$, of which the c_i^+, c_j^+ are members. The projector

$$P = \sum_{\mu\nu} |A_\nu)\, \chi^{-1}_{\nu\mu} (A_\mu|$$

$$\text{where} \qquad \chi_{\nu\mu} = (A_\nu | A_\mu) \tag{4.3.6}$$

projects onto R_o. The approximation within which $R_{ij}(z)$ is evaluated consists of neglecting R_1. This implies that Equation (4.3.5) is replaced by the approximate expression

$$\begin{aligned} R_{ij}(z) &= \left[Pc_i^+ \Omega \right| \frac{1}{z - P(L_o + H_1)P} Pc_j^+ \Omega \Big] \\ &= \sum_{\mu\nu} (c_i^+ \Omega | A_\nu)\, \chi^{-1}_{\nu\mu} \left[A_\mu \right| \frac{1}{z - P(L_o + H_1)} Pc_j^+ \Omega \Big] \\ &= \sum_{\mu\nu} (c_i^+ \Omega | A_\nu)\, \chi^{-1}_{\nu\mu}\, \mathfrak{R}_{\mu j}(z) \qquad . \end{aligned} \tag{4.3.7}$$

The new correlation functions

$$\mathfrak{R}_{\mu j}(z) = \left[A_\mu \right| \frac{1}{z - P(L_o + H_1)} Pc_j^+ \Omega \Big] \tag{4.3.8}$$

are computed from an equation of motion of the form

$$\sum_{\eta\mu} \left\{ z\, \chi_{\nu\eta} - \omega_{\nu\eta} \right\} \chi^{-1}_{\eta\mu}\, \mathfrak{R}_{\mu j}(z) = (A_\nu | c_j^+ \Omega) \tag{4.3.9}$$

where the frequency matrix $\omega_{\nu\eta}$ is given by

$$\omega_{\nu\eta} = (A_\mu | (L_o + H_1) A_\eta) \qquad . \tag{4.3.10}$$

Equation (4.3.9) follows from Equation (B17). Because R_1 is neglected, the memory matrix $M_{is}(z)$ vanishes here. In Appendix B the Mori scalar product was used instead of the bilinear form (4.2.20), but the derivation of Equation (4.3.9) is independent of that choice. The expectation values $(A_\nu | \, c_j^+ \, \Omega)$ can be evaluated similarly by using the approximation for Ω discussed in Section 4.2. Once the $\tilde{R}_{\mu j}(z)$ are known, the $R_{ij}(z)$ follow from Equation (4.3.7).

The above evaluation of the correlation function $R_{ij}(z)$ differs from the standard Mori–Zwanzig projection method. In order to understand these differences, consider Equation (4.3.3). Obviously correlations enter the function $R_{ij}(\tau)$ in two ways. One is through the time evolution operator $\exp[-\tau(L_o + H_1)]$ and the other is through the exact ground state represented by $|\Omega)$. Both types of correlations, i.e. those resulting from the **dynamics** and those from **static** properties (i.e. $|\Omega)$) are treated equivalently when Equation (4.3.5) is evaluated. By restricting the Liouville space to R_o one can treat the dynamics (see Equation (4.3.9)). But by the same method one can also calculate static quantities like $(A_\nu | \, c_j^+ \, \Omega)$ which appear in the dynamical equations. The operator $|\Omega)$ defined by Equation (4.2.23) is of a form which is suitable for projection techniques. This is not the case in the Mori–Zwanzig theory where static quantities must be obtained from somewhere else. One starts out in this theory from

$$R_{ij}(z) = \langle \psi_o^N | \, c_i \frac{1}{z-L} c_j^+ \, | \psi_o^N \rangle \qquad (4.3.11)$$

where L is the Liouville operator corresponding to H. This equation follows from (4.3.1) when, after a time shift by $-\tau$, the relation

$$c_j^+(-\tau) = e^{-L\tau} \, c_j^+ \qquad (4.3.12)$$

is used and the Laplace transform is taken. By applying the projection technique outlined in Appendix B, one finds the following equation of motion for $R_{ij}(z)$ (see Equation (B17)):

$$\sum_l \left[z\delta_{il} - \sum_s \left(L_{is} + M_{is}(z)\right) P_{sl}^{-1} \right] R_{lj}(z) = P_{ij} \qquad . \qquad (4.3.13)$$

The matrices L_{is}, $M_{is}(z)$ and P_{ij} are defined as follows

$$P_{ij} = \langle \psi_o^N | \, c_i \, c_j^+ \, | \psi_o^N \rangle$$

$$L_{is} = \langle \psi_o^N | \, c_i \, L \, c_j^+ \, | \psi_o^N \rangle$$

$$M_{is}(z) = <\psi_o^N| c_i L Q \frac{1}{z-QL} Q L c_j^+ |\psi_o^N> \quad . \quad (4.3.14)$$

The projector $Q = 1 - P$ equals

$$Q = 1 - \sum_{nm} c_n^+ |\psi_o^N> P_{nm}^{-1} <\psi_o^N| c_m \quad . \quad (4.3.15)$$

As mentioned above, the conventional projection method provides no recipe on how to compute matrices like P_{ij} and L_{is}, which are defined with respect to the **exact** ground state.

Often the following scalar product of two operators A and B is used:

$$(A|B)_a = <\psi_o^N| [A^+,B]_+ |\psi_o^N> \quad . \quad (4.3.16)$$

In this case the retarded Green's function $G_{ij}^R(\omega)$ can be written in the form

$$G_{ij}^R(\omega) = \left[c_i | \frac{i}{\omega - L + i\eta} c_j^+\right]_a \quad . \quad (4.3.17)$$

The proof is simple and follows immediately from the definition of $G_{ij}^R(t)$ and it's Fourier transform. The equations of Appendix B can be again applied in order to compute $G_{ij}^R(\omega)$.

Finally, it is worth pointing out the similarities and differences between the projection method and the equation of motion method. In the latter method the Green's function $<<A;B>>_\omega$ is coupled to $<<[A,H\}_-; B>>_\omega$. By setting up the equation of motion for the latter, one finds that it couples back to $<<A;B>>_\omega$ and contains terms of the form $<<[[A,H]_-,H]_-; B>>_\omega$ and so on. In the projection method the place of $<<[A,H]_-; B>>_\omega$ is taken by the memory matrix. But, in contrast to the latter, the equation of motion for the memory matrix does **not** contain coupling terms back to the original correlation function.

5. ELECTRON CORRELATIONS IN HIGH–T_c MATERIALS

5.1 Electronic Structure of the High–T_c Materials

The most important structural elements of the high–T_c superconductors are planes of $(Cu0_2)_x$. These planes are formed from octahedrons, pyramids or squares. In each case a Cu atom is surrounded by 0 atoms. This is indicated in Figure 5.1.1. The arrangement of these units into planes is shown in Figure 5.1.2. Examples are $La_{2-x}Sr_x$ CuO_4, where the planes are formed from octahedrons, $YBa_2Cu_3O_7$, where they consist of pyramids, and $Nd_{2-x}Ce_xCu$ O, where they are built from squares. The electron transport as well as the physical processes which lead to superconductivity are believed to take place within these planes. $YBa_2Cu_3O_7$ has another important structural element, viz. the chains formed from CuO_4 squares by sharing an O atom. It is believed that these $(CuO_3)_x$ chains may play a role in determining a number of physical properties, but that the superconducting electrons are predominantly in the planes.

In La_2CuO_4 the CuO_6 octahedrons are elongated (Jahn–Teller distortion). The Cu–O distances are 190 [pm] within the plane and 240 [pm] perpendicular to it. The distortion lifts the degeneracy of the Cu d–orbitals, which in octahedral symmetry are twofold ($e_g = \{d_{x^2-y^2}, d_{3z^2-r^2}\}$) and threefold ($t_{2g} = \{d_{xy}, d_{xz}, d_{yz}\}$). In counting valence electrons it is assumed that La is in a valence state of 3+ and O in one of 2–, so that the Cu atoms have a formal valency of Cu^{2+}. This implies a $3d^9$ configuration (remember that a Cu atom has a [Ar] $3d^{10}4s^1$ electron configuration). The hole in the 3d shell is placed into the highest antibonding Cu–O state, which has a predominantly $3d_{x^2-y^2}$ character. With one hole per unit CuO_4 one would expect La_2CuO_4 to be metallic with a half filled conduction band, provided the picture of independent electrons holds. In reality the material is an antiferromagnetic semiconductor with a Néel temperature of $T_N \simeq 280$ K. The latter depends sensitively on the stoichiometry. This is direct evidence for strong electron correlations. They lead to the opening of a gap at the Fermi energy and also to the observed magnetic properties.

In order to study the electronic structure of the high–T_c materials, LDA calculations have been performed. They neither reproduce the semiconducting behaviour of La_2CuO_4 nor do they predict the magnetic properties of the ground state. However, they produce well other properties of the systems, such as the electronic charge distributions which depend only slightly on the strong correlations. This can be explicitly checked by comparing the calculated charge densities with those which are obtained from quantum chemical calculations on negatively charged planar clusters of $(CuO_3)_n$ (n = 1,...,4). The latter employ a double–zeta basis set and an additional third set of Cu d functions. Correlations are included by the method of the Local Ansatz described in Section 4. The LDA as well as the quantum chemical calculations give a d electron number for Cu of $n_d \simeq 9.4$. It is interesting that the d electron count differs only slightly from that of a corresponding SCF calculation, which is $n_d^{SCF} = 9.6$. As in an intermediate valence system, the Cu electron configurations fluctuate between $3d^9$ and $3d^{10}$, i.e. the ground state wave function contains both configurations with almost equal weight. The strong correlations reduce the $3d^8$ configurations contained in the SCF ground state $|\Phi_{SCF}\rangle$ by 85%. Similarly, the $2p^4$

TOWARDS A THEORETICAL UNDERSTANDING OF THE HIGH–T_c SUPERCONDUCTING MATERIAL

Most important structural element element in $La_{2-x}Sr_xCuO_4$, $YBa_2Cu_3O_{8-\delta}$, $Bi_2Sr_2Ca_3Cu_3O_{10}$
are the Cu–O planes.
Hybridization between Cu : $3d_{x^2-y^2}$
O : O $2p_{x(y)}$ orbital

correlations are strong in planes

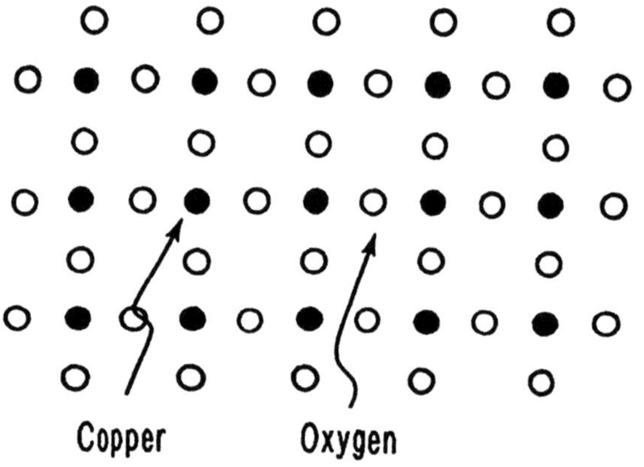

Figure 5.1.1 **The planes of $(CuO_2)_x$**

configurations of 0 ions are reduced by 75%. The correlations are strong not only on the Cu atoms, but also on the 0 atoms.

In order to study the electronic structure within the Cu–O planes in more detail a model Hamiltonian is introduced. It contains a number of parameters which are subsequently determined from a LDA or a quantum chemical calculation. It is useful to work within the hole instead of the electron picture. Consider first the following simple Hamiltonian:

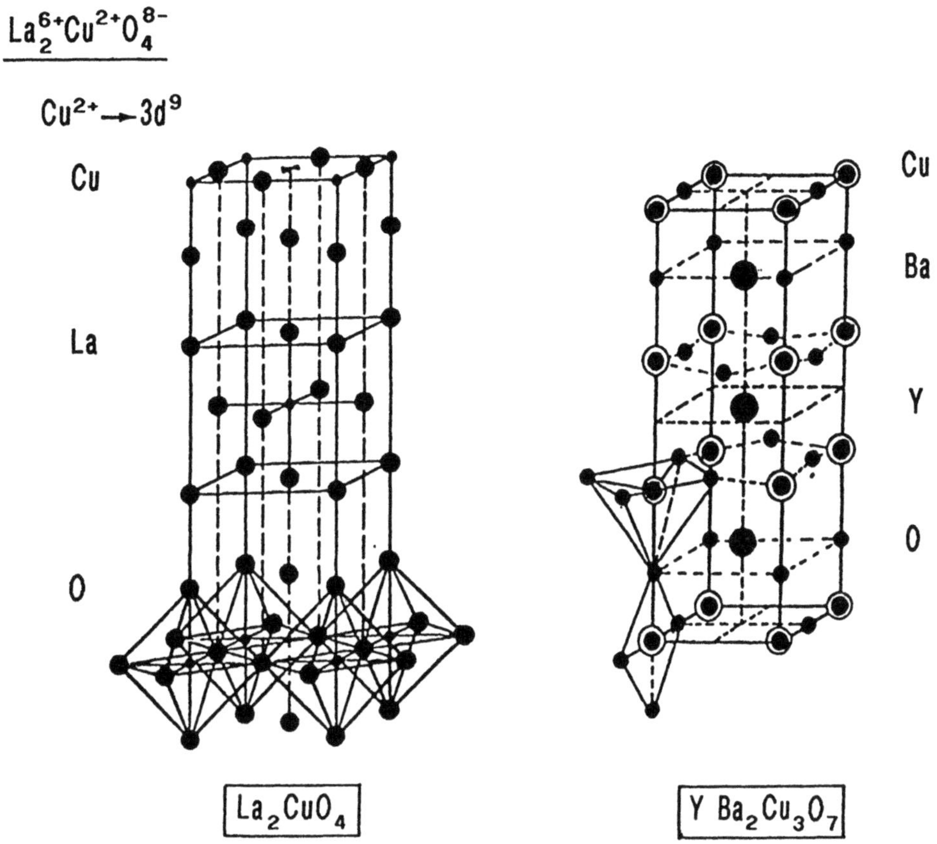

Figure 5.1.2 **Crystal structures of high T_c materials.**

$$H_o = \epsilon_d \sum_{i\sigma} n_{d\sigma}(i) + \epsilon_p \sum_{j\sigma} n_{p\sigma}(j) + \sum_{\langle ij\rangle} V_{ij}(d^+_{i\sigma} p_{j\sigma} + p^+_{j\sigma} d_{i\sigma}) . \tag{5.1.1}$$

The notation <ij> refers to taking pairs of nearest neighbours. The operator $d^+_{i\sigma}$ creates a hole with spin σ in the Cu $3d_{x^2-y^2}$ orbital at site i and the operator $p^+_{j\sigma}$ does the same with respect to the 0 $2p_x$ (or $2p_y$) orbital at site j. The corresponding number operators are $n_{d\sigma}(i)$ and $n_{p\sigma}(j)$, respectively. The hybridization between the two types of orbitals is given by the matrix elements V_{ij} of the form $V_{ij} = (-1)^{\alpha_{ij}} t_{pd}$ with $\alpha_{ij} = 0$. There are three orbitals per unit cell for which the Hamiltonian matrix is

$$H(k_x,k_y) = \begin{bmatrix} \epsilon_d & 2\, it_{pd} \sin \frac{k_x a}{2} & -2\, it_{pd} \sin \frac{k_y a}{2} \\ -2\, it_{pd} \sin \frac{k_x a}{2} & \epsilon_p & 0 \\ 2\, it_{pd} \sin \frac{k_y a}{2} & 0 & \epsilon_p \end{bmatrix} \tag{5.1.2}$$

Here $\underline{k} = (k_x,k_y)$ denotes the electron momentum. When the matrix is diagonalized, three bands are obtained. Their dispersion is given by

$$\epsilon_{1,2}(\underline{k}) = \frac{\epsilon_p + \epsilon_d}{2} \pm \sqrt{\left[\frac{\epsilon_p - \epsilon_d}{2}\right] + 4t^2_{pd}\left[\sin^2 \frac{k_x a}{2} + \sin^2 \frac{k_y a}{2}\right]}$$

$$\epsilon_3(\underline{k}) = \epsilon_p \tag{5.1.3}$$

The third solution ϵ_3 describes a nonbonding band. Within the one–particle Hamiltonian (5.1.1) La_2Cu0_4 would have a half–filled conduction band and be metallic. That this is not the case is due to the on–site Coulomb repulsion of the Cu 3d electrons which results in a Hubbard splitting of the conduction band. The Hamiltonian is therefore improved by adding the on–site repulsions of the 3d and 2p electrons to the Hamiltonian. Also included are the repulsions between electrons on nearest–neighbour sites and on neighbouring 0 sites and finally the direct 0–0 hybridization. This leads to the following form:

$$H = H_o + \sum_{\langle ij' \rangle} t_{pp}(j,j') \, (p^+_{j\sigma} \, p^+_{j'\sigma} + h.c.) + U_d \sum_i n_{d\uparrow}(i) \, n_{d\downarrow}(i)$$

$$+ U_p \sum_j n_{p\uparrow}(j) \, n_{p\downarrow}(j) + U_{pd} \sum_{\langle ij \rangle} n_d(i) \, n_p(j)$$

$$+ U_{pp} \sum_{\langle ij' \rangle} n_p(j) \, n_p(j') \tag{5.1.4}$$

As in Equation (5.1.1), the Hamiltonian includes Cu $3d_{x^2-y^2}$ and 0 $2p_y$, (Emery model). It has been written down in hole representation. This implies that two 3d holes on a Cu site equal each other with energy U_α and two 2p holes in a 0 site with energy U_p. Because of this, holes, which are introduced by doping, go predominantly into 0 2p orbitals otherwise we would have Cu sites doubly occupied with holes. Indeed, the quantum chemical calculations on planar $(CuO_3)_n0$ clusters find that added holes have 85% 0 2p and 15% Cu 3d character. The parameters of the Hamiltonian can be determined by either fitting them to a quantum chemical cluster calculation or to a constrained LDA calculation. In the latter case energy changes are calculated away from the ground state density distribution. For example, the d electron number at a Cu site i, $n_d(i)$, is constrained to a given value $n_d^c(i)$ and the ground state energy functional E[n] is minimized under this subsidiary condition, i.e.

$$E[n_d^c(i)] = \min \left\{ E[n] + \lambda \int_{MT} d^3r [n_d(\underline{r}) - n_d^c(\underline{r})] \right\} \quad . \tag{5.1.5}$$

Here λ denotes the Lagrange parameter and the integral extends over a muffin tin (MT), i.e. over the volume of the Cu site. The energy $E[n_d^c(i)]$ describes, within the LDA, the response of the system due to a local change of the charge distribution (charge fluctuation). In order to determine the parameters of the Hamiltonian (5.1.4), equivalent constrained calculations are done for its SCF part H_{SCF}. By bringing the results of the two calculations into agreement, the parameters of H are determined.

The ground state energy as well as the charge distributions, can be calculated analytically when the Hamiltonian (5.1.4) is somewhat simplified. For this purpose set $t_{pp}(j,j') = U_{pd} = U_{pp} = 0$ and furthermore assume that $U_d \longrightarrow \infty$ and that $U_p >> t_{pd}$, $\Delta = (\epsilon_p - \epsilon_d)$. The Coulomb repulsions U_d and U_p can be eliminated by going over to a reduced Hilbert space without doubly occupied Cu sites (i.e. $3d^8$ configurations) and 0 sites (i.e. $2p^4$ configurations). One finds the following effective Hamiltonian for the reduced space

$$H_{eff} = \tilde{H}_o + \tilde{H}_1 \tag{5.1.6}$$

with $\tilde{H}_o$ and $\tilde{H}_1$ given by

$$\tilde{H}_o = \epsilon_d \sum_{i\sigma} n_{d\sigma}(i) + \epsilon_p \sum_{j\sigma} n_{p\sigma}(j)$$

$$\tilde{H}_1 = \sum_{\langle ij\rangle\sigma} V_{ij} (\hat{d}^+_{i\sigma} \hat{p}_{j\sigma} + \hat{p}^+_{j\sigma} \hat{d}_{i\sigma}) -$$

$$-2 \sum_{\langle ij\rangle} \frac{V^2_{ij}}{U_p} \left\{ \frac{n_d(i)\, n_p(j)}{4} - \underline{S}_d(i)\, \underline{S}_p(j) \right\} -$$

$$- \sum_{i\neq m, j\sigma} \frac{V_{ij} V_{jm}}{U_p} \left\{ \hat{d}^+_{i\sigma} n_{p-\sigma}(j)\, \hat{d}_{m\sigma} - \hat{d}^+_{i\sigma} \underline{S}^{-\sigma}_p(j)\, \hat{d}_{m\sigma} \right\} . \tag{5.1.7}$$

The operators $\hat{d}^+_{i\sigma}$, $\hat{p}^+_{j\sigma}$ ensure the exclusion of double occupancies of sites. They are defined by

$$\hat{d}^+_{i\sigma} = d^+_{i\sigma}(1 - n_{d-\sigma}(i))$$

$$\hat{p}^+_{j\sigma} = p^+_{j\sigma}(1 - n_{p-\sigma}(j)) \tag{5.1.8}$$

The $\underline{S}_d(i)$ and $\underline{S}_p(j)$ are spin operators with respect to the d and p electrons. The last term of H_1 provides for a hopping from an 0 site to a neighbouring 0 site. The hopping process can have a spin flip associated with it. For the computation of the ground state energy consider the case of a half–filled band. Because $\Delta = (\epsilon_p - \epsilon_d) > 0$, the ground state of $\tilde{H}_o$ has one hole located on each Cu site. It is 2^{N_0} fold degenerate where N_0 is the number of Cu sites. In order to lift the degeneracy it is advantageous to add a term of the form

$$H' = -4 \sum_{i\neq j} J(i,j) \left\{ \frac{1}{4} n_d(i)\, n_d(j) - \underline{S}_d(i)\, \underline{S}_d(j) \right\} \tag{5.1.9}$$

to H_o and to subtract it again from H_1. The ground state of $\tilde{H}_o = \tilde{H}_o + H'$ is that of a two–dimensional Heisenberg antiferromagnet. The energy shift δE_o due to $H_1 = \tilde{H}_1 - H'$ is obtained by applying the projection technique described in Section 4.2. This implies that

Equation (4.2.29) must be evaluated. In order to do this, one must specify the relevant part of the Liouville space, i.e. the set of variables $\{A_\nu\}$ which span it. When in δE_o only terms of order U_p^{-1} are kept and higher order contributions are neglected. The natural choice for the operators A_ν is

$$A_1 = \sum_{\langle ij \rangle} V_{ij}\, \hat{p}^+_{j\sigma}\, \hat{d}_{i\sigma}$$

$$A_2 = (L_o + H_1)\, A_1$$

$$A_3 = (L_o + H_1)(L_o + H_1)\, A_1 \quad . \tag{5.1.10}$$

The Liouvillean L_o is defined as in Section 4.2. With these variables the energy shift δE_0 is obtained in the form of a continued fraction. To order $(t_{pd/\Delta})^6 = \tilde{t}^6$ the following result is found for the total energy E_o per Cu site

$$E_o = \epsilon_d + \frac{\delta E_o}{N_o}$$

$$= \epsilon_d - \cfrac{4\ \tilde{t}^2 \Delta}{(\omega_1/\Delta) + \cfrac{5\ \tilde{t}^2}{1 + 5.93\ \tilde{t}^2}} \tag{5.1.11}$$

The energy ω_1 is given by

$$\omega_1 = \Delta - \frac{t^2_{pd}}{U_p}\left(\frac{1}{4} - \langle \underset{\sim}{S}_d(i)\, \underset{\sim}{S}_d(j) \rangle\right) \tag{5.1.12}$$

where i and j are neighbouring Cu sites. The contribution $(t^2_{pd}/U_p)\ \langle \underset{\sim}{S}_d(i)\, \underset{\sim}{S}_d(j) \rangle$ is the ground state energy per Cu site of a two dimensional Heisenberg antiferromagnet with coupling constant $J = (t^2_{pd}/U_p)$. Its computation will be discussed below.

The continued fraction (5.1.11) terminates because the set $\{A_\nu\}$ consists of three operators only. One may estimate the resulting error in E_o by replacing the term $5.93\ \tilde{t}^2$ by $[5.93\ \tilde{t}^2/(1 + \delta)]$. In order to determine $\delta > 0$ one would have to include terms of higher order in $\tilde{t}$. For $\tilde{t} = 0.3$ the variation in $(E_o - \epsilon_d)$ is less than 20% when δ is varied between $0 \lesssim \delta < \infty$. One can obtain from $\langle n_p \rangle = (\partial E_o/\partial \epsilon_p)$ the average number of oxygen holes $\langle n_p \rangle$ due to the hybridization t_{pd}. This relation follows from the more general one $<n_p> = -(1/\beta)(\partial \ln Z/\partial \epsilon_p)$, which is easily verified by using the definition of the partition function Z. For $\tilde{t} = 0$ it is, of course, $<n_p> = 0$. But for $\tilde{t} = 0.3$ one finds the value of $<n_p> = 0.13$, in fair agreement with the results of the quantum chemical calculations.

5.2 2D Heisenberg Antiferromagnet

Neutron scattering experiments show that for many purposes the Cu–0 planes, in La_2Cu0_4 for example, can be considered as two dimensional (2D) Heisenberg antiferromagnets with spin $^1/_2$ attached to the Cu sites Also the expression (5.1.11) for the ground state energy of a Cu–0 plane with one hole per unit cell requires knowledge of the energy of a 2D Heisenberg antiferromagnet (AF). Therefore this system is considered in more detail. Especially interesting is a study of the motion of holes in a 2D Heisenberg AF. This is of help when in the next section the excitations of Cu–0 planes are investigated.

5.2.1 Ground State Energy

We begin with the computation of the ground state energy of the Hamiltonian

$$H = J \sum_{<ij>} \underline{S}_i \underline{S}_j \qquad ; \qquad J > 0 \qquad (5.2.1)$$

of N_0 sites on a square lattice. The bracket <ij> denotes nearest neighbour pairs. One immediate question is whether or not the ground state has long range order. From a general theorem due to Mermin and Wagner it is known that a 2D Heisenberg AF cannot have any long range order at any finite temperature. This does not exclude long range order in the ground state, however. A number of numerical studies, which are based on exact diagonalizations of small clusters (up to 16 sites) and on Quantum Monte Carlo calculations (up to 12×12 sites), suggest strongly that the ground state of the Hamiltonian (5.2.1) has indeed long range order. These numerical studies have also been applied in order to determine the ground state energy. We shall not discuss them in more detail because results of the same accuracy can be obtained from analytical calculations by using expansion methods.

Let us divide H into an Ising part H_0 and a spin flip part H_1, i.e.

$$H = H_0 + H_1$$

$$H_o = J \sum_{\langle ij \rangle} S_i^z S_j^z$$

$$H_1 = \frac{J}{2} \sum_{\langle ij \rangle} (S_i^+ S_j^- + S_i^- S_j^+) \qquad . \qquad (5.2.2)$$

The ground state of H_o is the Néel state $|\Phi_N\rangle$. It's energy is $E_o = 2J\, N_o/4$. In the Néel state the lattice is divided into two sub–lattices A and B with opposite spin polarization. In order to determine the effect of H_1 on E_o it is advantageous to rotate the spins on e.g. sub lattice B by π around the x axis. This corresponds to the transformation $S_j^z \to -S_j^z$, $S_j^y \to -S_j^y$ and $S_j^x \to S_j^x$. The Hamiltonian H_o changes sign under this transformation while H_1 goes over into

$$H_1 = \frac{J}{2} \sum_{\langle ij \rangle} (S_i^+ S_j^+ + S_i^- S_j^-)$$

$$= \quad H_1^+ + H_1^- \qquad . \qquad (5.2.3)$$

In the transformed ground state $|\Phi_N\rangle$ all spins point down. The change in the ground state energy δE_o due to H_1 is calculated by using Equation (4.2.24). Only terms up to order H_1^4 are considered here, although higher order terms can be included without major problems. One finds that

$$L_o H_1 |\Phi_N\rangle = L_o H_1^+ |\Phi_o\rangle$$

$$= 3J\, H_1^+ |\Phi_o\rangle \qquad (5.2.4)$$

The eigenvalue is 3J because one spin flip $S_i^+ S_j^+$ leads to six frustrated bonds of energy $J/2$ each. To lowest order one obtains therefore

$$\delta E_o^{(2)} = \lim_{z \to 0} \left[H_1^+ \middle| \frac{1}{z - L_o} H_1^+ \right]$$

$$= -\frac{2NJ^2}{4} \cdot \frac{1}{3J} = -\frac{NJ}{6} \qquad . \qquad (5.2.5)$$

The computation of the next higher order contribution is a nice little exercise in using cumulants. The term

$$\delta E_o^{(4)} = \lim_{z\to 0} \left[H_1 \middle| \left(\frac{1}{z - L_o} H_1\right)^3\right] \tag{5.2.6}$$

reduces to

$$\delta E_o^{(4)} = \frac{1}{9J^2} \lim_{z\to 0} M_o(z) \tag{5.2.7}$$

$$M_o(z) = \left[(H_1^+)^2 \middle| \frac{1}{z - L_o} (H_1^+)^2\right] \quad . \tag{5.2.8}$$

when $|(z - L_o)^{-1} H_1) = (z - 3J)^{-1}| H_1)$ is used.

Note that products of the form $H_1^- H_1^+$ do not contribute to the cumulant, because the corresponding expectation value factorizes into different parts. The operator $|(H_1^+)^2)$ generates two spin flips. The configurations which are created can be divided into four classes. The first three classes are denoted by v_ν ($\nu = 1,2,3$) and have 8, 10 and 12 frustrated bonds, respectively. The corresponding energy differences to the ground state are therefore 4J, 5J and 6J. The fourth class of configurations is denoted by v_4 and contributes only because one is dealing with cumulants and not ordinary expectation values. Applying the operator S_i^+ twice on $|\Phi_o\rangle$ would otherwise give zero. For the same reason the $|v_\nu\rangle$ are not eigenoperators of L_o. Where the cumulant (5.2.8) is expressed in terms of ordinary expectation values we obtain

$$M_o(z) = \langle\Phi_o| (H_1^-)^2 \frac{1}{z - L_o} (H_1^+)^2 |\Phi_o\rangle - \frac{\langle\Phi_o| H_1^- H_1^+ |\Phi_o\rangle^2}{z - 6J}$$

$$= \sum_{\nu=1}^{4} (v_\nu| \frac{1}{z - L_o} v_\nu) \tag{5.2.9}$$

The cumulants $(v_\nu|(z - L_o)^{-1} v_\nu)$ are of the form

$$\left[v_1 \middle| \frac{1}{z - L_o} \middle| v_1\right] = 2\, N_o 2 \left(\frac{J}{2}\right)^4 \left\{\frac{4}{z - 4J} - \frac{2}{z - 6J}\right\}$$

$$\left[v_2 \middle| \frac{1}{z - L_o} \middle| v_2\right] = 2\, N_o 14 \left(\frac{J}{2}\right)^4 \left\{\frac{2}{z - 5J} - \frac{2}{z - 6J}\right\}$$

$$\left[v_3 \middle| \frac{1}{z - L_o} \middle| v_3\right] = 0$$

$$\left[v_4 \left| \frac{1}{z - L_o} \right| v_4\right] \qquad = -2\, N_0 7 \left(\tfrac{J}{2}\right)^4 \frac{2}{z - 6J}\Big\} \ . \tag{5.2.10}$$

The result for $M_0(z)$ is therefore

$$M_0(z) = 2\, N_0 \left(\tfrac{J}{2}\right)^4 \left\{\frac{8}{z - 4J} + \frac{28}{z - 5J} - \frac{46}{z - 6J}\right\} \quad . \tag{5.2.11}$$

When this expression for $M_0(z)$ is set into Equation (5.2.7) one obtains

$$\delta E_0(4) = \frac{2N_0}{4} J \cdot 0.0019 \tag{5.2.12}$$

By adding up $\delta E_0(2)$ and $\delta E_0(4)$ one obtains for the ground state energy

$$E_0 = -2\, N_0\, H \cdot 0.3329 \tag{5.2.13}$$

When terms to sixth order in J are also included the factor 0.3329 is replaced by 0.3337. This can be compared with the value of 0.3344 of extensive Monte Carlo calculations.

One can also calculate the effect of H_1 on the staggered magnetization. For this purpose a Zeeman term of the form $2h \sum_i S_i^z$ is added to H_o (after the spin rotation of the sub–lattice B has been done) and the energy $E_o(h)$ is calculated. The magnetization m is obtained from $dE_o(h)/dh$. One finds that

$$\begin{aligned} m &= \mu_B (1 - 0.222 - 0.054 - \ldots.) \\ &\simeq 0.724\, \mu_B \end{aligned} \tag{5.2.14}$$

and the series seems to be rapidly convergent. This supports the conjecture that the ground state of the 2D Heisenberg AF has long–range order. The value for m is at variance with the one found by Monte Carlo calculations which is $m = 0.606\, \mu_b$. The origin of this discrepancy might be connected with the way in which m is calculated there. It is assumed that

$$m = \left(\frac{3}{N_0} \sum_i \langle S_i^z S_{i+L}^z \rangle\right)^{1/2} \tag{5.2.15}$$

where $L = \sqrt{N_0}$ is the linear dimension of the cluster for which the calculations are done. But there is no proof of the validity of this relation.

One might ask how the above calculations are modified when the Hamiltonian (5.2.1) is divided up differently into H_o and H_1. For example, one may choose for H_o the pairwise interactions. The remaining interactions are included in H_1. The ground state of H_o is a product of singlet pairs of the form

$$(ij) = \frac{1}{\sqrt{2}} (\alpha(i)\,\beta(j) - \beta(i)\,\alpha(j)) \tag{5.2.16}$$

where $\alpha(i)$ and $\beta(i)$ denote spin up and down states at site i. This division of H is within the spirit of the resonating valence bond description of a 2D Heisenberg AF. The ground state energy of H_o is

$$E_o^{(o)} = -2\,NJ \cdot \frac{3}{16} \quad . \tag{5.2.17}$$

When the lowest order correction in H_1 is included the energy becomes

$$E_o = E_o^{(o)} + \delta E_o^{(2)} = -2\,N_o J \cdot 0.3047 \quad . \tag{5.2.18}$$

One notices that the series is much less convergent than the previous one which started from the Néel state. Eventually, of course, the same energy is obtained independent of the particular choice of H_o. It is worth noticing that the magnetization remains zero to any order in H_1 when this division of H is used. This is due to the finite energy it takes to excite a triplet state. These findings suggest that a resonating valence bond picture is not a good starting point for describing the ground state of a 2D Heisenberg AF with nearest neighbour interactions on a square lattice.

5.2.2 Motion of a Hole

The motion of a hole in a 2D Heisenberg AF is a very interesting problem. In principle it should apply when the Cu–0 planes are doped with a small number of holes. However, in realistic systems imperfections may act like hole traps and change the overall picture. Our starting point is the so–called t–J Hamiltonian

$$H_{t-J} = -t \sum_{\langle ij\rangle\sigma} (\hat{a}^{+}_{i\sigma}\hat{a}_{j\sigma} + \text{h.c.}) + J \sum_{\langle ij\rangle} \left[\underline{S}_i\,\underline{S}_j - \frac{n_i\,n_j}{4} \right] . \tag{5.2.19}$$

The operators $\hat{a}^{+}_{i\sigma}$ stand for $a^{+}_{i\sigma}(1-n_{i-\sigma})$. This Hamiltonian is decomposed into a part H_o, which contains the Ising part of the spin interaction, and a part H_1 which contains the spin flips

$$H_{t-J} = H_o + H_1$$

$$H_o = -t \sum_{\langle ij\rangle\sigma} (\hat{a}^{+}_{i\sigma} \hat{a}_{j\sigma} + \text{h.c.}) + J \sum_{\langle ij\rangle} \left[S^z_i S^z_j - \frac{n_i n_j}{4}\right] .$$

$$H_1 = \frac{J}{2} \sum_{\langle ij\rangle} (S^+_i S^-_j + S^-_i S^+_j) \tag{5.2.20}$$

It is assumed that there are N_o lattice sites with $N_o - 1$ electrons, i.e. the system contains one mobile hole. In order to deal with this problem only H_o is considered first. This is the problem of a hole in an Ising AF. Its ground state in the absence of the hole is the Néel ordered state. Assume that a hole has been created at site j. As it moves away from this site it generates a string of disordered spins in the otherwise perfect Néel state. Each disordered pair of spins increases the energy by J/2. Therefore an effective potential is acting on the hole which increases with distance and binds it to site j. This argument can be made more quantitative. Let $|j, \nu_p\rangle$ denote a state which describes a hole created at site j after it has made ν hops. The symbol p denotes a set of numbers which parameterize the geometry of the path which the hole has taken. The following ansatz can then be made for the ground state of H_o with a localized hole centred at site j:

$$|\Phi_j\rangle = \sum_{\nu} \alpha_{\nu} \Big(\sum_{p} |j, \nu, p\rangle\Big) \quad . \tag{5.2.21}$$

The sum over p is over all paths of length ν and the coefficients α_ν are to be determined from the requirement that the total energy is a minimum. In evaluating this energy the following approximation is made

$$H_{Ising} \, |j, \nu, p\rangle = \frac{J}{2} \left((z-2)\,\nu + 1 - \delta_{\nu,o}\right) |j, \nu, p\rangle \tag{5.2.22}$$

where H_{Ising} is the Ising part of H_o and z is the number of nearest neighbour sites, i.e. z = 4 for a square lattice. The approximation (5.2.22) assumes that every hop of the hole creates (z − 2) new disordered or frustrated spin pairs, except for the first hop where this number is (z − 1). The energy depends in this case only on the length ν of the path the hole has taken. Furthermore, the zero point energy has been set equal to the Ising energy of the lattice with the hole at site j.

At this stage it should be pointed out that by applying Equation (5.2.22) the famous Trugman paths are neglected. As is seen from Figure 5.2.1, the hole can follow a spiral type of path which enables it to eliminate again the disordered spin pairs which it has generated. Paths of this form actually lead to an unbinding of the hole from site j. However, as it turns out, their effect is small compared with that of the spin flip processes contained in H_1. Therefore the Trugman paths are neglected here.

By minimizing the energy, thereby using Equation (5.2.22), one can derive the following set of difference equations for the coefficients α_ν

$$-zt\alpha_1 = E_B\,\alpha_o$$

$$-t\,((z-1)\,\alpha_{\nu+1} + \alpha_{\nu-1}) = (E_B - J\,(\nu + {}^1/_2))\,\alpha_\nu\ . \tag{5.2.23}$$

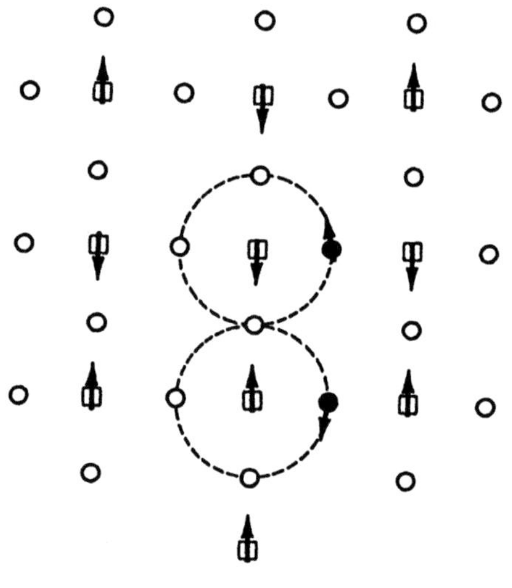

Figure 5.2.1 Coherent motion of a hole in AF Cu–O planes. Trugman paths.

E_B is the binding energy of the localized state $|\Phi_j>$. By the following change of variables

$$\beta_\nu = \frac{\alpha_\nu}{\sqrt{(z-1)^{\nu/2}}} \tag{5.2.24}$$

and $\tilde{t} = \sqrt{z-1}\, t$, these equations can be rewritten in the form

$$-\frac{z\tilde{t}}{z-1}\beta_1 = E_B\,\beta_o$$

$$-\tilde{t}\,(\beta_{\nu+1} + \beta_{\nu-1}) + J(\nu+1/2)\,\beta_\nu = E_B\,\beta_\nu \tag{5.2.25}$$

The last equation is just a one–dimensional Schrödinger equation with a linearly increasing potential. The solution of it can be expressed in terms of the Airy function but it is actually more convenient to solve it numerically.

The situation is drastically changed when the spin flip part H_1 of the Hamiltonian is included. These spin flip processes enable the hole to escape from the string potential. The Hamiltonian H_1 has matrix elements between different localized states $|\Phi_j>$. The matrix element connecting the two is therefore $(J/2)\ \alpha_o\ \alpha_2$. By means of the processes contained in H_1 the hole can tunnel from one localized state to another. This situation can be described by an effective tight–binding Hamiltonian. The following ansatz for the hole wave function is therefore suggested:

$$|\psi(\underline{k})> = \frac{1}{\sqrt{N_o}}\sum_j e^{i\underline{k}\,\underline{R}_j}\,|\Phi_j> \tag{5.2.26}$$

where $\underline{R}_j$ denotes the site position. The dispersion relation for the coherent motion of the hole is obtained from

$$E(\underline{k}) = \frac{<\psi(\underline{k})\ |H_{t-J}|\ \psi(\underline{k})>}{<\psi(\underline{k})\ |\ \psi(\underline{k})>} \quad . \tag{5.2.27}$$

In order to evaluate this expression one must know the matrix elements $<\Phi_i|H_{t-J}|\Phi_j>$. A string of any length $\nu \geq 2$ shortens by $\Delta\nu = 2$ when H_1 acts on the first two defects of the string. The starting point of the string is thereby shifted by 2 sites. Therefore one obtains an effective tight binding Hamiltonian with hopping elements $\tau_{1,1}$ and $\tau_{o,2}$ between second and third nearest neighbours. They are given by

$$\tau_{o,2} = J \sum_{\nu=o}^{\infty} (z-1)^{\nu} \alpha_{\nu} \alpha_{\nu+2}$$

$$= \frac{J}{(z-1)} \sum_{\nu=o}^{\infty} \beta_{\nu} \beta_{\nu+2}$$

$$\tau_{1,1} = 2\, \tau_{o,2} \quad . \qquad (5.2.28)$$

It is then easy to evaluate the dispersion relation for a square lattice and the following result is obtained

$$E(\underline{k}) = 4\, \tau_{1,1} (\cos k_x + \cos k_y)^2 + \text{const.} \qquad (5.2.29)$$

when the lattice constant is set equal to unity.

The bandwidth of the coherent motion of a hole is seen to be of order J and not t, which is the bare hopping matrix element. The effective mass of the quasiparticle is strongly renormalized by the strong interactions which the (5.2.19) Hamiltonian implicitly contains.

The above analysis can be extended by including the spin fluctuations which are present in the ground state of the system at half filling. Previously a Néel ordered state has been assumed for the ground state which is not the true ground state of the Heisenberg Hamiltonian. The net effect of including the ground state fluctuations is a renormalization of the bandwidth of the hole. The results of these calculations agree rather well with those based on evaluating the Green's function numerically and by using the Lanczos method for clusters of up to 18 sites.

5.3 Excited States of the Cu–0 Planes

Pure La_2Cu0_4 is an antiferromagnet below 300 [K]. At low temperatures the excitations of the weakly coupled Cu–0 planes are therefore antiferromagnetic spin waves. They have been studied in detail by inelastic neutron scattering and involve spin degrees of freedom of the electrons only. By partial substitution of La^{3+} by Ba^{2+} or Sr^{2+} (i.e. $La_{2-x}Ba_xCu0_4$ $La_{1-x}Sr_xCu0_4$) the Cu–0 planes are doped with holes. These holes are predominantly in the oxygen orbitals. This is in agreement with experiments based on electron energy loss spectroscopy (EELS). In these experiments a 1 s electron of an oxygen atom is excited into an empty valence state at the same site. Changes with doping in the p hole number of the oxygen sites are therefore directly measurable. Cu–0 planes doped with holes have also low energy excitations involving electronic charge degrees of freedom.

In the following, spin wave excitations are not considered because they are outside the scope of these lectures. Instead we concentrate on excitations which comprise the charge degrees of freedom. The simplest case is that of one hole in an otherwise half–filled band. The difference to the hole motion considered in Section 5.2.2 is that here a three band Hamiltonian is used instead of the simplified t–J model [Equation (5.2.19)]. Also we are interested here in calculating the spectral density of the excitations and not only in the coherent motion of the hole.

The excitations are determined by making use of the projection technique. The calculations serve as an example for the application of this method to excitations of strongly correlated systems. An alternative way is to compute the excitations by numerical studies of small clusters, thereby applying the Lanczos method. Calculations of this type are very helpful because they may serve as a kind of bench mark test for the analytic calculations.

The model Hamiltonian H of the system is divided into

$$
\begin{aligned}
H &= H_o + H_1 \\
H_o &= \epsilon_p \sum_{j\sigma} n_{p\sigma}(i) + \epsilon_d \sum_{i\sigma} \hat{d}^+_{i\sigma} \hat{d}_{i\sigma} + \sum_{\langle jj \rangle} t_{pp}(j,j') \, (p^+_{j\sigma} p_{j\sigma'} + \text{h.c.}) \\
H_1 &= \sum_{\langle jj \rangle} V_{ij} \, (\hat{d}^+_{i\sigma} p_{j\sigma} + p^+_{j\sigma} \hat{d}_{i\sigma}).
\end{aligned}
\tag{5.3.1}
$$

By using the operators $\hat{d}^+_{i\sigma} = d^+_{i\sigma} (1 - n_{d-\sigma}(i))$ one has excluded double occupancies of the d orbitals and therefore can dispose of the large Coulomb repulsion U_d, In contrast to the Hamiltonian (5.1.4) the interactions U_p and U_{pd} are neglected here. The matrix elements V_{ij} can be written as $V_{ij} = t_{pd} \, \gamma_j$ with $\gamma_1 = \gamma_2 = -\gamma_3 = -\gamma_4 = -1$.

The ground state of H_o has each Cu site occupied by one hole and is highly degenerate. Within the cumulant description presented in Section 4.2 it suffices to pick out of this manifold one specific state $|\Phi_o\rangle$. The excitation spectrum is found from the poles of the following correlation function

$$
C_{\underline{k}\alpha\sigma}(z) = \langle\psi_o| \, p_{\underline{k}\alpha\sigma} \frac{1}{z - L} p^+_{\underline{k}\alpha\sigma} \, |\psi_o\rangle \qquad . \tag{5.3.2}
$$

The exact ground state of H for half filling is denoted by $|\psi_o\rangle$. The Liouvillean L is defined with respect to H. The operators $p^+_{\underline{k}\alpha\sigma}$ ($p_{\underline{k}\alpha\sigma}$) create (destroy) a hole on oxygen sites. The indices $\underline{k}$ and σ refer to momentum and spin while the index α is sort of a band index. The numbering of the oxygen sites is $\alpha = 1,...4$. The Cu–0 plane is divided into unit cells with a Cu atom on sub lattice A (i.e. spin–up site), in the centre. The $p^+_{j\sigma}$ are relabelled $p^+_{i\alpha\sigma}$ in terms of which the $p^+_{\underline{k}\alpha\sigma}$ are expressed as

$$p^+_{\underline{k}\alpha\sigma} = \frac{1}{\sqrt{N_o}} \sum_i e^{-i\underline{k}(\underline{R}_i + \underline{r}_\alpha)} p^+_{i\alpha\sigma} \quad . \tag{5.3.3}$$

N_o is the number of unit cells and $\underline{R}_i$ and $(\underline{R}_i + \underline{r}_\alpha)$ are the positions of the Cu atom in the centre of cell i and of the 0 sites (i,α), respectively. By using the bilinear form (5.2.20) and adapting Equations (4.3.1 – 4.3.5) to the present case, the correlation function (5.3.2) is rewritten in terms of cumulants as

$$C_{\underline{k}\alpha\sigma}(z) = \left[\Omega |\, p_{\underline{k}\alpha\sigma} \frac{1}{z - L_o - H_1} p^+_{\underline{k}\alpha\sigma} \Omega\right] \quad . \tag{5.3.4}$$

The wave operator Ω is defined according to Equation (5.2.23).

In order to evaluate $C_{\underline{k}\alpha\sigma}(z)$ approximations have to be made. This is done by projecting all operators onto the relevant part R_o of the Liouville space. The latter is spanned by the set of elements $\{A_\nu\}$. The form of Equation (5.3.4) suggests the use of products of operators $\{\tilde{X}_\nu \Omega\}$ for the $\{A_\nu\}$. This ensures that the interplay between the hole which has been created and the changes in the ground state due to the presence of H_1 and represented by Ω is properly included in the theory. The projector P onto R_o is

$$P = \sum_{\nu\mu} |A_\nu)\, \chi^{-1}_{\nu\mu} (A_\mu|$$

$$\text{where} \qquad \chi_{\nu\mu} = (A_\nu | A_\mu) \tag{5.3.5}$$

and Equation (5.3.4) is replaced by

$$C_{\underline{k}\alpha\sigma}(z) = \left[\Omega |\, p_{\underline{k}\alpha\sigma} P \frac{1}{z - P\,(L_o + H_1)\,P} P\, p_{\underline{k}\alpha\sigma} \Omega\right] \quad . \tag{5.3.6}$$

The problem reduces therefore to calculating the matrix

$$G_{\nu\mu}(z) = \left[A_\nu |\, \frac{1}{z - P\,(L_o + H_1)\,P} P\, A_\mu\right] \tag{5.3.7}$$

in terms of which $C_{\underline{k}\alpha\sigma}(z)$ is easily expressed. This matrix is evaluated by applying the projection technique discussed in Appendix B. According to it, the following equation holds

$$\sum_{\nu} \left\{z\delta_{\eta\nu} - \omega_{\eta\nu} - \Sigma_{\eta\nu}(z)\right\} G_{\nu\mu}(z) = \chi_{\eta\mu} \quad . \tag{5.3.8}$$

Because all calculations are limited to R_o, the memory matrix $\Sigma_{\eta\nu}(z)$ vanishes in the present case. The frequency matrix $\omega_{\eta\nu}$ is defined according to

$$\omega_{\eta\nu} = \sum_{\lambda} (A_{\eta}| (L_o + H_1) A_{\lambda}) \chi^{-1}_{\lambda\mu} \quad . \tag{5.3.9}$$

The dimension of the matrix $G_{\nu\mu}(z)$ and therefore of the system of Equations (5.3.8) depends on the number of variables contained in the set $\{A_{\nu}\}$, i.e. on the dimension of R_o.

In order to find the most important operators $A_{\nu} = \tilde{A}_{\nu}\, \Omega$ which have to be included in R_o, the different microscopic processes must be analyzed. In doing this one has to take into account that the wave operator

$$\Omega = \lim_{z \to 0} \left[1 + \frac{1}{x - L_o} H_1 + \frac{1}{x - L_o} H_1 \frac{1}{x - L_o} H_1 + \ldots\right] \tag{5.3.10}$$

induces through H_1 charge fluctuations in the ground state by which holes hop from the Cu orbitals to the 0 orbitals and back. Only terms in Ω up to order H_1^2 will be considered. Clearly, the operators $p^+_{\underline{k}\alpha\sigma}$ $(\alpha = 1,\ldots,4)$ must be included in the set of $\tilde{A}_{\nu}$. They are labelled $\tilde{A}_1,\ldots,\tilde{A}_4$ and describe the hole motion due to direct 0–0 hopping when $t_{pp}(j,j') \neq 0$. Thereby a hole hops in a first step from a Cu to a 0 site. This can take place either through a ground state fluctuation contained in a Ω or through application of H_1 when $(z - L_o - H_1)^{-1}$ is expanded. In the latter case we speak of a **dynamical** Cu–0 charge fluctuation because it results from the dynamics or time evolution of the hole. In a second step the extra hole hops from an 0 site onto the empty Cu site so that as a net effect the hole has moved from one 0 site to another one. The final state is again described by the variables $\tilde{A}_{\nu}$ $(\nu = 1,\ldots,4)$, the operator describing the intermediate state as an independent variable because it is contained in the $\{A_{\nu}\} = \{A_{\nu}\, \Omega\}$ $\nu = 1,\ldots,4$. The above process is operative when the extra hole and the one on the Cu site have the same spin. But when the two holes have opposite spins, a spin exchange is associated with the corresponding two–step process. It gives rise to a new variable which is denoted by $\tilde{A}_5$. Assume that the extra hole $p^+_{\underline{k}\alpha\sigma}$ has $\sigma = +1$, i.e. spin up. The variable

$$\tilde{A}_5 = \frac{1}{2\sqrt{N_o}} \sum_{1\alpha} e^{-i\underline{k}(\underline{R}_1+2\underline{r}_1)} \gamma_\alpha p^+_{1'(\alpha)\,\alpha\downarrow} S^+_{1B} \qquad (5.3.11)$$

describes a spin exchange between the hole on the Cu B site belonging to unit cell 1 and the hole $p^+_{1'(\alpha)\,\alpha\downarrow}$ on the nearest–neighbour oxygen sites. These oxygen sites α belong to nearest neighbour unit cells $1'$. They have lattice vectors

$$\underline{R}_{1'(1)} = \underline{R}_1;\ \underline{R}_{1'(2)} = \underline{R}_1 + \underline{a}_2;\ \underline{R}_{1'(3)} = \underline{R}_1 + \underline{a}_1 + \underline{a}_2;\ \underline{R}_{1'(4)} = \underline{R}_1 + \underline{a}_1 \ . \qquad (5.3.12)$$

The next variables of importance are $\{\tilde{A}_6\}$ and $\{\tilde{A}_7\}$. The operator $\tilde{A}_6$ describes the dynamical Cu–0 charge fluctuation and is given by

$$\tilde{A}_6 = \frac{1}{4\sqrt{N_o}} \sum_{1\alpha\beta} e^{-i\underline{k}(\underline{R}_1+2\underline{r}_1)} \gamma_\alpha \gamma_\beta p^+_{1'(\alpha)\,\alpha\downarrow} p^+_{1''(\beta)\,\beta\downarrow} \hat{d}_{1B\downarrow} \ . \qquad (5.3.13)$$

It removes a hole from a Cu B site and generates an additional hole on a neighbouring 0 site.

The variable $\{\tilde{A}_7\}$ flips two neighbouring Cu spins. It is obtained from $\tilde{A}_5$ and is of the form

$$\tilde{A}_7 = \frac{1}{4\sqrt{N_o}} \sum_{1\alpha\beta} e^{-i\underline{k}(\underline{R}_1+2\underline{r}_1)} \gamma_\beta p^+_{1'(\alpha)\,\beta\uparrow} S^-_{1'(\alpha)A} S^+_{1B} \ . \qquad (5.3.14)$$

The inclusion of higher string variables would enable one to derive the energy dispersion of the coherent hole motion. This is not done here. Instead, the following considerations are limited to a Liouville subspace R_o spanned by $\{\tilde{A}\nu\,\Omega\}$ with $\nu = 1,...,7$.

The (7×7) matrix $G_{\nu\mu}(z)$ can be computed for any fixed value of $\underline{k}$ and from it $C_{\underline{k}\alpha\sigma}(z)$ can be determined according to Equation (5.3.6). The local spectral density $P(\omega)$ is related to the function $c_{\underline{k}\alpha\sigma}(z)$ through

$$P(\omega) = \frac{-1}{\pi\, N_o} \sum_{\underline{k}} \text{Im}\, C_{\underline{k}\alpha\sigma}(\omega + i\eta). \qquad (5.3.15)$$

where N_o = number of unit cells. Some calculated results are shown in Figure 5.3.1. The dominant structure near $\omega \simeq \epsilon_p$ corresponds to the non–bonding peak in a band structure calculation. It results from two combinations of $p^+_{1\alpha\uparrow}$ operators which would not couple with any of the other variables $\breve{A}_\nu$ in the case of $t_{pp} = 0$. The peak on the high energy side and the two–peak structure on the low energy side of the non–bonding peak have anti–bonding and bonding characters. They result predominantly from the two remaining combinations of the four operators $p^+_{1\alpha\uparrow}$ and from the variable $\breve{A}_5$. But the two bonding structure is also coupled appreciably to the variable $\breve{A}_7$. The anti–bonding structure can be interpreted by analyzing the eigenoperators in terms of singlet and triplet operators. The latter are formed by the hole on a Cu B site and that on a neighbouring 0 site (remember that we are working in Liouville space),

$$0_{s(\tau)} = \frac{1}{\sqrt{8\,N_o}} \sum_{1\alpha} e^{-i\underline{k}(\underline{R}_1+2\underline{r}_1)} \gamma_\alpha (p^+_{1(\alpha)\alpha\uparrow} \,(\mp)\, p^+_{1(\alpha)\alpha\downarrow} S^+_{1B}). \qquad (5.3.16)$$

Depending on the value of $\underline{k}$, the eigenoperator of the anti–bonding peak is found to have between 70–97% singlet character. To a good approximation the low lying bonding state can be

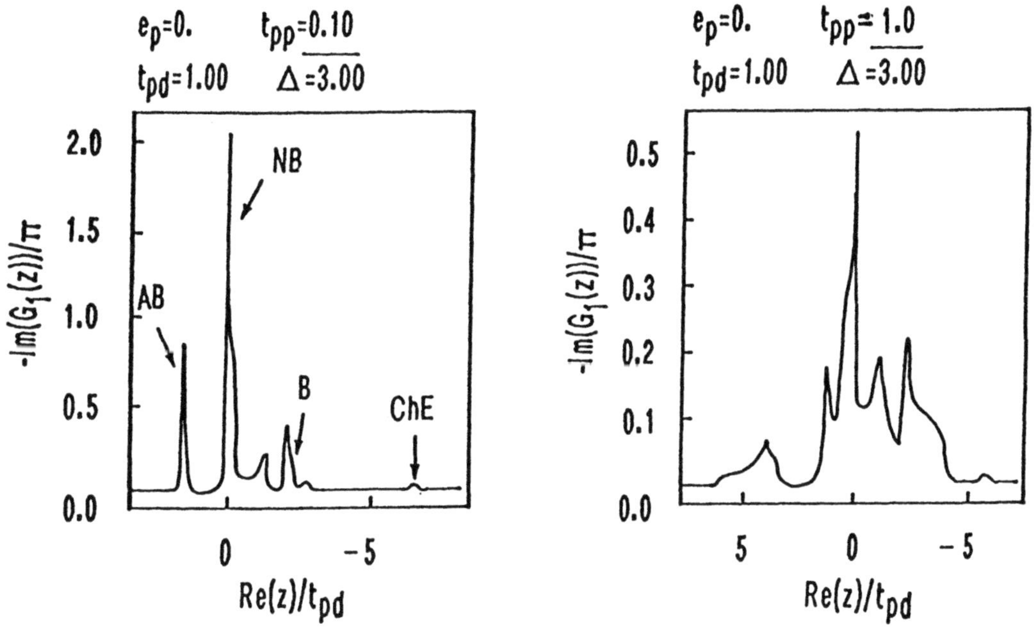

Figure 5.3.1 Spectral density $P(\omega)$ of Equation (5.3.15). The labels B, NB and AB refer to bonding, non–bonding and anti–bonding states. The Cu–O dynamic charge fluctuations give rise to the peak labelled ChE.

interpreted as a singlet of the form suggested by Zhang and Rice. The holes on a Cu site and on a neighbouring 0 site couple to form a singlet somewhat like electrons in a H_2 molecule at large interatomic distances where the Heitler–London theory applies. The high frequency structure in $P(\omega)$ at $\omega \simeq 6\, t_{pd}$ is due to the Cu–0 dynamical charge fluctuations, i.e. the variable $\tilde{X}_6$. The spectral weight turns out to be of order $(t_{pd}/\Delta)^6$ and therefore is very small. The above treatment requires two improvements.

One is the proper inclusion of the Cu spin interactions. This is an effect of order (t_{pd}^4/Δ^3). Until now only the Ising part of this interaction was used, which stabilizes the Néel state. A second improvement is the inclusion of higher string variables. This should give the quasiparticle excitation, which was the subject of Section 5.2.2 and has an energy dispersion of width J.

The above results can be compared with numerical calculations using small clusters of Cu–0. The hole spectral functions are thereby written in the form

$$P_{\underline{k}\alpha}(\omega) = -\frac{1}{\pi} \mathrm{Im} \langle \psi_0(N) | c^+_{\underline{k}\sigma} \frac{1}{\omega - E_0(N) + H + i\eta} c_{\underline{k}\sigma} | \psi_0(N) \rangle \tag{5.3.17}$$

where $|\psi_0(N)\rangle$ is the exact ground state of the N electron system with energy $E_0(N)$ and $c^+_{\underline{k}\sigma}$ is an electron, not hole, creation operator. The right hand side is evaluated with the help of the Lanczos method. In contrast to the previous analytic calculations, U_d is finite and U_p is also taken into account. Not only the hole spectrum is shown but also the electron spectrum, which can be measured by inverse photoelectron spectroscopy. In order to calculate the latter one must replace $c_{k\sigma}(c^+_{k\sigma})$ in Equation (5.3.17) by $c^+_{k\sigma}(c_{k\sigma})$, respectively. The dominant peak is again due to the nonbonding band. The structure far below the Fermi energy corresponds to various final states like $d^{10}p^4$ or d^8p^6, respectively. The inverse photoelectron spectrum is separated by a gap of order $(\epsilon_p - \epsilon_d)$ from the photoelectron spectrum (charge transfer gap). It represents the upper Hubbard band with a $d^{10}p^6$ final state (see Figure 5.3.2).

When the system is doped with holes, the inverse photoelectron spectrum is distinctly different from the one for the half filled band case. A new peak appears within the gap Δ as the hole concentration increases. Electrons are now injected not only into Cu 3d states but also into 0 2p states, giving rise to the additional peak. The above discussion shows that the excitations in Cu–0 planes have characteristic features which cannot be obtained from a conventional bandstructure calculation. In particular, the singlet–like excitation is a striking example of the strong correlations.

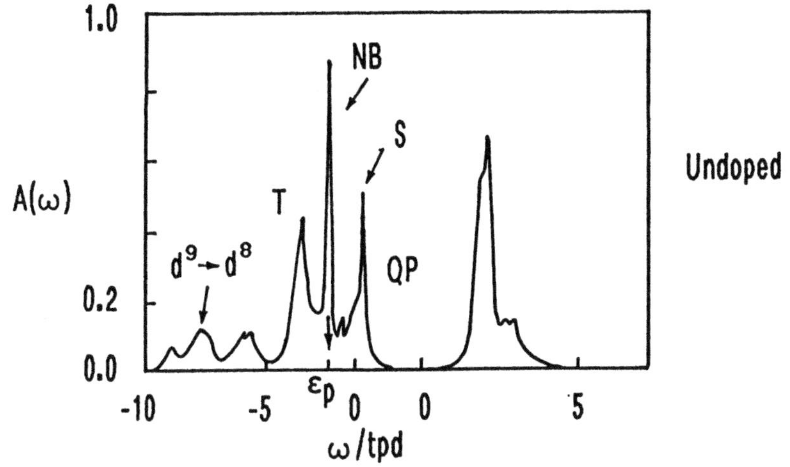

Heitler–London singlet and triplet (Zhang–Rice)

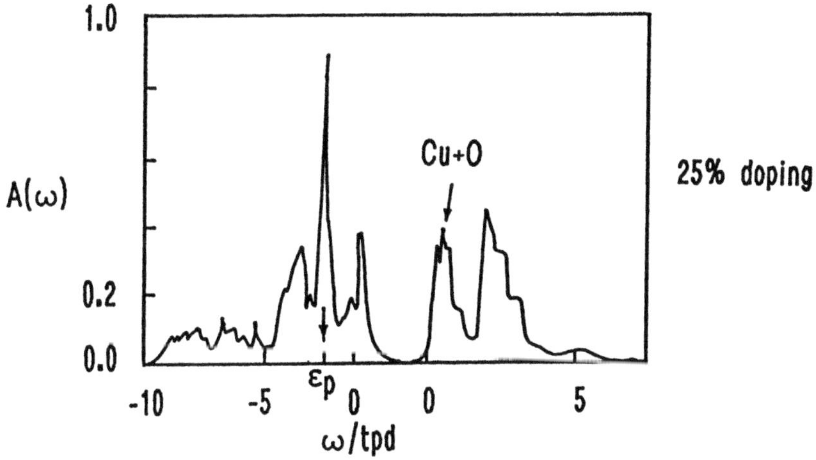

Figure 5.3.2 Photo–emission and inverse photo–emission for an undoped and doped system.

APPENDIX A

DERIVATION OF SOME RELATIONS INVOLVING CUMULANTS

In this Appendix the steps are indicated which are necessary to obtain some of the equations in Section 5.2.

First it is shown that Equation (4.2.19) follows from Equation (4.2.18). For that purpose write

$$I = <\Phi_o| \left[\frac{1}{z - L_o - H_1} - \frac{1}{z}\right] |\Phi_o>^c = <\Phi_o| \left[\frac{1}{z} \sum_{n=o}^{\infty} \frac{(L_o + H_1)^n}{z^n} - \frac{1}{z}\right] |\Phi_o>^c$$

$$= <\Phi_o| \frac{1}{z} \sum_{n=1}^{\infty} \frac{(L_o + H_1)^n}{z^n} |\Phi_o>^c \quad . \tag{A.1}$$

Because $L_o|\Phi_o> = 0$ and $<\Phi_o|L_o = 0$ the last expression equals

$$I = \frac{1}{z^2} <\Phi_o| \left[H_1 + H_1 \frac{1}{z - (L_o + H_1)} H_1\right] |\Phi_o>^c \tag{A.2}$$

which is Equation (4.2.19).

Next the equivalence of Equations (4.2.24) and (4.2.25) is shown. By expanding $(z - L_o)^{-1}$ in powers of L_o/z Equation (4.2.24) consists of a sum of terms of the form

$$(H_1|L_o^{m_1} H_1 L_o^{m_2} H_1 \,.....\, L_o^{m_n} H_1) \quad . \tag{A.3}$$

With the help of $L_o H_1 = [H_o, H_1]_-$ it is easily checked that e.g.

$$\begin{aligned}(H_1|L_o^m H_1) &= <\Phi_o|H_1 L_o^m H_1|\Phi_o>^c \\ &= (H_1|H_o^m H_1). \end{aligned} \tag{A.4}$$

This is so, because from the special properties of cumulants it follows that

$$<\Phi_o|H_1^2 H_o^n |\Phi_o>^c = 0 \,; \qquad n \geq 1. \tag{A.5}$$

A proof that the replacement $L_o \rightarrow H_o$ can be made everywhere in Equation (A.3) can be given by induction.

In order to prove the equivalence of Equations (4.2.25) and (4.2.26) one rewrites the latter equation as

$$\delta E_o = \lim_{z \to o} \sum_{n=0}^{\infty} \left(H_1 \Big| \left[\frac{1}{z} \sum_{\nu=0}^{\infty} \left[\frac{H_o - E_o 1^c}{z}\right]^{\nu} H_1\right]^n\right) \qquad . \qquad (A.6)$$

By expressing $(H_o - E_o\, 1^c)^{\nu}$ in terms of a binomial expansion this equation goes over into

$$\delta E_o = \lim_{z \to o} \sum_{n=0}^{\infty} \left(H_1 \Big| \left[\frac{1}{z} \sum_{\nu=0}^{\infty} \frac{1}{z}{}^{\nu} \left[\sum_{\mu=0}^{\infty} H_o^{\mu} (-E_o\, 1^c)^{\nu-\mu} \binom{\nu}{\mu}\right] H_1\right]^n\right) \quad .$$

$$(A.7)$$

Only terms with $\nu = \mu$ contribute because a factor 1^c or any power of it gives rise to a vanishing cumulant. This follows from the properties of cumulants, i.e. $(1^c | A) = 0$ for arbitrary operators A. From Equation (A.7) the expression (4.2.25) follows immediately.

An important relation is the equivalence of the two expressions in Equation (4.2.37) for the ground state energy, i.e.

$$(\Omega | H\Omega) = E_o + (H_1 | \Omega) \qquad (A.8)$$

provided $<H_1> = 0$. The energy E_o is the ground state energy of H_o. For the proof one uses the expression

$$|\Omega) = \lim_{z \to o} \sum_{n=0}^{\infty} \Big| \left[\frac{1}{z - L_o} H_1\right]^n \qquad (A.9)$$

and furthermore, the fact that, in expressions of the form of (A.3), L_o may be replaced by H_o.

Thus

$$(H_1 | \Omega) = \lim_{z \to o} \sum_{n=1}^{\infty} \left(H_1 \Big| \left[\frac{1}{z - H_o} H_1\right]^n\right) \qquad . \qquad (A.10)$$

On the other hand one obtains

$$(\Omega|H_1\Omega) = \lim_{z\to o}\left\{2\left[H_1|\frac{1}{z-H_o}H_1\right] + 3\left[H_1|\frac{1}{z-H_o}H_1\frac{1}{z-H_o}H_1\right]+....\right\}$$

$$= \sum_{n=1}^{\infty}(n+1)\left(H_1|\left[\frac{1}{z-H_o}H_1\right]^n\right)$$

$$= \sum_{n=1}^{\infty}n\left(H_1|\left[\frac{1}{z-H_o}H_1\right]^n\right) + (H_1|\Omega) \qquad . \qquad (A.11)$$

Similarly one finds that

$$(\Omega|H_o\Omega) = E_o + \lim_{z\to o}\left\{\left[H_1|\frac{1}{z-H_o}H_o\frac{1}{z-H_o}H_1\right] + ...\right\}$$

$$= E_o - \lim_{z\to o}\left\{\left[H_1|\frac{1}{z-H_o}H_1\right] + 2\left[H_1|\frac{1}{z-H_o}H_1\cdot\frac{1}{z-H_o}H_1\right]+....\right\}$$

$$= E_o - \sum_{n=1}^{\infty}n\left(H_1|\left[\frac{1}{z-H_o}H_1\right]^n\right) \qquad . \qquad (A.12)$$

By adding up Equations (A.11) and (A.12) one recovers the right hand side of Equation (A.8).

APPENDIX B

PROJECTION METHOD OF MORI AND ZWANZIG

The projection method of Mori and Zwanzig enables one to compute resolvents. In quantum chemistry it is often called the superoperator method.

Consider the Liouville– or operator–space. There are various ways of defining an operator product in it. One particular way was chosen in Section 5.2 and specified by Equation (4.2.20). A different form is

$$(A|B) = \langle\psi_o|[A^+,B]_+|\psi_o\rangle \qquad . \qquad (B.1)$$

Note that the expectation value is with respect to the exact ground state $|\psi_0>$ and that in contrast to (4.2.20) no cumulants are used. Which choice for $(A|B)$ is most appropriate depends on the problem one wants to consider. The above choice is particularly suitable when Green's functions are calculated (see Equation (4.3.17)). Most of the following calculations are independent of the choice of the operator product. They also apply when the form (4.2.20) is used instead.

The Liouville operator L corresponding to H is defined by analogy to Equation (4.2.4) as

$$LA = [H,A]_- \tag{B.2}$$

where A is an arbitrary operator.

Since $[H,A] = -i\, dA/dt$ it follows that

$$LA = -i \frac{dA}{dt} \quad . \tag{B.3}$$

This equation has the formal solution

$$A(t) = e^{iLt} A(0). \tag{B.4}$$

A set of operators $|A_i)$ is introduced which are called dynamical variables in the following. The aim is to evaluate the matrix $\underline{R}(z)$ with matrix elements

$$R_{ij}(z) = (A_i | \frac{1}{z - L} A_j) \tag{B.5}$$

It is assumed that $z = \omega + i\eta$ where η is a positive infinitesimal number. It serves to specify the analytic properties of $R_{ij}(\omega)$. For example, when the $|A_i)$ are the electron– creation operators $|c^+_{p\sigma})$ then it can be shown that R^σ_{pp} is the retarded Green's function $G^R_\sigma(\underline{p},\omega)$ of the electrons.

In order to evaluate Equation (B.5) a projection operator

$$P = \sum_{ij} |A_i)\, \chi^{-1}_{ij}\, (A_j|$$

where $$\chi_{ij} = (A_i|A_j) \tag{B.6}$$

is introduced. It has the property that, when it acts on any operator $|B)$, it extracts from it all the components which are proportional to the variables $|A)$. The operator $Q = 1 - P$ does the same for the Liouville space, which is orthogonal to the space spanned by the $|A_i)$,

Consider again the matrix $\underline{R}$ with matrix elements (B.5). By making use of the identity

$$\frac{1}{a+b} = \frac{1}{a} - \frac{1}{a} b \frac{1}{a+b} \tag{B.7}$$

which is verified by expanding both sides, and by using the relationship $L = PL + QL$, one can rewrite R_{ij} in the form

$$R_{ij} = (A_i| \left[\frac{1}{z - LQ} + \frac{1}{z - LQ} LP \frac{1}{z - L}) A_j\right] . \tag{B.8}$$

One convinces oneself that

$$\begin{aligned} (A_i| \frac{1}{z - LQ} A_j) &= \frac{1}{z} (A_i|A_j) \\ &= \frac{1}{z} \chi_{ij} \quad . \end{aligned} \tag{B.9}$$

This is so since $Q|A_j) = 0$. Therefore Equation (B.8) goes over into

$$R_{ij} = \frac{1}{z} \chi_{ij} + \sum_{\ell m} (A_i| \frac{1}{z - LQ} L |A_l) \chi^{-1}_{\ell m} R_{mj} \tag{B.10}$$

or in matrix notation

$$(\underline{z} - \underline{K}\underline{\chi}^{-1}) \underline{R} = \underline{\chi} \tag{B.11}$$

where the matrix $\underline{K}$ has the matrix elements

$$K_{i\ell} = (A_i| \frac{z}{z - LQ} L |A_l) \quad . \tag{B.12}$$

These elements can be decomposed into

$$K_{i\ell} = (A_l| L | A_l) + (A_i| L Q \frac{1}{z - LQ} L |A_l) \tag{B.13}$$

and consist of two parts. One part consists of the elements L_{ij} of <u>frequency</u> matrix

$$L_{i\ell} = (A_i \mid L \mid A_l) \quad . \tag{B.14}$$

The second part defines the memory matrix

$$M_{i\ell}(z) = (A_i \mid LQ \frac{1}{z - QLQ} QL \mid A_l) \quad . \tag{B.15}$$

Here $Q^2 = Q$ has been used. It should be noticed that $M_{i\ell}(z)$ is again of the form of Equation (B.5) but with $|A_j)$ replaced by $QL|A_j)$ and L replaced by QLQ. Therefore the same procedure can be repeated over and over again, and at each step the space within which one is working is orthogonal to the former one.

When Equations (B.14 and B.15) are substituted into Equation (B.10) it takes the form

$$\underline{\underline{R}}(z) = \frac{1}{z - (\underline{\underline{L}} + \underline{\underline{M}}(z))\ \underline{\underline{\chi}}^{-1}}\ \underline{\underline{\chi}} \tag{B.16}$$

or alternatively

$$\sum_{\ell} [z\delta_{i\ell} - \sum_{s} (L_{is} + M_{is}(z))\ \chi^{-1}_{s\ell}]\ R_{\ell j}(z) = \chi_{ij} \quad . \tag{B.17}$$

It is seen from Equation (B.16) that, when $\underline{M}(z)$ is written in the same form as $\underline{\underline{R}}(z)$, one obtains a continued fraction. Each order contains a new frequency– and memory matrix. When $\underline{\underline{R}}(z)$ is evaluated in this way, it contains a number of static quantities such as the "susceptibility" matrices χ_{ij} or the frequency matrix $L_{i\ell}$. Because of the form (B.1) they are unknown since the exact ground state $|\psi_0>$ is not known. Therefore independent ways of approximating these quantities have to be found. In conclusion it is found that the conventional projection method of Mori and Zwanzig expresses dynamical correlation functions in terms of static quantities. The latter have to be determined independently. Note that the projection method presented in Section 4.2 allows for computing a static quantity i.e. the ground state energy shift. Therefore it can be looked upon as a generalization of the Mori–Zwanzig method presented here.

Superfluid ^{3}He – Flow, Textures and Rotation

John Hook

Physics Department,Manchester University,Manchester M13 9PL,UK

1 Introduction

The A and B phases of liquid ^{3}He which exist at temperatures below about 2.5 mK were discovered by Osheroff, Richardson and Lee (1972). It is impossible in 5 lectures to do justice to the large amount of experimental work that has been done on the new phases since then. My choice of topics has been guided by my own interests. For further information on the theory and experimental properties of superfluid ^{3}He the reader is recommended to consult Leggett (1975), Wheatley (1975), Vollhardt and Wölfle (1990), Halperin and Pitaevskii (1990) and the many excellent review articles that have appeared in *Progress in Low Temperature Physics* (ed. by D. F. Brewer, published by North-Holland, Amsterdam, Vol. VIIa onwards).

There is a natural progression in the lectures. I begin by discussing experiments which investigate the extent to which the new phases show superfluid behaviour, analogous to that of the other superfluids that have been investigated experimentally, namely superfluid ^{4}He and superconducting metals. I then progress to experiments which show the differences in behaviour between ^{3}He and other superfluids, resulting from the more complicated order parameter of superfluid ^{3}He. In particular, experiments are described which indicate that the new phases have properties similar to those of liquid crystals. Finally I discuss the interaction between the liquid-crystal-like and superfluid properties of the A phase. This interaction, which has not yet been extensively investigated experimentally, raises doubts as to the extent that the A phase satisfies the normal criteria for defining a superfluid system.

As well as teaching some of the properties of superfluid ^{3}He my intention is also to demonstrate the beauty and the difficulty of the experiments. The difficulties arise not only because of the need to maintain temperatures of around 1 mK, often in a rotating cryostat, but also in the design and interpretation of the experiments themselves. The theory of the experimental method is often as difficult as the more-fundamental theory used to calculate the properties of ^{3}He. I think it is important that such points should be made to the theoretical students attending the summer school.

In Sect. 2 an experiment is described which shows that the superflow around a wire immersed in superfluid ^{3}He–B is quantized. In Sect. 3 I discuss experiments which show that it is possible to establish 'persistent' currents around a toroidal channel in both phases. These two sections establish the credentials of ^{3}He to be described as a superfluid. In Sect. 4 the beautiful experiments of Avenel and Varoquaux are described, which show that both isotopes of helium exhibit Josephson effects in their superfluid states; the widely-held belief that such effects should be present in the helium superfluids as well as in superconductors has led to many experimental attempts to find them but only Avenel and Varoquaux have so far succeeded with certainty.

In Sect. 5 the behaviour of superfluids under rotation is discussed and the differences in this respect between superfluid ^{4}He and ^{3}He–B are described. In Sect. 6 I describe an experiment which illustrates the liquid crystal properties of superfluid ^{3}He–A. In Sect. 7 I explain how superflow in ^{3}He–A differs from that in ^{3}He–B, ^{4}He and superconductors, how it interacts with the liquid crystal properties and how the interaction manifests itself in the properties of ^{3}He–A under rotation.

2 Quantization of Circulation in ^{3}He–B

The most convincing demonstration of dissipationless flow in a superfluid is the persistence of flow around a toroidal channel for indefinite periods of time. The stability of the persistent current can be explained if the superfluid is described by a quantum mechanical wave function $\Psi = |\Psi| \exp(i\theta)$, which remains phase coherent on a macroscopic length scale and in particular around the channel. This description implies that the circulation,

$$\kappa = \oint \boldsymbol{v}_s . d\boldsymbol{r}, \tag{1}$$

of the superfluid velocity $\boldsymbol{v}_s$ around the channel should be quantized, as we now demonstrate. If the wave function Ψ is identified with particles of mass M then the superfluid velocity is related to the phase θ by

$$\boldsymbol{v}_s = \frac{\hbar}{M} \nabla \theta. \tag{2}$$

If one imposes the requirement that the phase of a single-valued wave function can change only by $2\pi n$ on going around the channel, then integration of (2) around the channel gives

$$\kappa = n \frac{h}{M}, \tag{3}$$

where n is an integer.

Quantization of circulation in superfluid ^{4}He was first demonstrated experimentally by Vinen (1961). He measured the circulation around a wire immersed in the fluid through its effect on transverse vibrations of the wire and found that the quantum of circulation was h/m_4 indicating that the mass M of the superfluid particles in ^{4}He is just the atomic mass m_4. Vinen's technique was subsequently refined by Whitmore and Zimmermann (1968) and has recently been applied to ^{3}He–B by Davis et al. (1991) at Berkeley.

In the Berkeley experiment a superconducting Nb-Ti wire of diameter 16 μm was stretched along the axis of a 2.8 mm-inner-diameter, 50 mm-long brass cylinder containing the ^{3}He. A 50 mT magnetic field applied perpendicular to the wire enabled the wire to be "plucked" by passing a pulse of electric current through it. The decay of the resulting transverse vibrations of the wire at the resonant frequency of order 347 Hz was then observed by monitoring the voltage induced across the wire due to its motion in the applied field.

A simplified theory of the wire dynamics will suffice to show how the circulation can be deduced. We assume that the displacement $\boldsymbol{r}$ of the wire at any point along its length satisfies the equation of motion

$$\mu\ddot{\boldsymbol{r}} + \rho_s\kappa\hat{\boldsymbol{z}} \times \dot{\boldsymbol{r}} + \mu\omega_0^2\boldsymbol{r} = 0, \tag{4}$$

where μ is the mass per unit length of the wire, ρ_s the superfluid density and ω_0 is the resonant angular frequency. The third term on the left hand side represents the restoring force on the wire, due ultimately to the fact that it is fixed at each end. The second term shows that the force on the wire due to the circulation κ is at right angles to the wire direction $\hat{\boldsymbol{z}}$ and also to its velocity; the origin of the force is the same as that of the lift force on an aeroplane wing (see, for example, Lamb 1932).

We take the components of (4) along the x and y directions in the plane perpendicular to the wire:

$$\begin{aligned} \mu\ddot{x} - \rho_s\kappa\dot{y} + \mu\omega_x^2 x &= 0, \\ \mu\ddot{y} + \rho_s\kappa\dot{x} + \mu\omega_y^2 y &= 0, \end{aligned} \tag{5}$$

where the introduction of ω_x and ω_y in place of ω_0 takes account of the fact that the wire will not have a perfectly circular cross section. The two transverse normal modes of a wire of elliptical cross section are plane polarized in perpendicular directions in the absence of circulation; ω_x and ω_y are the frequencies of these modes and the x and y axes are chosen along the polarization directions. The plucking of the wire causes an initial displacement in the direction perpendicular to the field; since this does not coincide with either the x or y axis, both normal modes are excited in this process. Beating is therefore observed in the decay of the oscillations at a beat frequency $\omega_x - \omega_y$ in the absence of circulation as shown in Fig. 1(a).

We see from (5) that the effect of circulation is to couple the vibrations in the x and y directions; the normal modes are then elliptically polarized. To calculate the frequencies of the elliptical modes we substitute a solution of the form

$$x = x_0 \exp(\mathrm{i}\omega t), \quad y = y_0 \exp(\mathrm{i}\omega t)$$

into (5) to obtain

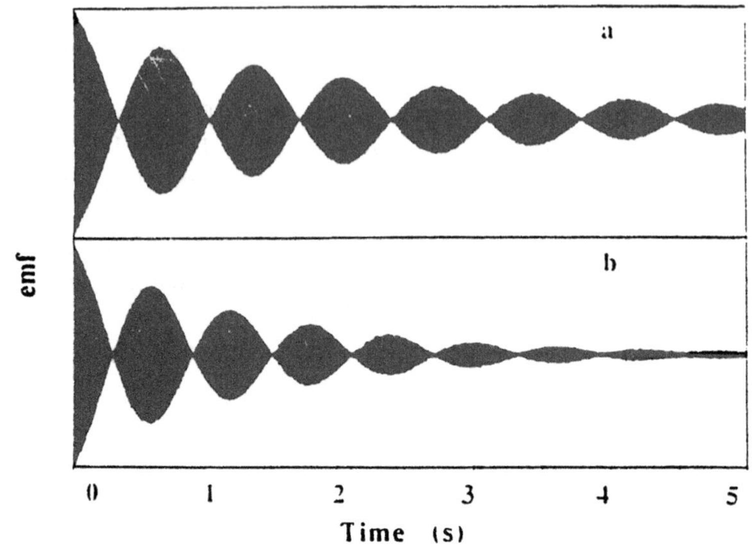

Fig. 1. (a) Beats observed in the free decay of oscillations in the absence of circulation. (b) The beats are at a higher frequency when there is one quantum of circulation around the wire.

$$\mu(\omega_x^2 - \omega^2)x_0 - \mathrm{i}\omega\rho_s\kappa y_0 = 0,$$
$$\mathrm{i}\omega\rho_s\kappa x_0 + \mu(\omega_y^2 - \omega^2)y_0 = 0.$$

It is possible to simplify these equations to

$$2\mu(\omega_x - \omega)x_0 - \mathrm{i}\rho_s\kappa y_0 = 0,$$
$$\mathrm{i}\rho_s\kappa x_0 + 2\mu(\omega_y - \omega)y_0 = 0,$$

by assuming that $\omega \simeq \omega_x \simeq \omega_y$. The resulting homogeneous equations have a non-trivial solution only if the determinant of coefficients vanishes and this determines the frequencies of the two modes as

$$\omega_\pm = \frac{\omega_x + \omega_y}{2} \pm \frac{\sqrt{(\omega_x - \omega_y)^2 + \kappa^2\rho_s^2/\mu^2}}{2}$$

ω_+ (ω_-) is the frequency of the elliptically polarized mode rotating in the same (opposite) direction as the circulation. In the presence of circulation the beat frequency observed in the decay of transverse vibrations is $\omega_+ - \omega_-$ and is therefore given by

$$(\omega_+ - \omega_-)^2 = (\omega_x - \omega_y)^2 + \frac{\kappa^2\rho_s^2}{\mu^2}. \tag{6}$$

Figure 1(b) clearly shows that the beat frequency is different in the presence of circulation and from Figs. 1 it is possible to deduce a value for κ by use of (6).

It was not possible to measure κ at temperatures above 400 μK because damping of the motion of the wire by thermal excitations prevented observation of beats above this temperature (for simplicity we have not included this damping in our analysis). At temperatures below 400 μK, ρ_s in (6) can be replaced by the total fluid density ρ.

States of finite circulation could be created by rotation of the cryostat containing the experiment and then bringing the cryostat to rest. Figure 2 shows the measured circulation as a function of the maximum rotation rate of the cryostat. Quantization of circulation in ^{3}He–B in steps of $h/2m_3$ is clearly indicated. The appearance of $2m_3$ in the denominator indicates that the mass of the superfluid particles is twice the atomic mass and supports the theoretical picture of superfluidity in ^{3}He as arising from Cooper pairing of the atoms. The circulation measurement was calibrated by putting superfluid ^{4}He in the apparatus and assuming that the quantum of circulation is h/m_4 in this case; this essentially determines a value for μ in (6). The values of κ between the quantized values on Fig. 2 were unstable and probably correspond to a situation in which a quantized vortex line (see Sect. 5) detaches itself from the wire at some point along its length.

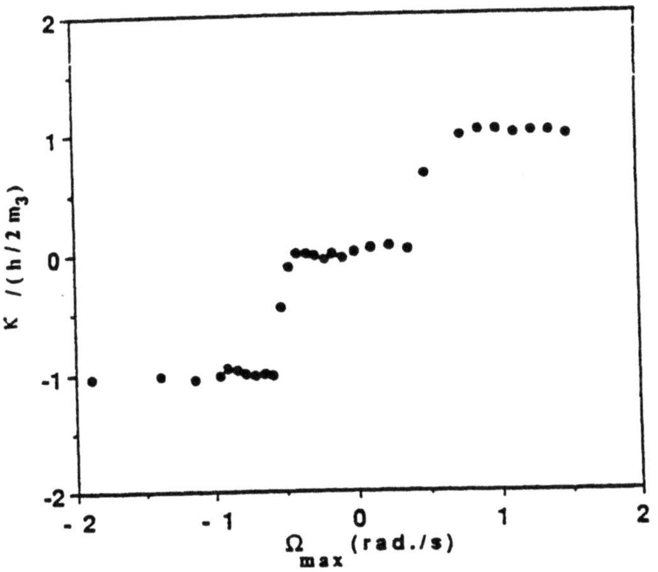

Fig. 2. The average value of the circulation (in units of $h/2m_3$) measured for 10 minutes after the cryostat has been ramped to Ω_{max} and back to 0. The direction of the circulation was established by noting the effect on the beat pattern of a rotation of the applied magnetic field.

It would be interesting to perform this experiment for the A phase also. As we shall see in Sect. 7 there are reasons to doubt that the circulation would be quantized in this case. This would however be a more difficult experiment since the A phase only exists down to low temperatures in magnetic fields $\geq 0.4\,\mathrm{T}$.

3 Persistent Currents

Persistent current experiments on the A and B phases of ^{3}He have been performed at Cornell (Gammel, Hall and Reppy 1984; Gammel, Ho and Reppy 1985) and Helsinki (Pekola and Simola 1985). The principle of both experiments is the same and has evolved from the pioneering dc gyroscope experiments of Reppy in ^{4}He (see e.g. Langer and Reppy 1970). In these experiments the persistent current was measured by monitoring the torque required to cause precession of the associated angular momentum. This is illustrated in Fig. 3. A toroidal channel is mounted in a vertical plane so that the angular momentum $\boldsymbol{L}_\mathrm{p}$ of a persistent current around it is in the horizontal plane as shown; if the toroid is rotated about the vertical axis at angular velocity $\boldsymbol{\Omega}$ then the precession of $\boldsymbol{L}_\mathrm{p}$ requires a torque $d\boldsymbol{L}_\mathrm{p}/dt$, which is equal to $\boldsymbol{\Omega} \times \boldsymbol{L}_\mathrm{p}$ and thus perpendicular to both $\boldsymbol{L}_\mathrm{p}$ and the rotation axis. The torque is therefore parallel to the tungsten fibre on which the torus is mounted and is provided by a slight twist of this fibre, through an angle θ, given by

$$\Omega L_\mathrm{p} = c\theta, \tag{7}$$

where c is the torsion constant of the fibre. The twisting angle θ is measured using a capacitive deflection sensor and hence L_p determined using (7). The toroid was packed with a fine powder in order to increase the critical current for superflow.

To perform this experiment in ^{3}He is more difficult because of the need to fill the torus and maintain thermal contact without reducing the sensitivity. The sensitivity can be enhanced by using the ac gyroscopic technique employed at Cornell and Helsinki. Suppose the equilibrium position of the torus is in the horizontal plane but that small oscillations are possible about an axis $\hat{\boldsymbol{\theta}}$ in this plane as shown in Fig. 4. The torque required to produce the resulting oscillation of the persistent angular momentum affects the motion of the oscillator in a manner that we now calculate.

The total angular momentum of the torus is $\boldsymbol{L} = I_\theta \dot{\boldsymbol{\theta}} + \boldsymbol{L}_\mathrm{p}$, where $\dot{\boldsymbol{\theta}}$ is the angular velocity and I_θ is the moment of inertia about the $\hat{\boldsymbol{\theta}}$ axis. We suppose that, when the torus is rotated through $\boldsymbol{\theta}$, it is subject to a restoring torque $-c\boldsymbol{\theta}$. The equation of motion of the torus is therefore $d\boldsymbol{L}/dt = I_\theta \ddot{\boldsymbol{\theta}} + d\boldsymbol{L}_\mathrm{p}/dt = -c\boldsymbol{\theta}$. Because of the oscillatory motion of $\boldsymbol{L}_\mathrm{p}$ we have $d\boldsymbol{L}_\mathrm{p}/dt = \dot{\boldsymbol{\theta}} \times \boldsymbol{L}_\mathrm{p}$ and the equation of motion becomes

$$I\ddot{\boldsymbol{\theta}} + \dot{\boldsymbol{\theta}} \times \boldsymbol{L}_\mathrm{p} = -c\boldsymbol{\theta}. \tag{8}$$

The similarity in structure of (8) and (4) means that the effect of the persistent angular momentum on the motion of the torus is similar to that of circulation on the motion of a vibrating wire. We take the components of (8) along orthogonal x and y axes in the plane of the torus

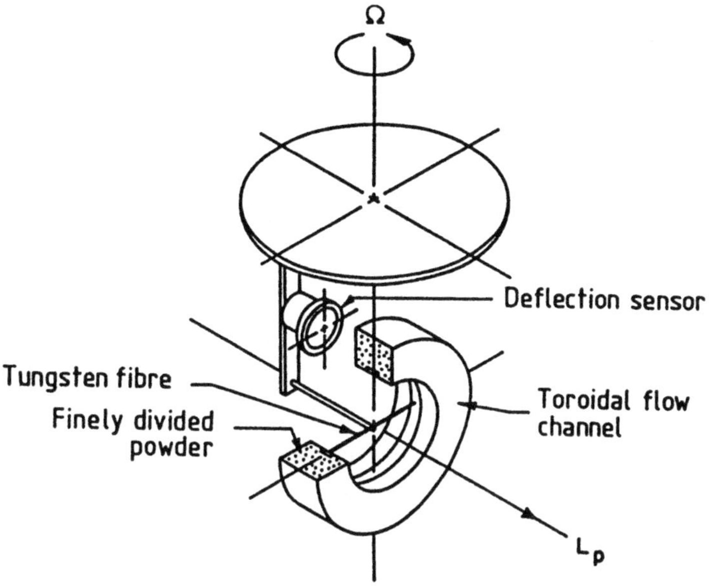

Fig. 3. DC gyroscope technique invented by John Reppy for measuring the angular momentum $\boldsymbol{L}_\mathbf{p}$ of a persistent current.

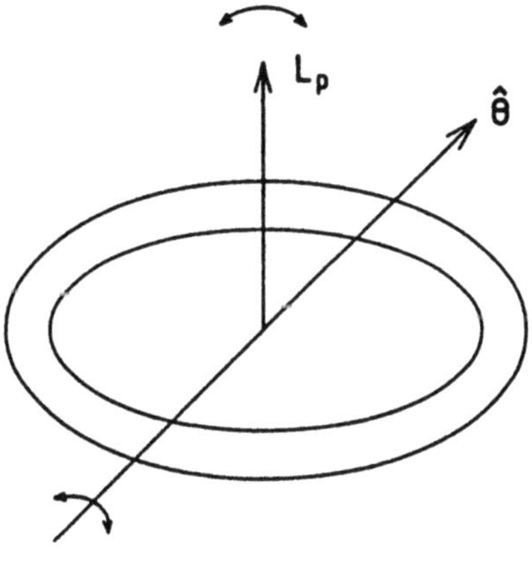

Fig. 4. Vibrations of the torus about the horizontal axis $\hat{\boldsymbol{\theta}}$ cause oscillations of the angular momentum of the persistent current.

$$\begin{aligned} I_x\ddot{\theta}_x + I_x\gamma_x\dot{\theta}_x + L_\mathrm{p}\dot{\theta}_y + c_x\theta_x &= 0, \\ I_y\ddot{\theta}_y + I_y g_y\dot{\theta}_y - L_\mathrm{p}\dot{\theta}_x + c_y\theta_y &= 0. \end{aligned} \tag{9}$$

We have allowed for asymmetries in the oscillator construction by introducing different moments of inertia (I_x and I_y) and different restoring forces ($-c_x\theta_x$ and $-c_y\theta_y$) for rotations about the x and y axes; we have also introduced damping forces ($I_x\gamma_x\dot{\theta}_x$ and $I_y\gamma_y\dot{\theta}_y$) into the equations. Although this latter step was unnecessary for the vibrating wire calculation it is necessary here since the oscillator was driven continuously rather than by an initial pulse.

In the absence of persistent angular momentum the two normal modes of the oscillator are oscillations about the x and y axes at frequencies $\omega_x = \sqrt{c_x/I_x}$ and $\omega_y = \sqrt{c_y/I_y}$ respectively; the x and y axes are of course chosen to correspond to the normal mode directions. The effect of the persistent angular momentum is to couple together the x and y motions so that the normal modes are elliptically polarized. The persistent angular momentum can therefore be detected by, for example, driving the oscillation about the x axis at the resonant frequency ω_y of the 'y' normal mode; if the oscillations about the x axis are given by $\theta_x = \theta_{x0}\exp(\mathrm{i}\omega_y t)$ then, by substitution in the second of Eqs. (9), we find that the persistent angular momentum causes a response about the y axis, $\theta_y = \theta_{y0}\exp(\mathrm{i}\omega_y t)$, where

$$\theta_{y0} = \frac{L_\mathrm{p}\theta_{x0}}{I_y\gamma_y} = \frac{Q_y L_\mathrm{p}\omega_y\theta_{x0}}{c_y}. \tag{10}$$

To write θ_{y0} in the second form we have introduced the quality factor, $Q_y = c_y/\omega_y I_y\gamma_y$, of the '$y$' normal mode; comparison with (7) and identifying $\omega_y\theta_{x0}$ as the angular velocity at which L_p is oscillated shows that the sensitivity of the ac gyroscope is enhanced over that of the dc gyroscope by the quality factor Q_y, which can be very large.

The oscillator used in the Cornell experiments is shown in Fig. 5. The restoring force for oscillation about the x and y axes is provided by bending of the thin walled tube on which the torus is mounted; this tube also acts as the fill line for the cell. The centre of mass of the torus must be situated at the mid point of the tube if the modes are to be pure rotations. The motions were driven and detected electrostatically. The frequency of the x and y modes was about 321 Hz and by careful preparation of the oscillator the frequency difference between them was reduced to a mere 0.15 Hz; this enhanced the sensitivity even further since the 'x' oscillation was also being driven at a frequency close to resonance and hence only a small drive was required to produce a large oscillation amplitude θ_{x0}. The oscillator used in the Helsinki experiment was very different in detail although the principle of the measurement was the same; in particular the symmetry of the Helsinki oscillator was such that ω_x and ω_y were very different. Although this meant that the Helsinki apparatus was less sensitive, the sensitivity was more stable since it was less affected by small drifts in the normal mode frequencies.

Although (10) provides a method for determining L_p it is not used directly since it is difficult to measure the values of θ_{x0} and θ_{y0}; it is changes in inter-electrode capacitances that are actually measured and the sensitivity factors relating these to the actual displacement of the electrodes may be different for the x and y axes. Fortunately nature provides a convenient method for calibrating the measurement. Measurements of the coupling between the modes are made with the cryostat

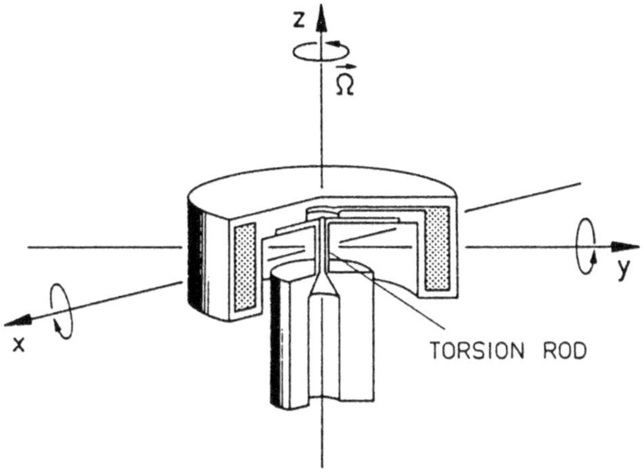

Fig. 5. The Cornell ac gyroscope.

rotating about a vertical axis at an angular velocity Ω. In this case coupling due to a persistent current is completely dwarfed by coupling produced by the Coriolis force $2\delta m \boldsymbol{v} \times \boldsymbol{\Omega}$ acting on the elements δm of the oscillator. Calculation of the oscillation amplitude $\theta_{y\mathrm{c}}$ due to the Coriolis force gives

$$\theta_{y\mathrm{c}} = \frac{(I_z - I_x)\Omega\theta_{x0}}{I_y\gamma_y}, \tag{11}$$

where I_z is the moment of inertia for rotations about the vertical axis. Division of (10) by (11) gives

$$\frac{\theta_{y0}}{\theta_{y\mathrm{c}}} = \frac{L_\mathrm{p}}{(I_z - I_x)\Omega}, \tag{12}$$

in which only the ratio $\theta_{y0}/\theta_{y\mathrm{c}}$ appears so that L_p can be determined from easily measured quantities.

Persistent currents were created by rotating the cryostat about a vertical axis. The persistent angular momentum was measured in the Helsinki apparatus with the cryostat at rest following a period of rotation at a "preparation" angular velocity Ω_p. L_p is shown as a function of Ω_p in Fig. 6 for ^{3}He–B. We now explain this behaviour. Rotation of the cryostat at angular frequency Ω causes the normal fluid component to undergo solid body rotation, $v_\mathrm{n} = \Omega R$, where R is the radius of the torus; for simplicity we ignore the finite width of the channel and assume that R has a unique value. Even a perfectly irrotational superfluid component ($\mathrm{curl}\boldsymbol{v}_\mathrm{s} = 0$) does not remain at rest but is dragged along by the geometrical effect of the fine powder with which the channel is packed in order to increase the critical velocity. We can write the averaged superfluid velocity

$$\bar{v}_\mathrm{s} = \chi v_\mathrm{n} = \chi\Omega R, \tag{13}$$

where χ, the "dragging factor", can in principle be calculated by solving the equations for the motion of an ideal fluid within the complicated geometry of the powder-packed channel. For small Ω, up to $\Omega_p = \Omega_c$ in Fig. 6, this situation is reversible so that there is no persistent current when the cryostat is brought to rest. With increasing Ω, the counterflow velocity $v_n - \bar{v}_s$ eventually reaches a critical value v_c at which the superfluid component ceases to behave as an ideal fluid; this is the point $\Omega_p = \Omega_c$ in Fig. 6. At this point

$$v_c = v_n - \bar{v}_s = (1 - \chi)R\Omega_c. \tag{14}$$

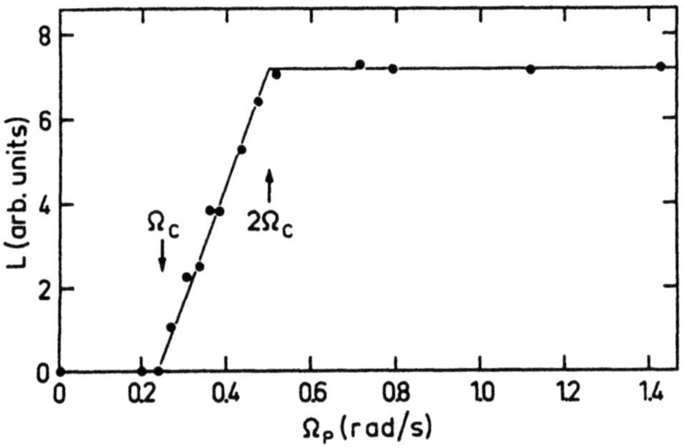

Fig. 6. Persistent angular momentum L as a function of Ω_p for ^{3}He–B at 8 bars and $T/T_c = 0.6$.

As Ω is increased above Ω_c the superfluid velocity increases so that the counterflow velocity remains at its critical value to a good approximation. Thus at $\Omega = \Omega_p$,

$$\bar{v}_s^{\max} = R\Omega_p - v_c. \tag{15}$$

When Ω is reduced the counterflow velocity becomes sub-critical and the superfluid component exhibits ideal behaviour. Thus, from (13),

$$\Delta\bar{v}_s = \chi\Delta v_n. \tag{16}$$

When the cryostat is brought to rest therefore there will be a persistent superfluid velocity,

$$v_{sp} = \bar{v}_s^{\max} + \Delta\bar{v}_s, \tag{17}$$

given by (15) and (16), corresponding to a persistent angular momentum

$$L_p = \rho_s V_L R v_{sp} = \rho_s V_L R^2 (1 - \chi)(\Omega_p - \Omega_c), \tag{18}$$

where V_L is the volume of liquid in the channel. Equation (18) explains why L_p varies linearly with $(\Omega_p - \Omega_c)$ above $\Omega_p = \Omega_c$ and shows that an experimental value for the dragging factor χ can be obtained from the slope of the graph.For values of $\Omega_p > 2\Omega_c$ the value of $|v_{sp}|$ predicted by (17) exceeds v_c. This of course cannot happen and the persistent angular momentum reaches the limiting value

$$L_p^{max} = \rho_s V_L R v_c$$

for $\Omega_p > 2\Omega_c$ as is observed on Fig. 6.

Fig. 7 shows measured values of the persistent angular momentum as a function of temperature for ^{3}He–B for two different values of Ω_p and two different magnetic fields. A discontinuity is apparent in the values of L_p at a reduced temperature of about 0.62. We will return to this discontinuity in Sect. 5.

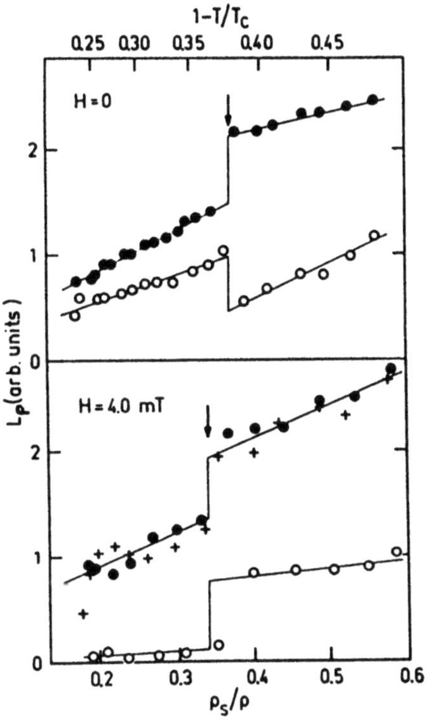

Fig. 7. Persistent angular momentum L_p in ^{3}He–B as function of ρ_s/ρ (or temperature) at 23 bars at the magnetic fields indicated. The preparation angular velocities were $\Omega_p = 0.40$ rad/s (open circles) and $\Omega_p = 0.86$ rad/s (filled circles and crosses).

No persistent currents were observed in the A phase by the method we have described above but an alternative method was used successfully to observe persistent currents in both phases at Cornell. This method exploited the fact that the Cornell oscillator could undergo torsional oscillations about a vertical axis in addition to the oscillations about the x and y axes that we have already discussed. Gammel et al.(1985) observed the torsional oscillations with the cryostat rotating and discovered that the quality factor decreased when the cryostat was rotated in either direction for both the A and B phases as shown in Fig. 8(a). As we have explained above the effect of the rotation is to produce a finite dc counterflow $v_n - \bar{v}_s$ so that this behaviour indicates that the dissipation of the ac torsional oscillations is increased by the presence of the dc counterflow. In the A phase the extra dissipation can be associated with the motion of the orbital vector $\boldsymbol{l}$; dissipation from this source had earlier been observed in a torsional oscillator by Gay et al. (1983). The existence of the effect in the B phase was quite unexpected and is still not understood. If there is a persistent current present in the torus then the counterflow is non-zero with the cryostat stationary; the minimum dissipation (maximum Q) then occurs with the cryostat rotating at a rate Ω_m such that the counterflow velocity is reduced to zero (Fig. 8(b)); the counterflow velocity of the persistent current can therefore be deduced. The magnitudes of the measured critical velocities for the A phase persistent currents were such as to explain why the persistent angular momentum had not been observed by the earlier method. For measurements close to T_c, where ρ_s is small, the second method has the advantage that it is the velocity of the persistent current $\bar{v}_{sp}$ rather than its angular momentum ($\propto \rho_s \bar{v}_{sp}$) that is measured.

It is worth commenting on the link between the stability of persistent currents and the quantization of circulation discussed in the previous section. Quantized states of circulation occur because they represent local minima in the energy. A persistent current corresponds to a state of quantized circulation and its decay therefore requires a transition to a state with a smaller number of quanta. The energy barrier between the corresponding local minima in energy determines the probability that the transition will occur. For T not too close to T_c and persistent currents that are not too large, the lifetime of the persistent current is effectively infinite (see Langer and Reppy 1970 for further discussion). That it is $\boldsymbol{v}_s$ that is fixed by circulation quantization rather than the angular momentum is strikingly demonstrated by cooling and warming a torus in which a persistent current has been "frozen". The persistent angular momentum increases and decreases reversibly in this process reflecting the temperature dependence of ρ_s.

4 Josephson Effects

Josephson effects are further indications that superfluids and superconductors can be described by a quantum mechanical wave function that remains phase coherent on a macroscopic length scale. I begin by reviewing briefly Josephson effects in superconductors. Consider first 2 isolated samples of a superconductor as shown in Fig. 9. If the samples are at the same temperature the magnitudes of their wave functions are equal but the phases are unrelated. Suppose that the superconductors are now connected by a weak link (sometimes called a Josephson junction) which allows Cooper pairs to pass through it from one superconductor to the other.

The weakly coupled superconductors will have minimum energy when the phases of the wave function on the two sides of the junction are equal; also, since

(a)

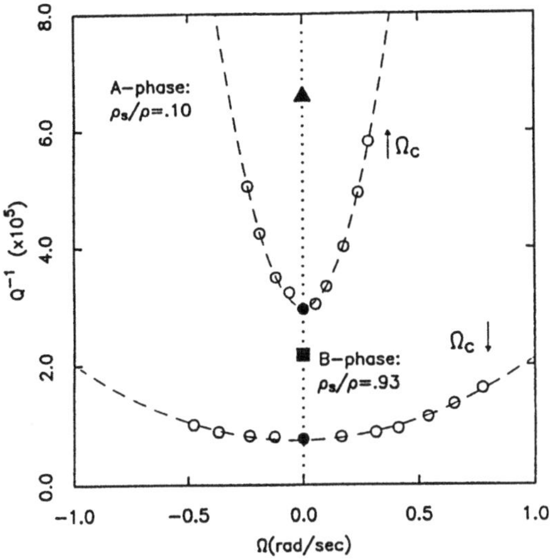

(b)

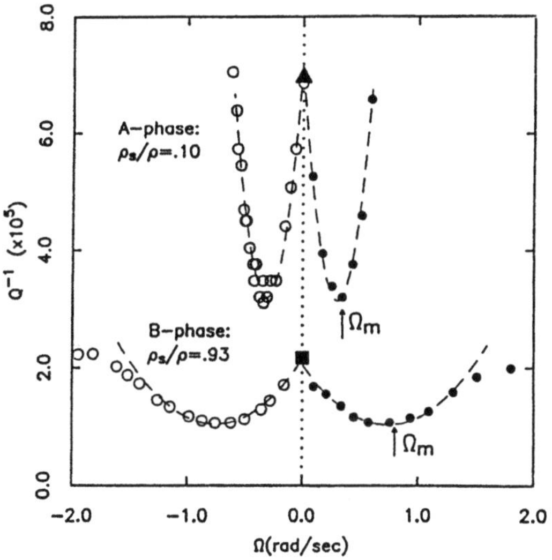

Fig. 8. (a) Dissipation Q^{-1} as a function of the cryostat angular velocity Ω for the case where no persistent current is present with the cryostat at rest. The solid triangle and solid square represent A and B phase dissipation states with large persistent currents present. The dashed lines are quadratic fits to the data.
(b) Dissipation as a function of cryostat angular velocity for the case where there is a persistent current at zero Ω. The solid (open) circles are for cases where the persistent current is created by positive (negative) rotation.

Fig. 9. In two isolated samples of a superconductor at the same temperature the amplitudes of the order parameter will be the same but the phases will be unrelated.

a change in the value of the phase by 2π leaves the wave function unchanged, the energy must be a periodic function of the phase difference $\Delta\theta = \theta_1 - \theta_2$ with period 2π. The simplest function with these properties is

$$E_\theta = -A\cos(\theta_1 - \theta_2), \tag{18}$$

where A is positive; this form of energy can be justified theoretically in the limit of a very weak link. It is possible to generate a phase difference across a weak link by passing a small current through it or by applying a potential difference across it. To allow for the presence of a potential difference we add to (18) the potential energy,

$$E_V = -2N_1eV_1 - 2N_2eV_2, \tag{19}$$

associated with the N_1 Cooper pairs in superconductor 1 at potential V_1 and the N_2 Cooper pairs in superconductor 2 with potential V_2.

The dynamics of the weakly coupled superconductors can now be deduced by using the fact that N and $\hbar\theta$ are conjugate variables, just like position and momentum, and applying Hamilton's equations of motion:

$$\begin{aligned} \frac{\partial N_1}{\partial t} &= \frac{1}{\hbar}\frac{\partial E}{\partial \theta_1} = \frac{A}{\hbar}\sin\Delta\theta, \\ \hbar\frac{\partial \theta_1}{\partial t} &= -\frac{\partial E}{\partial N_1} = 2eV_1, \end{aligned} \tag{20}$$

where E is the total energy obtained by adding (18) and (19). The first of these equations gives the current flowing through the weak link as

$$I = \frac{2eA}{\hbar}\sin\Delta\theta. \tag{21}$$

The second of Eqs. (20), on subtracting the corresponding equation for superconductor 2, gives

$$\hbar\frac{\partial \Delta\theta}{\partial t} = 2e(V_1 - V_2). \tag{22}$$

Equations (21) and (22) are the fundamental equations which describe the behaviour of the Josephson junction.

Two experimentally observable effects which follow simply from these equations are:

1. **Flow of a dc supercurrent at zero applied potential difference.** With zero applied potential difference, (22) integrates to give $\Delta\theta = \text{constant}$ and thus, from (21) a steady supercurrent of Cooper pairs occurs. In practice the value of $\Delta\theta$ adjusts itself so that the supercurrent through the weak link, as given by (21), matches the current fed in by the external circuit. Clearly the maximum current that can be matched in this way is $I_c = 2eA/\hbar$ when $\Delta\theta = \pi/2$. If a current in excess of this is passed through the junction then it will make a transition to a resistive state.
2. **Flow of ac supercurrent at finite dc potential difference.** If there is a finite constant potential difference ΔV across the junction then (22) integrates to give $\Delta\theta = 2e\Delta Vt/\hbar + \theta_0$, where θ_0 is the constant of integration. Substitution of this in (21) shows that there is an ac supercurrent of Cooper pairs $I = I_c \sin(2e\Delta Vt/\hbar + \theta_0)$ at a frequency
$$\omega_J = \frac{2e\Delta V}{\hbar}. \tag{23}$$
Measurement of the frequency of the supercurrent can be used to determine the ratio $e/\hbar$. The limitation on the accuracy of the method is the measurement of ΔV and National Standards Laboratories now use (23) to establish the voltage standard. This is achieved by defining a value of $e/\hbar$, which is of course consistent with the best known value, and then measuring ω_J to establish ΔV. The ac supercurrent occurs in parallel with the dc normal current due to the passage of excitations through the weak link at finite voltage difference.

Amongst the many types of weak link that have been successfully used in superconductivity is the Dayem Bridge (Fig. 10). Most attempts to find Josephson effects in superfluid helium have used a structure equivalent to this, namely a small hole connecting two volumes of bulk liquid; early indications of success for superfluid ^{4}He (Richards and Anderson 1965) proved to be erroneous and it was not until 1985 that Avenel and Varoquaux finally succeeded in seeing Josephson-like phenomena in superfluid ^{4}He. Successful observation for ^{3}He–B followed shortly after (Avenel and Varoquaux 1987).

The apparatus used by Avenel and Varoquaux was the superfluid analogue of the superconducting rf SQUID (Superconducting Quantum Interference Device). In the rf SQUID the weak link interrupts a superconducting ring (Fig. 11(a)), which is inductively coupled to the inductance of a parallel LC circuit (a tank circuit in the jargon). The tank circuit is fed by an ac current source at its resonant frequency (typically 20 MHz) and the amplitude of the ac voltage across the circuit is measured. Fig. 11(b) shows the detected voltage as a function of the applied current; because of the steps in the figure it is referred to as *the staircase pattern.* We discuss later the extent to which the appearance of a staircase pattern indicates the existence of Josephson effects within a weak link.

Avenel and Varoquaux's apparatus is shown schematically in Fig. 12(a); the orifice, which is the weak link, was milled using an ion beam into a nickel film of

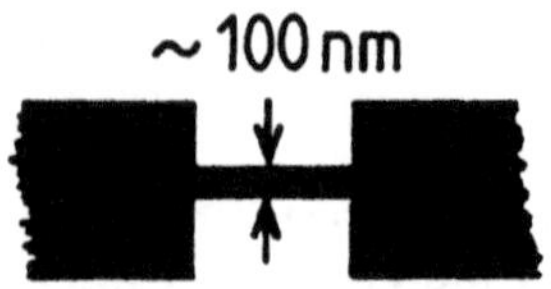

Fig. 10. The Dayem Bridge geometry. In a superconductor the length and width of the bridge are typically of order 100 nm.

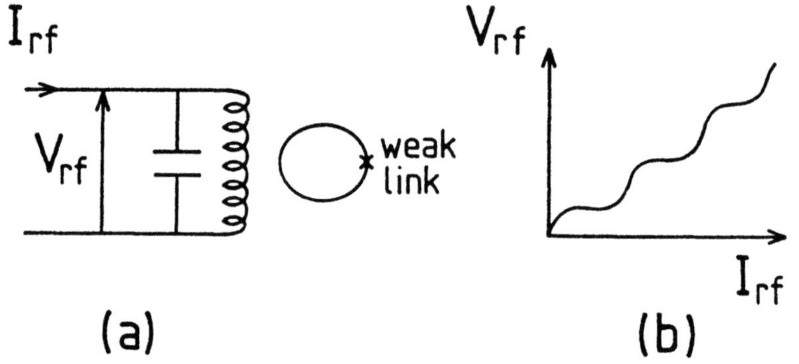

Fig. 11. (a) Parallel *LC* resonant circuit inductively linked to a superconducting ring containing a weak link. (b) Detected rf voltage as a function of the amplitude of the driving rf current showing the staircase pattern.

thickness 0.2 μm; it has the approximately rectangular cross section 0.3 μm×5 μm shown in Fig. 12(b). The long channel in Fig. 12(a) completes the superfluid ring and the topology is thus the same as that of the rf SQUID. Oscillations of the aluminized kapton diaphragm at its resonant frequency (~1–10 Hz) are excited by applying a voltage to the capacitor plate. The motion of the diaphragm is detected inductively; a persistent supercurrent I_0 through the flat coil applies a magnetic field at the diaphragm and displacement of the superconducting Al film therefore changes the flux linked with the flat coil. The changes in flux were measured using a superconducting rf SQUID; by this means a diaphragm displacement as small as 10^{-13} m could be observed.

Fig. 13 shows the amplitude of the oscillations as a function of the driving voltage with superfluid ^{4}He in the apparatus and the similarity of the staircase pattern with that for a superconducting SQUID is very striking.

I will now explain the behaviour observed for ^{4}He. Since the orifice dimensions are much bigger than the coherence length it should be possible, at least at small flow rates, to describe the superflow within the orifice by (2), $\boldsymbol{v}_s = \hbar\nabla\theta/m_4$. This is a very different relation between the current through the orifice and the phase difference across it than the Josephson equation (21); we will discuss the connection between these equations later. The analogue of the second of Eqs. (20) and thus of (22) for a neutral superfluid is readily obtained by noting that the chemical potential μ per particle fulfils the role of the Cooper pair potential energy $2eV$. Thus

$$\hbar \frac{\partial \theta}{\partial t} = -\mu, \tag{24}$$

and

$$\hbar \frac{\partial \Delta\theta}{\partial t} = -\Delta\mu. \tag{25}$$

Equations (2) and (24) together give the Landau superfluid acceleration equation

$$m_4 \frac{\partial \boldsymbol{v}_s}{\partial t} = -\nabla\mu. \tag{26}$$

The motion of the diaphragm creates a pressure difference Δp and hence a chemical potential difference $\Delta\mu/m_4 = \Delta p/\rho$ between the ends of the orifice. The resulting superfluid acceleration, according to (26), is

$$\frac{\partial \boldsymbol{v}_s}{\partial t} = -\frac{\Delta p}{\rho l}, \tag{27}$$

for an orifice of length l. The associated mass flow rate is $\rho_s A \boldsymbol{v}_s$, where A is the cross-sectional area of the orifice. To obtain the equation of motion of the diaphragm we must relate the mass flow and the pressure difference to the diaphragm displacement. A straightforward calculation shows that the displacement ψ_0 of the centre of a circular diaphragm of radius a and isotropic tension T subject to a uniform static pressure difference is $\psi_0 = \Delta p a^2/4T$ (the inertia of the diaphragm can be ignored). The volume of fluid displaced is $\Delta V = \pi a^2 \psi_0/2$; since the fluid can be regarded as incompressible under the conditions of this experiment the mass displaced by the diaphragm $\rho\Delta V$ must flow through the long channel and orifice in parallel.

Putting these results together gives

$$\frac{\partial^2 \psi_0}{\partial t^2} + \omega_H^2 \psi_0 = 0 \tag{28}$$

for the equation of motion of the diaphragm, where

$$\omega_H^2 = \frac{8T\rho_s}{\rho^2 \pi a^4}\left(\frac{A_1}{l_1} + \frac{A_2}{l_2}\right);$$

ω_H is the resonant frequency and A_1 (A_2) and l_1 (l_2) are the length and area of the orifice (long channel). No dissipation has been included in (28) and the experimental results of Fig. 13 indicate that the damping is very small before the first step of the staircase pattern is reached. The first step indicates the onset of a dissipation mechanism which limits the oscillation amplitude at an almost constant value for a range of driving force. The onset is associated with the flow velocity exceeding a critical value within the orifice. That a critical value is reached within

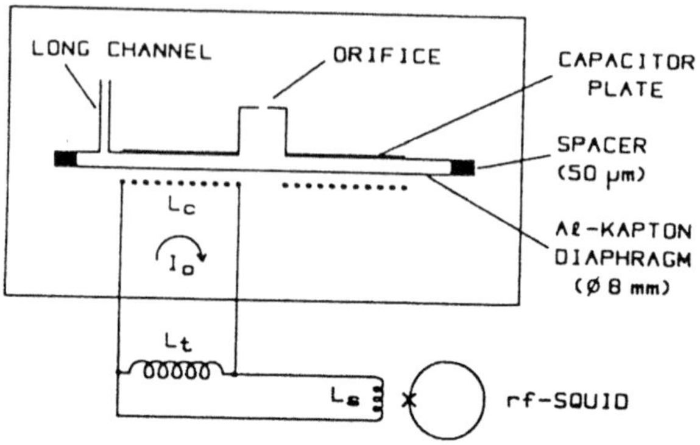

(a)

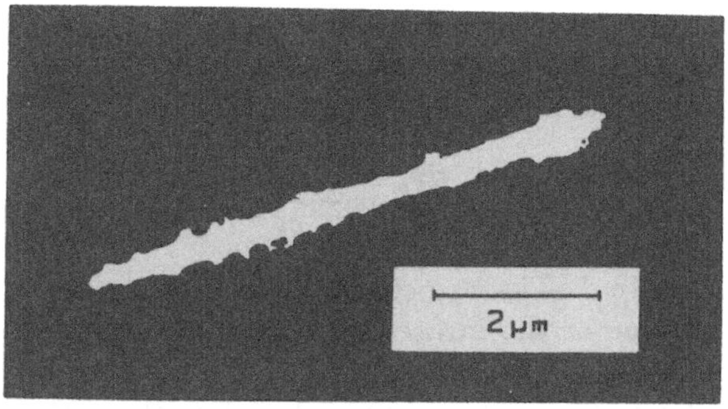

(b)

Fig. 12. (a) Schematic diagram of the experimental cell used by Avenel and Varoquaux. (b) Micrograph obtained from a transmission electron microscope of the aperture.

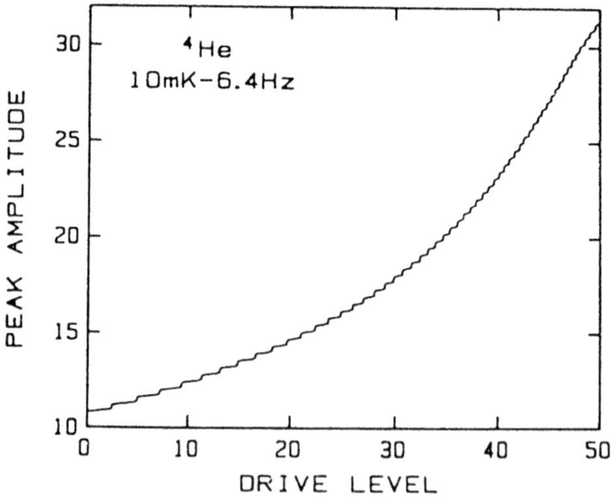

Fig. 13. A staircase pattern for superfluid ^{4}He, showing more than 50 steps.

the orifice before the long channel can be seen by comparing the flow velocity amplitudes, v_{s1} and v_{s2}, predicted by (27)

$$\frac{v_{s1}}{v_{s2}} = \frac{l_2}{l_1}; \tag{29}$$

the velocity is thus higher in the shorter channel.

Further information concerning the nature of the dissipation is revealed in Fig. 14 where the peak amplitude of oscillation at constant drive voltage is plotted as a function of time. The discontinuities indicate that whenever the critical flow velocity is exceeded a dissipative event occurs in which the same amount of energy is always extracted from the oscillator; although there is a small random variation in the amplitude at which the dissipation events occur the decrease in amplitude at each event is always the same. The measured loss of energy at each event is 1.21×10^{-17} J.

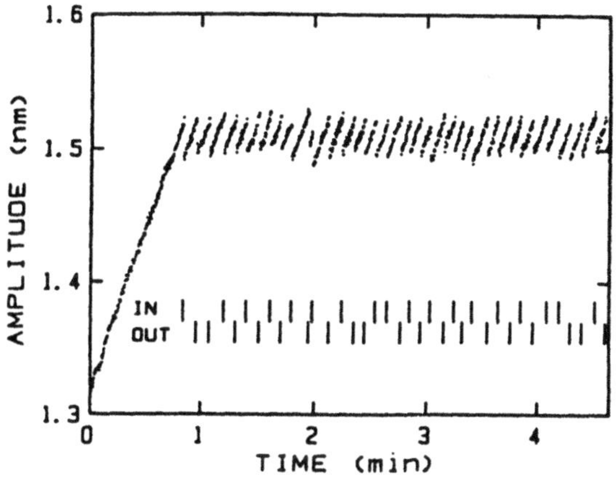

Fig. 14. Peak amplitude of the membrane oscillation as a function of time for a constant drive power. Each dot represents a measurement and the time between dots is half a period. The lines labelled 'in' and 'out' indicate whether the jump occurs with fluid flowing in or out of the cell.

The effect of changing the drive level is shown in Fig. 15. The average amplitude of oscillation is the same in each case thus explaining the horizontal portion of the first step of the staircase pattern on Fig. 13; the dissipative events just occur more frequently at the higher drive levels on this step. The end of the first step occurs when the drive is big enough to cause one dissipative event per cycle. The second step corresponds to between one and two dissipative events per cycle and so on.

The constancy of the energy change in the dissipative events implies that some kind of quantization condition is important and this turns out to be the quantization of circulation (see Sect. 2). Thus when the critical value v_c is exceeded the velocity within the orifice decreases rapidly (on the time scale of the oscillation) by $\Delta v_{s1} = h/m_4 l_1$, corresponding to change in circulation of one quantum around

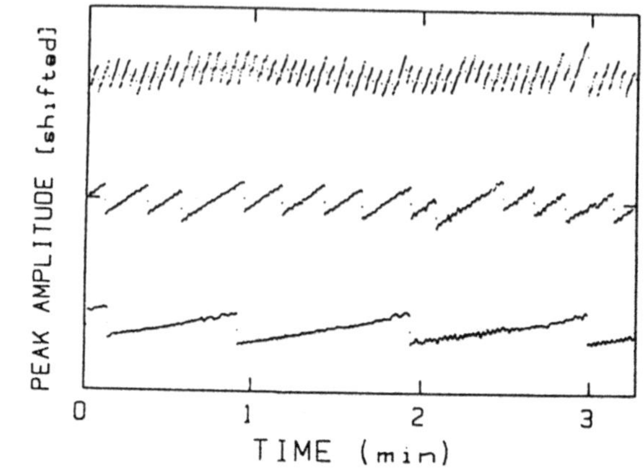

Fig. 15. Peak amplitude jumps, as in Fig. 14, at three different drive powers.

a curve threading the superfluid ring. The associated change in the kinetic energy of the fluid in the orifice is (for $|\Delta v_{s1}| \ll v_c$)

$$\Delta E = \frac{\rho_s A_1 v_c h}{m_4};$$

this value of ΔE agrees with the measured energy loss at each dissipative event.

A possible mechanism for a change in circulation by one quantum is the passage of a singly quantized vortex line (see Sect. 5) across the full cross-section of the orifice. In such a process the phase difference $\Delta\theta$ between the ends of the orifice decreases by 2π; the values of the superfluid order parameter at the ends of the orifice are unchanged by 2π phase slips of this kind so that the rest of the superfluid ring is unaware that anything has happened. By assuming the scenario that we have outlined Avenel and Varoquaux were able to fit the entire behaviour of their oscillator to within the experimental uncertainty. The critical velocity for the staircase pattern of Fig. 13 corresponds to a phase difference of order 52π across the orifice, indicating that the change in velocity at a phase slip event is indeed small compared to the critical velocity.

The successful explanation of the staircase pattern using the linear current-phase relation (2) demonstrates that the sinusoidal current-phase relation (21) is not necessary to produce this pattern. Aslamasow and Larkin (1969) used Ginzburg-Landau theory to investigate the current-phase relation of a superconducting Dayem bridge. Departures from (2) are caused by depression of the magnitude of the superconducting wave function within the bridge by the high current density there; such effects are larger for smaller bridges and become important when the dimensions of the bridge are comparable to the temperature-dependent Ginzburg-Landau coherence length. In the limit of a very small bridge the current-phase relation tends towards the "Josephson form" (21).

The way in which a smooth transition is achieved between the two very different relations (2) and (21) is illustrated in Fig. 16. Fig. 16(a) shows that (2) leads to a multi-branched current-phase relation because the superfluid wave function at either end of the channel is unchanged by a change in phase of 2π; the vertical arrow between two of the branches illustrates the 2π phase slip events needed to explain the ^{4}He experiments of Avenel and Varoquaux. Fig. 16(b) shows the sinusoidal current-phase relation (21). Figs. 16(c) and 16(d) show the intermediate current-phase relations that interpolate between these two extremes. As the size of an initially very small bridge increases the sinusoidal phase relation distorts in the manner shown in Fig 16(c) and eventually becomes multivalued as indicated in Fig. 16(d). The dashed sections of the multivalued curves are unstable and are therefore not included in the extreme limit of a large bridge as shown in Fig. 16(a).

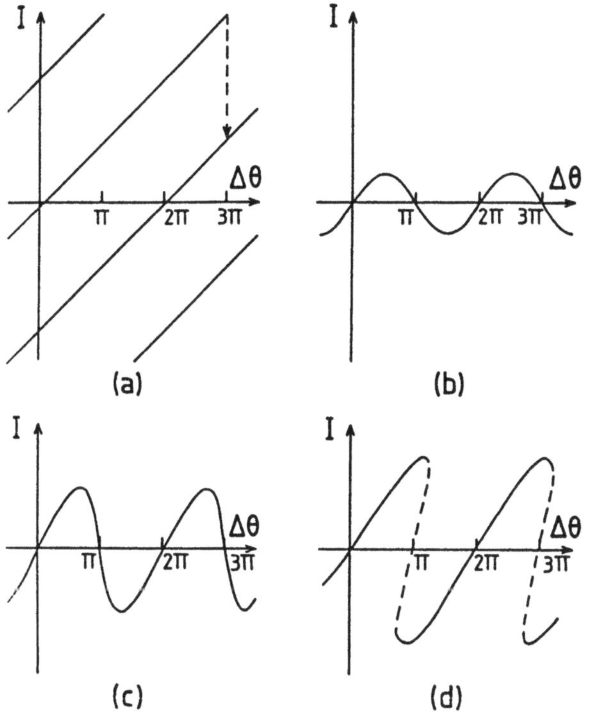

Fig. 16. Explanation of how a linear-current phase relation (a) transforms smoothly into a sinusoidal current phase relation (b). The intermediate stages are shown in (c) and (d).

The orifice used in the Avenel-Varoquaux experiments is a large Dayem bridge for ^{4}He since the coherence length is of order atomic dimensions; this is not so for ^{3}He where the coherence length is typically hundreds of angstroms. A staircase pattern observed for ^{3}He–B in the same apparatus is shown in Fig. 17; at first sight it is qualitatively very similar to that for ^{4}He.

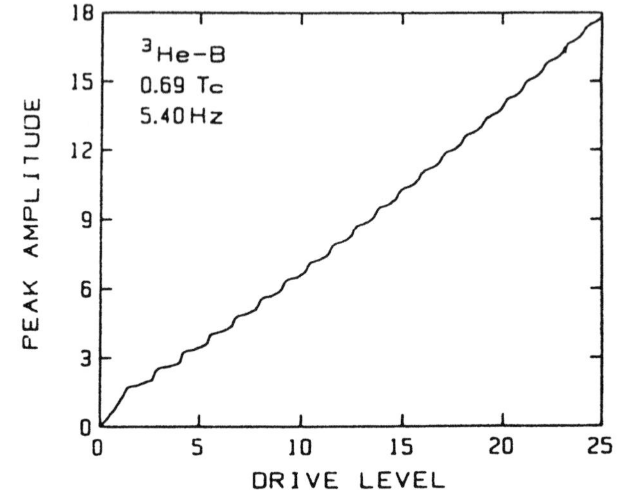

Fig. 17. Staircase pattern for ^{3}He-B showing more than 20 steps.

It is interesting to discuss the extent to which the experimental results can be used to investigate the current-phase relation for ^{3}He–B. For an arbitrary current-phase relation, $I = I_c f(\Delta\theta)$, the equations describing the dynamics of the diaphragm for ^{3}He–B are:

$$\frac{\pi a^2 \rho}{2}\frac{\partial \psi_0}{\partial t} = I_c f(\Delta\theta) + \frac{\rho_s A_2 \hbar \Delta\theta}{2m_3 l_2}, \tag{30}$$

$$\frac{\hbar}{2m_3}\frac{\partial \Delta\theta}{\partial t} = -\Delta\mu = -\frac{\Delta p}{\rho}, \tag{31}$$

$$\frac{4T\psi_0}{a^2} + \gamma\frac{\partial \psi_0}{\partial t} = \Delta p + P_0 \cos\omega t. \tag{32}$$

Equation (30) states that the rate at which fluid mass is displaced by the diaphragm equals the rate of flow through the orifice and long channel in parallel; (2) is used to obtain the flow through the long channel. Equation (31) is obtained from (25). Equation (32) expresses the balance of forces on the diaphragm; $P_0 \cos\omega t$ is the driving force and the second term on the left hand side represents the damping of the oscillations. The intrinsic damping in ^{3}He–B is much bigger than in ^{4}He as is apparent from comparing the slopes of Figs. 13 and 17 at drive levels below that at which the first step occurs.

Avenel and Varoquaux assumed a current-phase relation for ^{3}He–B of the form

$$I = I_c \sin\zeta, \tag{33}$$

where ζ is related to the phase difference $\Delta\theta$ by

$$\zeta + \alpha \sin\zeta = \Delta\theta. \tag{34}$$

This model form of current-phase relation, with a suitable choice of the parameter α fits curves of the form of Fig. 16; $\alpha = 0$ corresponds to the Josephson limit (21), $\alpha = \infty$ corresponds to the large bridge limit (2) and at $\alpha = 1$ the current-phase relation first becomes multivalued. By solving (30)–(32) numerically with the current-phase relation (33) Avenel and Varoquaux were able to explain the observed motions of their diaphragm for ^{3}He–B at not too low reduced temperature. Fig. 18 shows a typical fit to the staircase pattern obtained by using a value of α of 0.5 at $T = 0.89T_c$; calculated curves for $\alpha = 0$ and $\alpha = 1$ are shown for comparison. The current-phase relation thus seems to be closer to (21) than to (2) for ^{3}He–B at this temperature. With decreasing temperature the value of α required to fit the experimental data increases; this is what one should expect since the effective size of the orifice increases as the temperature-dependent coherence length decreases. For temperatures below 0.7 mK the behaviour is not understood and this may reflect the transition to a multivalued current-phase relation.

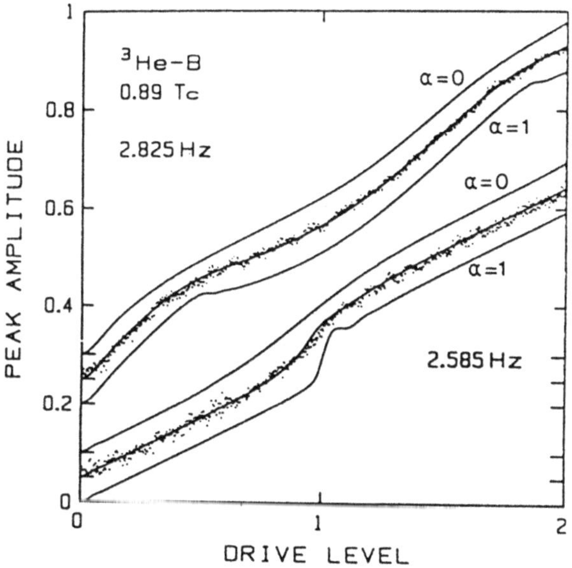

Fig. 18. Fits to the first step of the staircase pattern for $\alpha = 0.5$ (superposed) and $\alpha = 0$ and 1 (shifted).

Monien and Tewordt (1987) have performed calculations, using Ginzburg-Landau equations, of the current-phase relation in various Dayem bridge geometries for ^{3}He–B. There are two major differences to the superconductivity case: it is necessary to take account of the full complexity of the order parameter (see Sect. 5) and of the depression of the order parameter at the surfaces of the orifice. An example of the rather complicated results of Monien and Tewordt is shown in Fig. 19. The vertical arrow indicates the possible behaviour as the phase difference increases into a region where there are no solutions on the same branch of the current-phase relation; the time-independent equations used by Monien and Tewordt could not be used to predict such time-dependent phenomena. Kopnin and Salomaa (1990) have considered time-dependent phase slip phenomena but using a highly simplified approach. An attempt to consider both space and time-dependent phenomena in sufficient generality for direct comparison with experiment has yet to be made.

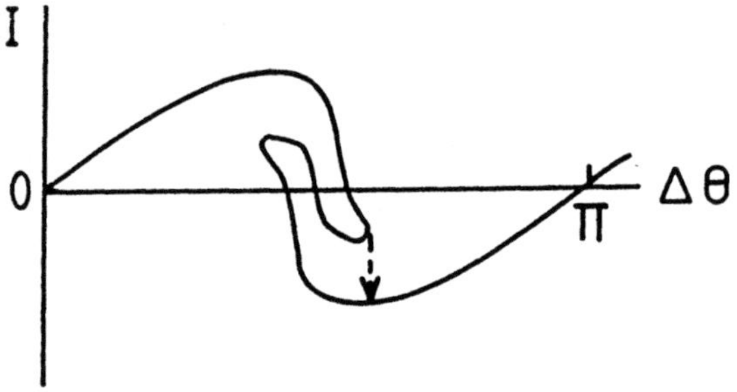

Fig. 19. Current-phase relation calculated by Monien and Tewordt for an orifice of rectangular cross-section $7\xi \times 175\xi$ and length 7ξ.

5 Rotation of Superfluid ^{3}He–B

5.1 Quantized Vortex Lines

The equilibrium state of a normal liquid in a bucket rotating with angular frequency Ω is solid-body rotation with the fluid velocity v equal to Ωr at distance r from the rotation axis. This is apparent from the shape of the meniscus, determined by constancy of the gravitational plus centrifugal potential at the liquid surface (normally surface tension can be ignored). Thus

$$gh - \frac{1}{2}\Omega^2 r^2 = \text{constant}, \tag{35}$$

giving a parobolic meniscus. In superfluid ^{4}He, the superfluid velocity satisfies (2) and hence is irrotational,

$$\text{curl}\boldsymbol{v}_s = 0. \tag{36}$$

Solid-body rotation is thus impossible since it would correspond to $\text{curl}\boldsymbol{v}_s = 2\Omega$. The only solution for $\boldsymbol{v}_s$ that is irrotational everywhere inside a rotating cylindrically symmetric bucket is $\boldsymbol{v}_s = 0$ which would give a horizontal meniscus at $T = 0\,\text{K}$ where the normal fluid component goes to zero. Nevertheless when a bucket of superfluid ^{4}He is rotated the meniscus adopts the same shape to a good approximation as for a normal liquid.

Closer inspection shows that the superfluid simulates solid-body rotation by having an array of vortex lines threading it (Fig. 20). The circulation κ around each line is quantized according to (3) and the total circulation surrounding N lines is $N\kappa$; to simulate solid-body rotation within a cylinder of radius R requires $\boldsymbol{v}_s = \Omega R$ at the circumference and hence there must be N vortex lines within it, where

$$\oint \boldsymbol{v}_s . dr = \Omega R . 2\pi R = N\kappa .$$

The vortex density n_v is thus

$$n_v = \frac{N}{\pi R^2} = \frac{2\Omega}{\kappa}. \tag{37}$$

For singly quantized vortex lines ($\kappa = h/m_4$) and for a rotation rate of 1 radian s^{-1} about 2000 vortex lines cm^{-2} are required. In equilibrium the vortices rotate with the container (see Ho and Mermin 1980).

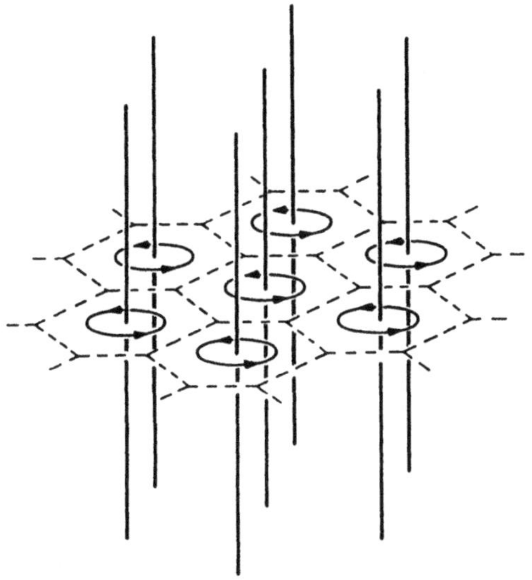

Fig. 20. Hexagonal array of quantized vortex lines.

It will facilitate our consideration of the behaviour of superfluid ^{3}He–B under rotation if we discuss in more detail the structure of an isolated vortex line in ^{4}He. We use cylindrical polar coordinates (r, φ, z). We assume that the vortex line is at $r = 0$, that it is cylindrically symmetric and that $\boldsymbol{v}_s = \hbar \nabla \theta$ everywhere except at $r = 0$. The superfluid velocity can then be determined in terms of the circulation $\kappa = nh/m_4$ around the vortex using

$$\oint \boldsymbol{v}_s . d\boldsymbol{r} = v_s 2\pi r = \frac{nh}{m_4}.$$

Thus

$$v_s = \hat{\varphi}\frac{n\hbar}{m_4 r}. \tag{38}$$

The requirement of Stokes' theorem that $\oint v_s.dr = \int \mathrm{curl}v_s.dS$ is then satisfied by assuming a δ-function of $\mathrm{curl}v_s$ at $r = 0$, namely

$$\mathrm{curl}v_s = \frac{nh}{m_4}\delta(r). \tag{39}$$

It is clear from (38) and (39) that the vortex axis $r = 0$ is a rather special place. The divergence of the superfluid velocity implies an infinite kinetic energy density ($\frac{1}{2}\rho_s v_s^2$) at this point and also a multivalued wave function since the phase θ must take all values between 0 and $2\pi n$ there. These problems are avoided if the magnitude of the wave function $|\Psi|$ goes to zero at $r = 0$, i.e. if the vortex acquires a normal core, $\rho_s \to 0$ at $r = 0$ (Fig. 21). The radius r_c of the core is determined by the kinetic energy density of the superflow becoming comparable with the condensation energy that stabilises the superfluid state relative to the normal state. This means that r_c is comparable to the coherence length and thus of order an atomic spacing in ^4He. As we have already mentioned in Sect. 4, the coherence length in ^3He is much longer and the structure of the core can therefore be explored in much greater detail.

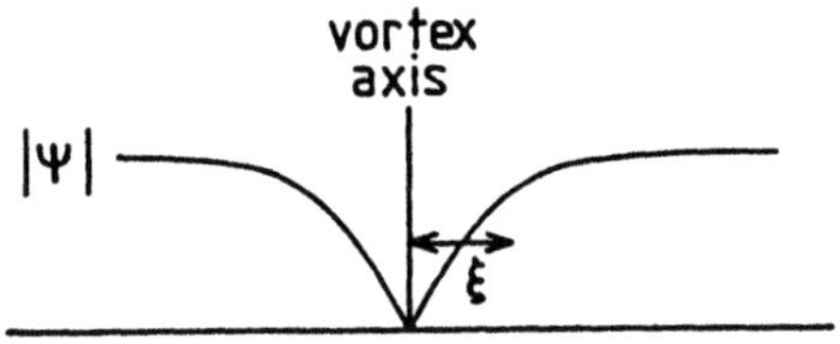

Fig. 21. The magnitude of the superconducting wave function goes to zero along the axis of a quantized vortex.

To see why singly quantized vortex lines are energetically most favourable it is necessary to calculate the energy per unit length of the vortex line. To a good approximation this is given by the kinetic energy

$$\epsilon_v = \oint \frac{1}{2}\rho_s v_s^2 2\pi r dr = \frac{\pi \rho_s n^2 \hbar^2}{m_4^2} \ln\left(\frac{r_{\mathrm{max}}}{r_c}\right) \tag{40}$$

of the circulating superflow. We have used (38) and have inserted maximum and minimum values, r_{max} and r_c of r to avoid the logarithmic divergence of the integral. The upper cut-off is provided by the radius of the container or the vortex line spacing; the core radius provides the lower cut-off. From (40) the energy of a single vortex varies as n^2 and from (37) the vortex density for a given rotation rate varies as n^{-1}; the energy of the vortex lattice as a whole is thus proportional to n and singly quantized lines are thus favoured.

So far in my lectures, I have been able to interpret the experimental properties of superfluid ^{3}He by using a wave function which is a complex scalar function of position. The wave function (more usually referred to as the order parameter) of ^{3}He is in reality more complicated than this. To understand the nature of quantized vortex lines in ^{3}He–B and also the behaviour of ^{3}He–A described in Sects. 6 and 7 we will need to know about these complications. In the next subsection we digress briefly from the main subject of Sect. 5 in order to describe in more detail the order parameter of the A and B phases of superfluid ^{3}He.

5.2 The Order Parameter of ^{3}He–A and ^{3}He–B

It will be adequate for our purposes to regard the order parameter of superfluid ^{3}He as the wave function of the Cooper pairs. As such it is a function of two position variables, which we will take to be the centre of mass coordinate $\boldsymbol{r} = (\boldsymbol{r}_1 + \boldsymbol{r}_2)/2$ and the relative coordinate $\boldsymbol{\rho} = \boldsymbol{r}_1 - \boldsymbol{r}_2$ of the atoms forming the pair. The pairing is into a state with orbital angular momentum quantum number 1. In constructing the order parameter we can therefore use the three $L = 1$ spherical harmonics:

$$Y_{1\,1} \propto \sin\theta \exp(\mathrm{i}\varphi), \quad Y_{1\,0} \propto \cos\theta \quad \text{and} \quad Y_{1\,-1} \propto \sin\theta \exp(-\mathrm{i}\varphi), \tag{41}$$

where θ and φ are the spherical polar coordinates of the direction of $\boldsymbol{\rho}$. For a pair wave function with an odd orbital angular momentum quantum number to be antisymmetric under interchange of the particles only the eigenstates of spin corresponding to $S = 1$ can be used; thus we will construct the pair wave function by using the above spherical harmonics in conjunction with the three spin triplet states:

$$\chi_{1\,1} = |\uparrow\uparrow\rangle, \quad \chi_{1\,0} = |\uparrow\downarrow + \downarrow\uparrow\rangle \quad \text{and} \quad \chi_{1\,-1} = |\downarrow\downarrow\rangle. \tag{42}$$

It is convenient for us to assume that the coordinate axes used to define the orbital states (41) and the spin states (42) are differently aligned as shown in Fig. 22.

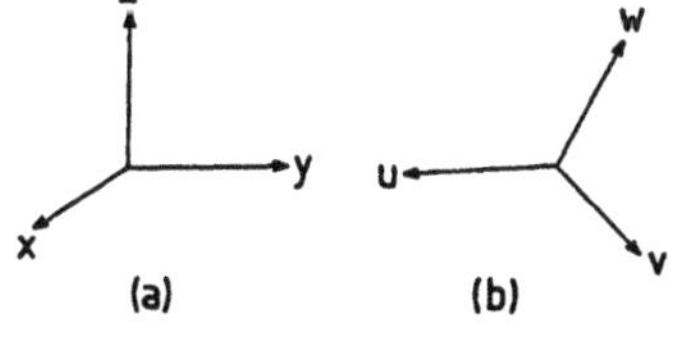

Fig. 22. (a) orbital axes and (b) spin axes used to define the orbital and spin states, (41) and (42) respectively, of the Cooper pair.The spin quantisation axis is w.

For a spatially homogeneous superfluid with no superflow the order parameter does not depend on the centre of mass coordinate $\boldsymbol{r}$ and can therefore, for any $L = 1$ state, be written as

$$\Psi = \sum_{\nu=1}^{3} \sum_{\mu=1}^{3} a_{\nu\mu} \chi_{1\nu} Y_{1\mu}, \tag{43}$$

where the $a_{\nu\mu}$ are constant coefficients. For the B phase we have

$$a_{1\,-1} = a_{0\,0} = a_{-1\,1}, \tag{44}$$

and all the other $a_{\nu\mu}$'s are zero. If the spin and orbital axes of Fig. 22 were aligned this would correspond to the spherically symmetric $J = 0$ combination of angular momentum eigenfunctions. The B phase order parameter is completely specified by specifying the rotation matrix $\boldsymbol{R}$ which relates the two sets of axes. The magnetic interaction between the nuclear dipole moments is minimized if the axes are rotated with respect to each other by an angle $\cos^{-1}(-1/4) \simeq 104°$. The axis of this rotation can point in any direction; the direction is normally specified by a unit vector $\boldsymbol{n}$. The existence of this axis breaks the spherical symmetry of the B phase and gives it liquid-crystal-like properties; however, many of the properties, including the energy gap, remain isotropic to a very good approximation.

For the A phase the only non-zero $a_{\nu\mu}$ in (43) is $a_{0\,1}$ and the order parameter takes the form

$$\Psi = Y_{1\,1}\chi_{1\,0} = Y_{11}|\uparrow\downarrow + \downarrow\uparrow\rangle. \tag{45}$$

The quantum numbers of the pair, with reference to Fig. 22, are thus $L_z = 1$, $S_w = 0$. The A phase order parameter is therefore completely specified by giving the directions of the x, y and z axes in Fig. 22 (we use unit vectors $\boldsymbol{m}$, $\boldsymbol{n}$ and $\boldsymbol{l}$ for this purpose) and the direction of the w axis (we use the unit vector $\boldsymbol{d}$ for this purpose) along which the spin definitely has component zero. The energy gap of the A phase has a $\sin^2\theta$ dependence reflecting the angular dependence of $|\Psi|^2 \propto |Y_{1\,1}|^2$. The magnetic interaction between the nuclear moments is minimised for the order parameter (45) by aligning the x, y, z axes with the u, v, w axes. Note that sometimes the order parameter is written in the form

$$\Psi = Y_{1\,1}\frac{1}{\sqrt{2}}(|\uparrow\uparrow\rangle - |\downarrow\downarrow\rangle).$$

This is just the form (45) takes if the u axis rather than the w axis is used as the spin quantisation axis.

5.3 Vortices in ^{3}He–B

Superflow corresponds to centre of mass motion of the Cooper pairs and is therefore described in ^{3}He–B by an order parameter obtained from that given by (43) and (44) by multiplication by a phase factor $\exp(i\theta)$, where the phase angle θ depends on the centre of mass coordinate $\boldsymbol{r}$. The superfluid velocity is then given by (2) with the mass M of the superfluid particles equal to $2m_3$; thus

$$\boldsymbol{v}_s = \frac{\hbar}{2m_3}\nabla\theta. \tag{46}$$

The similarity of (46) to $\boldsymbol{v}_s = \hbar\nabla\theta/m_4$ for ^{4}He implies that the behaviour of ^{3}He–B under rotation is very similar to that of ^{4}He. The discussion of Sect. 5.1 can be taken over merely by replacing m_4 by $2m_3$; thus, for example, ^{3}He–B simulates solid body rotation by the formation of a lattice of singly quantised vortices each with circulation $h/2m_3$. As already mentioned in Sect. 5.1 the vortex core is much larger in ^{3}He–B and it differs from the core of a ^{4}He vortex in other important ways.

Outside the vortex core the order parameter, apart from the phase factor associated with the superflow, has the form appropriate to the bulk homogeneous superfluid; thus in ^{4}He the modulus of the order parameter is constant and in ^{3}He–B the $a_{\nu\mu}$'s take the form given by (44). The size of the core is determined by comparability of the kinetic energy density of the circulating superflow with the condensation energy. When this occurs the order parameter is forced to vary from its 'bulk' form. In ^{4}He the only degree of freedom is the modulus of the order parameter, which is depressed to zero at the centre of the vortex (Fig.21). In ^{3}He the order parameter is described by the 9 complex amplitudes $a_{\nu\mu}$; these can depart from their 'bulk' B-phase values in various ways to give cores with different symmetries. Salomaa and Volovik (1983, 1985) show that, even if only vortices of cylindrical symmetry are considered, there are still five possible types of B-phase vortex core; in order to explain the experimental results it is necessary also to consider non-cylindrically symmetric vortices.

Since the order parameter differs from the B phase form of (44) within the core, (46) need no longer hold there. Some $L = 1$ phases of ^{3}He have curl$\boldsymbol{v}_s \neq 0$ and this allows the intriguing possibility of a vortex with a superfluid core; in such a vortex there would be no line along which curl$\boldsymbol{v}_s$ has a δ-function singularity of the form of (39) and along which all the components $a_{\mu\nu}$ must vanish. Instead curl$\boldsymbol{v}_s$ remains finite throughout the core in such a way that Stokes' theorem gives the correct quantised circulation around the vortex. An example of a vortex with a superfluid core is shown in Fig. 23; the c coefficients plotted here give the radial dependence of the a coefficients of (43) with $c_{++} \propto a_{11}$, etc. At large distances from the centre of the vortex the order parameter tends to the bulk B phase form of (44), $c_{+-} = c_{00} = c{-+}$. All nine components of the order parameter are finite somewhere within the core region, but the only non-zero components at the centre are c_{0+} and c_{+0}; the first of these is the biggest and this means that the centre of the vortex is A-phase-like. As we shall see in Sect. 7 the A phase permits a non-zero value of curl$\boldsymbol{v}_s$.

The finite magnetic moment of the ^{3}He nucleus allows another property of the ^{3}He vortices which is not possible in ^{4}He. A vortex has a sense of rotation and as a result the energy of spins pointing parallel and anti-parallel to the rotation axis will not in general be the same. The vortex core is thus ferromagnetic; the spontaneous moment is small but leads to observable effects.

Most of the experimental information on B phase vortices has come from NMR measurements on the rotating cryostats at Helsinki; Hakonen et al (1989) give an excellent review of the experiments. The ^{3}He was contained in a cylinder of diameter 5 mm parallel to the rotation axis. The vortex separation at the maximum rotation rate is about 100 μm to be compared with a vortex core size of order 50 nm; the vortex cores therefore occupy only a small fraction of the volume of the cylinder and cannot be observed directly; instead the vortices are detected through

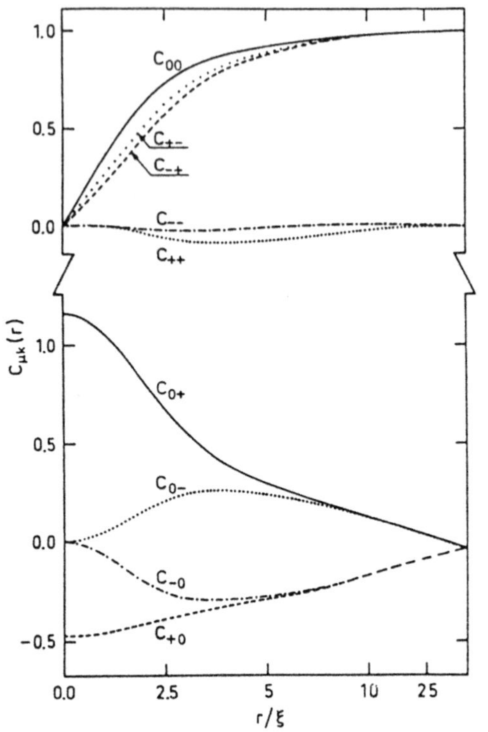

Fig. 23. The calculated structure of the axisymmetric v-vortex of Salomaa and Volovik (1983, 1985). ξ is the temperature-dependent Ginzburg-Landau coherence length.

their effect on the texture within the whole cylinder of the $\boldsymbol{n}$ axis which defines the rotation of the spin axes relative to the orbital axes in ^{3}He–B.

In the absence of vortices the equilibrium $\boldsymbol{n}$ texture in the cylinder is determined by competition between magnetic field alignment, the gradient energy associated with spatially varying $\boldsymbol{n}$ and the boundary condition on $\boldsymbol{n}$ at the surface of a cylinder. For a full discussion of these the reader is recommended to consult Brinkman and Cross 1978; my description will of necessity be briefer. The magnetic field orientation energy density is of the form

$$f_{\mathrm{m}} = -a(\boldsymbol{n}.\boldsymbol{H})^2, \tag{47}$$

and thus favours $\boldsymbol{n} \parallel \boldsymbol{H}$; the gradient energy favours $\boldsymbol{n}$ to vary as slowly with position as possible. The boundary condition at the walls in a magnetic field comparable to those used in the experiments strongly anchors $\boldsymbol{n}$ in a direction such that a rotation by the 'magic' angle of 104° about it takes the surface normal into the magnetic field direction. For a field parallel to the cylinder axis the appropriate direction is obtained by setting the angles α and β in Fig. 24 to 60° and 63° respectively. Fig. 24 shows the 'flare-out' texture in the cylinder that minimises the energy subject to this boundary condition (Spencer and Ihas 1982). The competition between the magnetic field alignment and bending energies leads to a characteristic bending length for $\boldsymbol{n}$

$$\xi_H = \sqrt{\frac{\rho_s}{a}} \frac{\hbar}{2m_3 H}, \tag{48}$$

where a is defined by (47); ξ_H is typically 0.5 mm in a field of 30 mT.

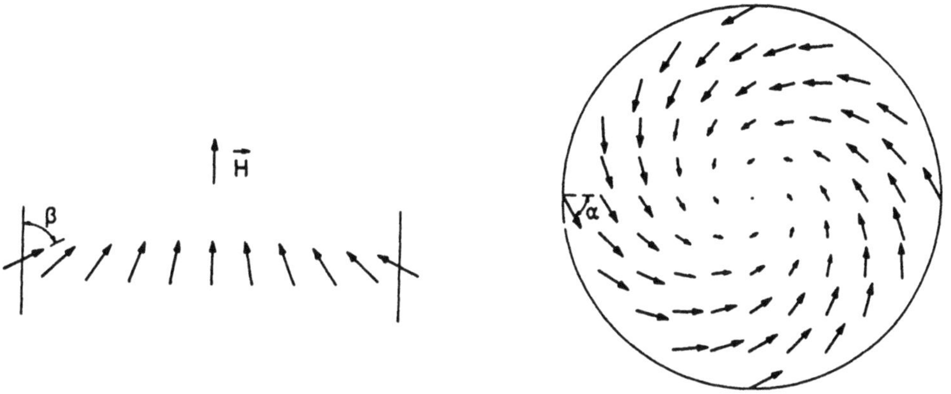

Fig. 24. Flare-out texture of $\boldsymbol{n}$ in a cylinder, shown in side view on the left and plan view on the right. The azimuthal angle α and inclination angle β are used to define the direction of $\boldsymbol{n}$.

An example of the NMR absorption signal from the flare-out texture and the effect of rotation on it are shown in Fig. 25. The simplest approach to understanding the spectrum is the 'local oscillator' model which assumes that each region of the texture contributes to the spectrum at the frequency and amplitude that it would if the texture were spatially uniform at the orientation within that region. This approach predicts correctly the envelope of the spectrum but fails to explain the sharp peaks that are apparent in Fig. 25. These suggest the existence of a series of discrete energy levels which turn out to be spin-wave modes, trapped in a potential well provided by the spatial variation of $\sin^2 \beta$, where β is the inclination angle defined in Fig. 24.

The effect of rotation is to increase the separation of the spin-wave modes; this is examined in more detail in Fig. 26 in which the frequency differences between the spin wave resonances are plotted as a function of temperature for zero rotation rate and for $\Omega = 1.40\,\mathrm{rad\,s^{-1}}$. As well as the increased splitting, the other notable features of the results obtained while rotating are the discontinuity in the frequency differences at the reduced temperature of 0.6 and their increased dependence on the relative directions of $\boldsymbol{\Omega}$ and $\boldsymbol{H}$ at temperatures below this.

The increased spacing of the spin wave modes on rotation suggests that vortices decrease the width of the potential well formed by the flare-out texture. Since the vortex separation is much less than the bending length ξ_H of $\boldsymbol{n}$, the orienting effect of the vortices can be averaged over many vortices and this results in two additional contributions to the energy density, normally written as

$$f_{v1} + f_{v2} = \frac{2}{5}a\lambda(\hat{\boldsymbol{\Omega}}.\boldsymbol{R}.\boldsymbol{H})^2 + \frac{4}{5}a\kappa\hat{\boldsymbol{\Omega}}.\boldsymbol{R}.\boldsymbol{H}, \tag{49}$$

where a is defined by (47) and λ and κ are coefficients, proportional to the vortex density n_v, which contain information on the structure of the vortices; $\boldsymbol{R}$ is the rotation matrix which achieves rotation about $\boldsymbol{n}$ by an angle of 104° and is given explicitly by

$$R_{\mu i} = -\frac{1}{4}\delta_{\mu i} + \frac{5}{4}n_\mu n_i - \frac{\sqrt{15}}{4}\epsilon_{\mu ij}n_j$$

The energy f_{v1} arises because of the superflow around the vortices and the magnetic anisotropy of the vortex core and λ is proportional to the integral over the vortex core of

$$\sum_k(|c_{0k}|^2 - \frac{1}{2}|c_{+k}|^2 - \frac{1}{2}|c_{-k}|^2);$$

f_{v2} arises because of the ferromagnetism of the vortex cores and is proportional to the integral over the core of

$$\sum_k(|c_{+k}|^2 - |c_{-k}|^2).$$

From the experimental results such as those of Fig. 26 it is possible to deduce values for λ and κ; since $f_{v1} \propto H^2$ and $f_{v2} \propto H$ these parameters can be separately determined. From the form of f_{v2}, we see that it is this term which is responsible for the difference between the experimental results with $\boldsymbol{\Omega}$ parallel and anti-parallel to $\boldsymbol{H}$ in Fig. 26. In principle the nature of the vortex core can be determined from the measured values of λ and κ, but in practice the parameters of the theory are not sufficiently well determined to enable this.

The discontinuity in the experimental results of Fig. 26 seems to indicate a change in vortex core type; the locus of this transition on the ^{3}He–B phase diagram, shown in Fig. 27, is close to that at which the critical velocity shows a discontinuity in the persistent current measurements (see Fig. 7), but it is not yet clear whether these phenomena are related. It does not seem to be possible to explain the experimental results using only cylindrically symmetric vortices. Although the V1 vortex in Fig. 27 could correspond to the axisymmetric v-vortex of Fig. 23 of Salomaa and Volovik (1983, 1985), Thuneberg 1986 has proposed that the V2 vortex is non-axisymmetric and has a double core structure; the double core is A-phase-like with a half quantum of circulation about each part. Support for this proposal has been provided by recent experiments of Kondo et al. 1992 which are interpreted as providing evidence for twisting oscillations of the double core structure of the V2 vortex.

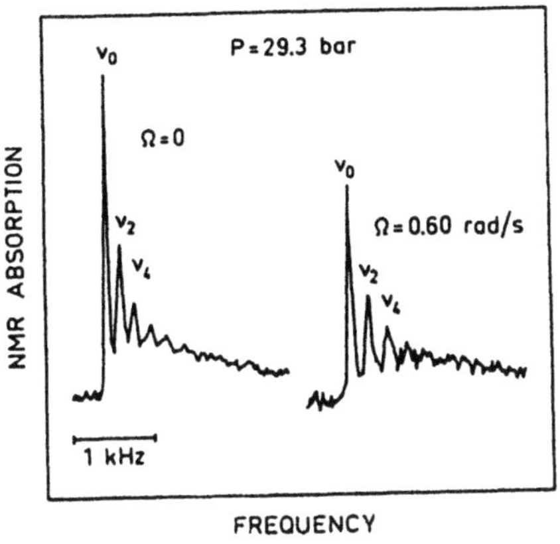

Fig. 25. NMR spectrum of ^{3}He–B measured at H=28.4 mT and 29.3 bar with the cryostat stationary and with it rotating at $\Omega = 0.60\,\mathrm{rad\,s^{-1}}$.

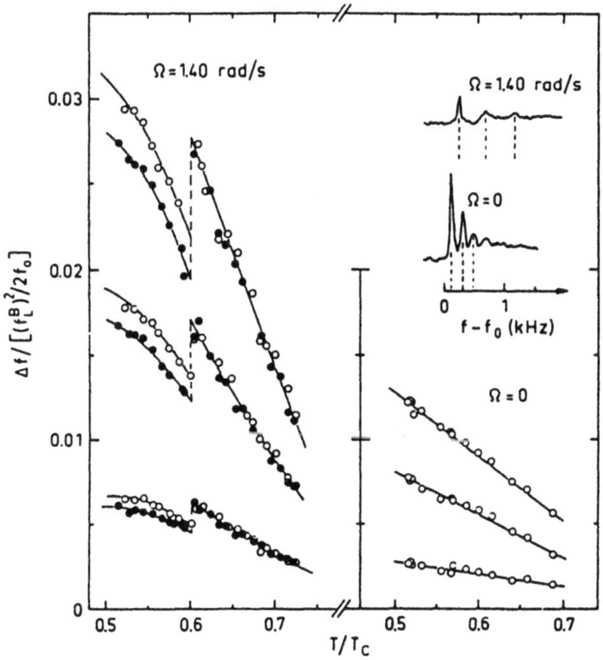

Fig. 26. Measured frequency differences $\Delta f = f - f_0$ of the spin-wave resonances relative to the Larmor frequency at 29.3 bar and 28.4 mT, normalised to $(f_L^B)^2/2f_0$ (f_L^B is the B-phase longitudinal ringing frequency) for $\Omega = 0$ and $\Omega = 1.40\,\mathrm{rad\,s^{-1}}$. The open and filled symbols correspond to rotation with $\boldsymbol{\Omega}$ parallel to $\boldsymbol{H}$ and $\boldsymbol{\Omega}$ anti-parallel to $\boldsymbol{H}$. The actual NMR spectra at $T = 0.51T_c$ are illustrated in the upper right corner of the figure.

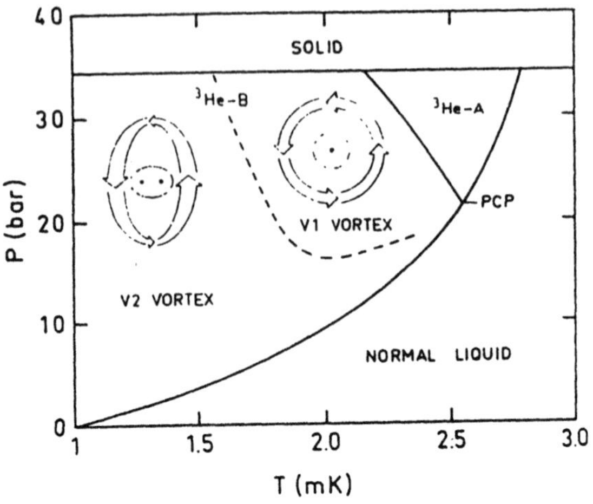

Fig. 27. Phase diagram of ^{3}He below 3 mK showing the locus of the vortex core transition observed in the Helsinki NMR measurements. As schematically illustrated, the V1 vortices are believed to be axisymmetric whereas the V2 vortices are non-axisymmetric with a double-core structure.

6 Liquid-Crystal-Like Behaviour of Superfluid ^{3}He–A

In this section I will describe an experiment performed at Manchester University (Hook 1989, Hook et al 1989) which illustrates the liquid-crystal-like properties of ^{3}He–A. The order parameter describing the local pairing state in ^{3}He–A is given by (45) and is completely specified by the triad of orthogonal unit vectors $\boldsymbol{m}$, $\boldsymbol{n}$ and $\boldsymbol{l}$ along the axes of the spherical harmonic Y_{11}, which defines the orbital state of the pair, and the unit vector $\boldsymbol{d}$ along which the pair has spin component zero. In infinite bulk ^{3}He–A, the orientation of $\boldsymbol{d}$ with respect to the $\boldsymbol{m},\boldsymbol{n},\boldsymbol{l}$ triad is determined by the interaction energy

$$E_{\mathrm{D}} = -\frac{1}{2}\lambda_{\mathrm{D}}(\boldsymbol{l}.\boldsymbol{d})^2, \tag{50}$$

per unit volume, between the nuclear magnetic dipole moments; thus $\boldsymbol{l} \parallel \boldsymbol{d}$ is favoured. The orientation of the $\boldsymbol{m}$ and $\boldsymbol{n}$ vectors does not play a role in this section and we do not discuss them further here; we shall see in Sect. 7 that they play a crucial role in superflow in ^{3}He–A.

The liquid-crystal-like properties of ^{3}He–A result from the fact that the magnetic properties of the fluid, such as the susceptibility, are anisotropic with respect to the $\boldsymbol{d}$ direction and the mechanical properties such as the superfluid density are anisotropic with respect to the $\boldsymbol{l}$ direction. The anisotropy of the magnetic susceptibility means that a magnetic field will tend to orient $\boldsymbol{d}$ to maximise the susceptibility; the associated energy density is

$$E_H = \frac{1}{2}\Delta\chi(\boldsymbol{H}.\boldsymbol{d})^2, \tag{51}$$

where

$$\Delta\chi = \chi_\perp - \chi_\parallel, \tag{52}$$

and $\chi_\perp$ and $\chi_\parallel$ are the susceptibilities measured with $\boldsymbol{H}$ perpendicular and parallel to $\boldsymbol{d}$, respectively. Since $\Delta\chi > 0$ the effect of a magnetic field is to tend to align $\boldsymbol{d}$ perpendicular to it.

6.1 The Fréedericksz Transition

In the Manchester experiment the texture of the $\boldsymbol{l}$ and $\boldsymbol{d}$ vectors in a thin-slab geometry was investigated. The boundary conditions on $\boldsymbol{l}$ and $\boldsymbol{d}$ at the surfaces of the slab are (Brinkman and Cross 1978)

$$\begin{aligned} \boldsymbol{l}.\hat{\boldsymbol{s}} &= \pm 1, \\ (\hat{\boldsymbol{s}}.\nabla)\boldsymbol{d} &= 0, \end{aligned} \tag{53}$$

where $\hat{\boldsymbol{s}}$ is the surface normal. Thus $\boldsymbol{l}$ is perpendicular to the surfaces and the component of $\nabla\boldsymbol{d}$ perpendicular to the walls is zero; the physical meaning of the boundary condition on $\boldsymbol{d}$ is that the component of spin supercurrent normal to non-magnetic walls should vanish. A magnetic field applied perpendicular to the slab will tend to align $\boldsymbol{d}$ parallel to the slab and the walls tend to align $\boldsymbol{l}$ perpendicular to the slab. These requirements are incompatible with the parallel orientation of $\boldsymbol{l}$ and $\boldsymbol{d}$ required to minimise the dipolar energy (50). To calculate the resulting texture it is necessary to add the bending energies, E_l and E_d, of $\boldsymbol{l}$ and $\boldsymbol{d}$ to the dipolar energy (50) and magnetic anisotropy energy (51); these take the form

$$E_l = \left(\frac{\hbar}{2m}\right)^2 [\frac{1}{2}K_\mathrm{S}(\mathrm{div}\boldsymbol{l})^2 + \frac{1}{2}K_\mathrm{T}(\boldsymbol{l}.\mathrm{curl}\boldsymbol{l})^2 + \frac{1}{2}K_\mathrm{B}(\boldsymbol{l}\times\mathrm{curl}\boldsymbol{l})^2], \tag{54}$$

and

$$E_d = \left(\frac{\hbar}{2m}\right)^2 [\frac{1}{2}\rho_{\mathrm{sp}\parallel}(\boldsymbol{l}.\nabla d_\alpha)^2 + \frac{1}{2}\rho_{\mathrm{sp}\perp}(\boldsymbol{l}\times\nabla d_\alpha)^2] \tag{55}$$

per unit volume. E_l is of the same form as the bending energy associated with spatial variation of the director vector of a nematic liquid crystal (de Gennes 1974); K_S, K_T and K_B are the splay, twist and bend coefficients. E_d can be interpreted as the kinetic energy of the spin supercurrents; $\rho_{\mathrm{sp}\parallel}$ and $\rho_{\mathrm{sp}\perp}$ are the spin superfluid densities for spin currents parallel and perpendicular to $\boldsymbol{l}$.

The factors of $(\hbar/2m)^2$ in (54) and (55) mean that all the coefficients are densities; the coefficients are comparable in magnitude and this enables us to choose one of them $\rho_{\mathrm{sp}\parallel}$ to define a characteristic length L_D, the dipolar length, by comparing a bending energy with the dipolar energy (50). Since $\boldsymbol{l}$ and $\boldsymbol{d}$ are unit vectors, bending on a length scale x gives rise to an energy density $\sim (\hbar/2m)^2\rho_{\mathrm{sp}\parallel}/x^2$ (by setting $\nabla \sim 1/x$ in (55)) which is comparable to the dipolar energy ($\sim \lambda_\mathrm{D}$) if

$$x \sim L_{\mathrm{D}} = \frac{\hbar}{2m}\sqrt{\frac{\rho_{\mathrm{sp}\|}}{\lambda_{\mathrm{D}}}}. \tag{56}$$

The physical significance of L_{D}, which is around 10 μm and varies only weakly with temperature, is that if $\boldsymbol{l}$ is forced to vary on a length scale short compared to L_{D} then $\boldsymbol{d}$ does not remain parallel to $\boldsymbol{l}$ since the cost in bending energy exceeds the energy gain from the dipolar energy. A characteristic magnetic field H_{D}, the dipolar field, can be defined by comparing the dipolar energy (50) with the magnetic anisotropy energy (51). These are comparable at a field

$$H \sim H_{\mathrm{D}} = \sqrt{\frac{\lambda_{\mathrm{D}}}{\Delta\chi}}. \tag{57}$$

At fields greater than H_{D}, which is around 2.5 mT and varies only weakly with temperature, the orientation of $\boldsymbol{d}$ is determined by the field rather than the dipolar energy.

For simplicity we will calculate the texture for a thick slab (thickness $D \gg L_{\mathrm{D}}$) in a small field ($H \ll H_{\mathrm{D}}$); these conditions are approximately true for the Manchester experiment. In this case the dominant energy is the dipolar energy (50) which maintains $\boldsymbol{l} \parallel \boldsymbol{d}$ to a good approximation, a situation known as dipole locking. The replacement of $\boldsymbol{d}$ by $\boldsymbol{l}$ in (55) and adding the result to (54) shows that the bending energy of the $\boldsymbol{l}, \boldsymbol{d}$ combination in the dipole-locked limit can be written in the form of (54) by redefining the three coefficients as:

$$K'_{\mathrm{S}} = K_{\mathrm{S}} + \rho_{\mathrm{sp}\perp}, \quad K'_{\mathrm{T}} = K_{\mathrm{T}} + \rho_{\mathrm{sp}\perp}, \quad K'_{\mathrm{B}} = K_{\mathrm{B}} + \rho_{\mathrm{sp}\|}.$$

In zero magnetic field the equilibrium $\boldsymbol{l}, \boldsymbol{d}$ texture is uniform and perpendicular to the slab as shown in Fig. 28(a). Application of a field perpendicular to the slab tends to destabilise this texture because of the magnetic anisotropy energy (51). If we assume that the dipole-locked $\boldsymbol{l}$ and $\boldsymbol{d}$ make an angle $\theta(z)$ to the slab normal so that

$$\boldsymbol{l} = \boldsymbol{d} = \hat{\boldsymbol{y}} \sin\theta + \hat{\boldsymbol{z}} \cos\theta, \tag{58}$$

then the texture is determined by minimising the sum of the magnetic anisotropy energy and bending energies integrated over the slab

$$\begin{aligned} \int_0^D E dz &= \int_0^D (E_H + E_l + E_d) dz \\ &= \int_0^D \left(\frac{1}{2}(K'_{\mathrm{S}} \sin^2\theta + K'_{\mathrm{B}} \cos^2\theta)\theta'^2 + \frac{1}{2}\Delta\chi H^2 \cos^2\theta \right) dz, \end{aligned} \tag{59}$$

subject to the boundary condition (53) on $\boldsymbol{l}$, which can be written

$$\theta = 0 \quad \text{at} \quad z = 0 \quad \text{and} \quad z = D. \tag{60}$$

This problem is easily solved using the calculus of variations (Hook 1990). The Euler equation

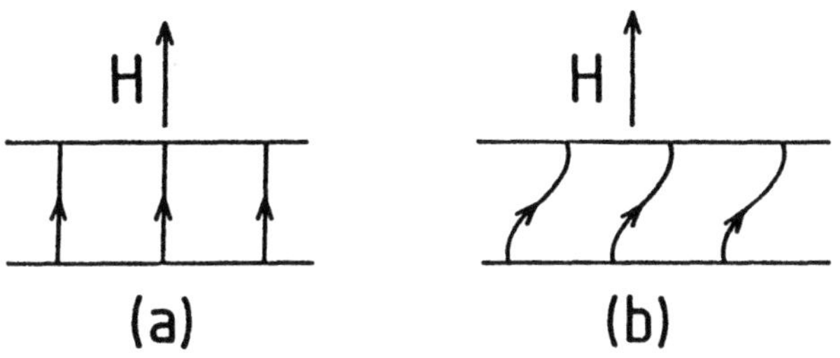

Fig. 28. (a) Uniform texture in a slab geometry for $H < H_F$. (b) Distorted texture for $H > H_F$.

$$\frac{\partial E}{\partial \theta} - \frac{d}{dz}\frac{\partial E}{\partial \theta'} = 0,$$

where E is defined by (59), has only the solution $\theta = 0$ for values of applied magnetic field perpendicular to the slab less than a threshold value H_F, where

$$H_F = \frac{\pi \hbar}{2mD}\sqrt{\frac{K'_B}{\Delta\chi}}; \tag{61}$$

$\theta = 0$ corresponds to the uniform texture of Fig. 28(a) with $\boldsymbol{l}$ and $\boldsymbol{d}$ perpendicular to the slab. Above H_F the texture is distorted in the manner indicated in Fig. 28(b) and for $H \gg H_F$ the texture has $\boldsymbol{l}$ and $\boldsymbol{d}$ approximately perpendicular to the field except in narrow regions close to each wall where bending is necessary for $\boldsymbol{l}$ to satisfy the boundary condition (60). The textural transition we have described is an example of a class of transitions which are well-known in liquid crystal physics and were first discovered by Fréedericksz and Repiewa 1927.

The Manchester apparatus used to observe the Fréedericksz transition in ^{3}He–A is shown in Fig. 29. The slab containing the ^{3}He–A has a thickness of about 100 μm and a diameter of 8.5 mm and is contained within a torsional oscillator. Because of the anisotropy of the viscosity and superfluid density, changes in the $\boldsymbol{l}$ texture affect the inertia of the fluid coupling to the oscillator motion and also the level of dissipation within the fluid; the textural changes can therefore be investigated by measuring the resonant frequency and bandwidth of the oscillator. Figure 30 shows a Fréedericksz transition in ^{3}He–A at 29.3 bar and $T/T_c = 0.834$; the resonant frequency ν and bandwidth ν_B of the oscillator are plotted as a function of the magnetic field applied perpendicular to the slab. H_F is identified as the field at which ν and ν_B first depart from the values for the uniform texture. The change is most striking in ν; the value of ν_B decreases slightly at H_F before increasing in higher fields.

Figure 31 compares the observed threshold field for the transition as a function of reduced temperature with the theoretical value obtained for slab thicknesses 105 μm and 115 μm; these are the limits of uncertainty in the value of D. Dipole-unlocking has been taken into account in calculating the theoretical value; the

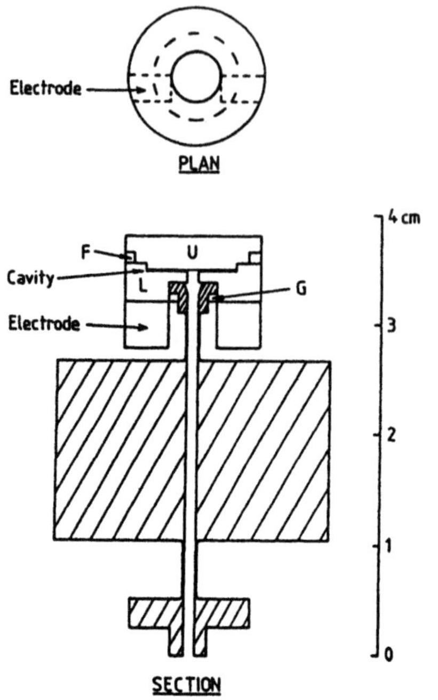

Fig. 29. Manchester torsional oscillator containing a thin slab of ^{3}He–A.

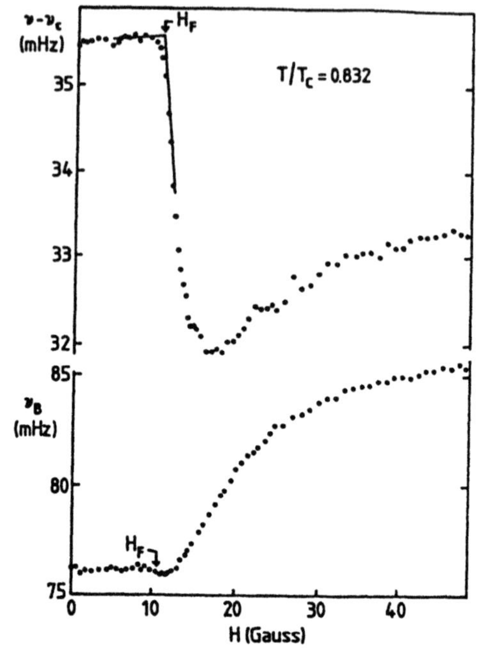

Fig. 30. Resonant frequency and bandwidth of the oscillator as functions of the magnetic field applied perpendicular to the slab; ν_c is the resonant frequency ($\sim$ 1333 Hz) at T_c.

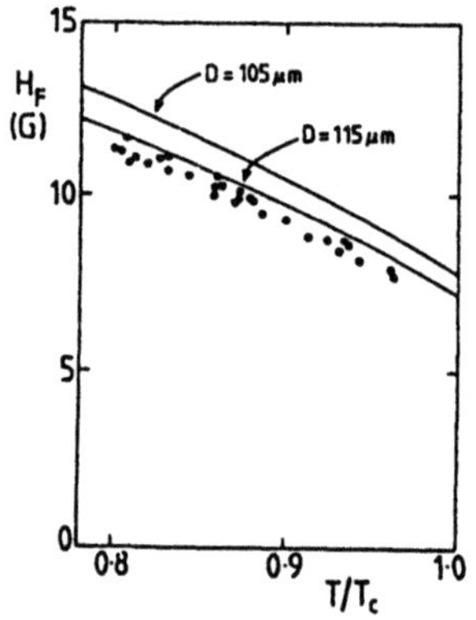

Fig. 31. Comparison of experimental and theoretical values of the Fréedericksz transition field. The theoretical curves were calculated for the values of D indicated.

parameters which enter the calculation are not sufficiently well known to decide whether the deviations between theory and experiment are significant.

By making measurements of the resonant frequency and bandwidth as a function of magnetic field applied at various angles to the slab it was possible to study the textural variations within the slab in more detail and also to investigate the anisotropy of the viscosity and superfluid density of ^{3}He–A. Figure 32 shows the variation of ν and ν_B with fields at different angles to the slab at $T/T_c = 0.832$; the continuous curves are theoretical fits in which the viscosity and superfluid density are the fitting parameters.

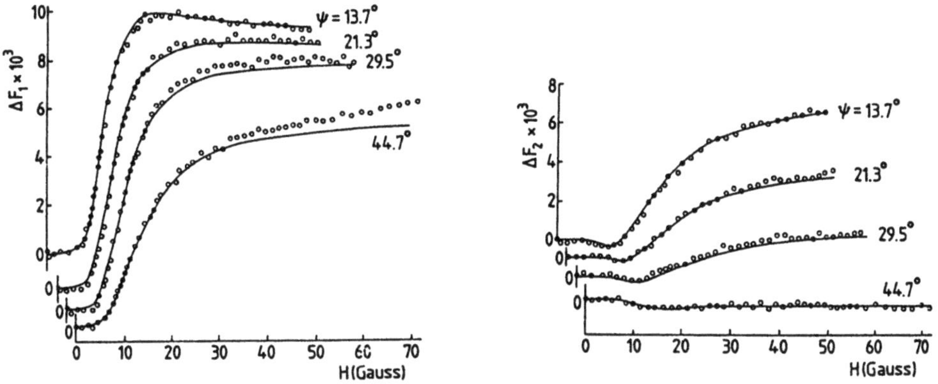

Fig. 32. Resonant frequency (left) and bandwidth (right) as a function of magnetic field applied at the angles ψ to the slab normal indicated. F_1 and F_2 are dimensionless measures of ν and ν_B obtained by dividing by the resonant frequency difference between states in which the fluid is uncoupled and perfectly coupled to the oscillator motion.

For an insight into the kind of complications associated with calculating the theoretical curves in Fig. 32 we indicate in Fig. 33 the geometry of the flow associated with the oscillator motion. The magnetic field is assumed to be in the y, z plane and the textural energy is then minimised if $\boldsymbol{l}$ and $\boldsymbol{d}$ also distort in the y, z plane as in (58); $\boldsymbol{l}$ thus acquires a y component as indicated. To a first approximation the normal fluid is clamped to the oscillator by its viscosity so that $\boldsymbol{v}_n$ is approximately uniform and azimuthal; because of the anisotropy of the normal fluid density with respect to the $\boldsymbol{l}$ direction, however, the normal mass current density $\boldsymbol{j}_n$ is larger where $\boldsymbol{l}$ has a component parallel to $\boldsymbol{v}_n$ (points A and C on Fig. 33) than where $\boldsymbol{l}$ and $\boldsymbol{v}_n$ are orthogonal (points B and D on Fig. 33). Under the conditions of the experiment however the fluid is incompressible so that the variation in normal current density must be compensated by superflow; this is in complete contrast to the situation for an isotropic superfluid in a torsional oscillator where the superfluid stays at rest. The additional superflow has the geometry indicated in Fig. 33 in order to give an essentially azimuthal total mass current density as required; for this texture the superflow must satisfy $\mathrm{curl}\,\boldsymbol{v}_s = 0$, even in the A phase. The details of the hydrodynamics can be found in Hook (1989) and the experimental information obtained on the 2 coefficients ($\rho_{s\perp}$ and $\rho_{s\parallel}$) required to specify the superfluid density and the 5 coefficients required to specify the viscosity of the uniaxially anisotropic ^{3}He–A can be found in Hook et al (1989).

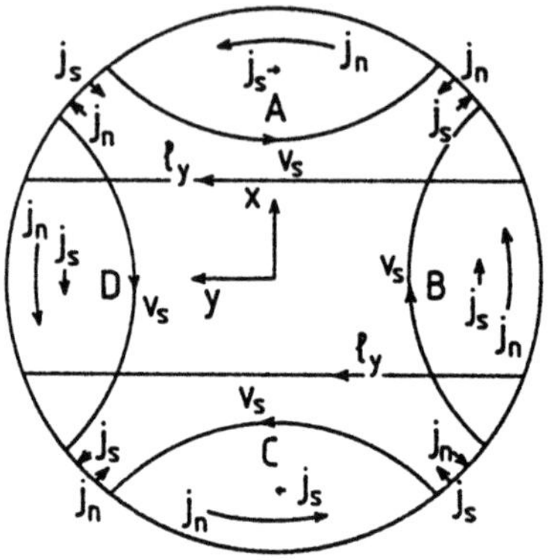

Fig. 33. Geometry of the flows generated by the torsional oscillations of the slab when the $\boldsymbol{l}$ vector is in the y, z plane.

7 Superflow in Superfluid ^{3}He–A

From (45) the order parameter of spatially homogeneous ^{3}He–A can be written

$$\Psi \propto \sin\theta \exp(\mathrm{i}\varphi)\chi_{1\,0}, \tag{62}$$

where θ and φ are the spherical polar angles with respect to the orbital axes of the relative coordinate $\boldsymbol{\rho} = \boldsymbol{r}_1 - \boldsymbol{r}_2$ of the pair. Superflow corresponds to centre-of-mass motion of the pair and is therefore associated with a phase factor $\exp(\mathrm{i}\gamma)$, where γ is a function of the centre-of-mass coordinate $\boldsymbol{r} = (\boldsymbol{r}_1 + \boldsymbol{r}_2)/2$ of the pair (we use γ rather than θ for the phase variable here in order to avoid confusion with the spherical polar coordinate). Multiplying (62) by this phase factor gives an order parameter

$$\Psi \propto \sin\theta \exp[\mathrm{i}(\varphi + \gamma)]\chi_{1\,0}. \tag{63}$$

The appearance of the factor $(\varphi + \gamma)$ indicates that in ^{3}He–A the centre-of-mass and internal orbital motion of the Cooper pair are inextricably linked and this has many important consequences. More precisely we see that multiplying the order parameter by a phase factor $\exp(\mathrm{i}\gamma)$ is equivalent to specifying the internal motion of the pair with respect to orbital x and y axes rotated by $-\gamma$ with respect to the original set; in terms of the $\boldsymbol{m}, \boldsymbol{n}, \boldsymbol{l}$ unit vectors introduced in Sect. 5.2 to define the local directions of the x, y, z axes, the effect of the multiplication is to rotate $\boldsymbol{m}$ and $\boldsymbol{n}$ about $\boldsymbol{l}$ by an angle $-\gamma$. It follows that it is not possible or desirable to factorise from the wave function a phase factor associated with centre-of-mass motion; the superflow is contained within an order parameter of the form of (62) through spatial variation of the $\boldsymbol{m}$ and $\boldsymbol{n}$ axes, with respect to which the spherical harmonic Y_{11} is specified. The superfluid velocity is related to the gradient of the 'rotation angle' α of $\boldsymbol{m}$ and $\boldsymbol{n}$ about $\boldsymbol{l}$ by

$$v_S = -\frac{\hbar}{2m_3}\nabla\alpha, \qquad (64)$$

where the $-$ sign is needed because anti-clockwise rotation of $\boldsymbol{m}$ and $\boldsymbol{n}$ corresponds to an increase in $\varphi+\gamma$ in (63). Equation (64) provides the link between the liquid-crystal-like and superfluid properties of ^{3}He–A; Fig. 34 shows the spatial variation of $\boldsymbol{m}$ and $\boldsymbol{n}$ corresponding to superflow parallel and perpendicular to a uniform $\boldsymbol{l}$ texture.

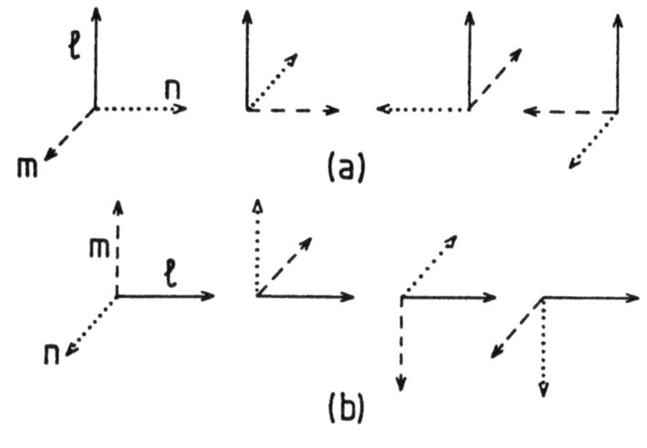

Fig. 34. (a) A gradient of the rotation angle α perpendicular to $\boldsymbol{l}$ generates $\boldsymbol{v}_s \perp \boldsymbol{l}$. (b) A gradient of the rotation angle α parallel to $\boldsymbol{l}$ generates $\boldsymbol{v}_s \parallel \boldsymbol{l}$.

The most interesting situations, however, are those in which $\boldsymbol{l}$ varies spatially and under these circumstances the similarity of (64) to the corresponding equations, (2) and (46), for ^{4}He and ^{3}He–B is misleading as we will now demonstrate. Let us suppose that the texture at four points A, B, C and D within the fluid is as shown in Fig. 35. In going directly from A to D a rotation of $\boldsymbol{m}$ and $\boldsymbol{n}$ about $\boldsymbol{l}$ by $\pi/2$ is involved so that one could say that α has increased by $\pi/2$ in this process. Going from A to D via B and C however is achieved by successive rotations of $\pi/2$ about $\boldsymbol{m}$, $\boldsymbol{n}$ and $-\boldsymbol{m}$ so that α has apparently not changed in this process. The conclusion to be drawn is that it is not possible to generate a well-defined scalar function of position α by integration of $\nabla\alpha$; the answer obtained depends on the path taken. The results that we derived for ^{4}He and ^{3}He–B, such as $\text{curl}\,\boldsymbol{v}_s = 0$ and $\partial \boldsymbol{v}_s/\partial t = -\nabla\mu/M$, that depended on $\boldsymbol{v}_s$ being proportional to the gradient of a well-defined scalar function of position will therefore no longer be valid for ^{3}He–A.

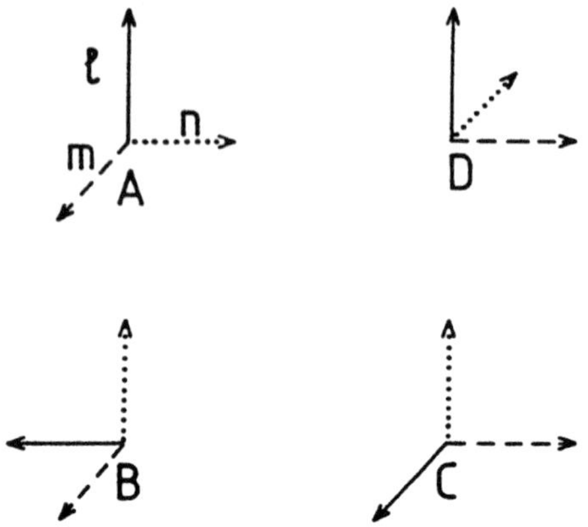

Fig. 35. Illustration that $\nabla\alpha$ cannot be integrated to give a well-defined scalar function of position.

To obtain the amended versions of these results for ^{3}He–A, we notice from Fig. 35 that the reason why α is not well defined is that finite rotations do not commute in three dimensions. The non-commutativity of two rotations $\boldsymbol{\theta}_1$ and $\boldsymbol{\theta}_2$ can be neatly expressed in the form

$$\boldsymbol{\theta}_1\boldsymbol{\theta}_2 - \boldsymbol{\theta}_2\boldsymbol{\theta}_1 = \boldsymbol{\theta}_1 \times \boldsymbol{\theta}_2. \tag{65}$$

It follows directly from this that two successive differentiations of α do not commute; explicitly

$$\partial_1\partial_2\alpha - \partial_2\partial_1\alpha = \boldsymbol{l}.(\partial_2\boldsymbol{l} \times \partial_1\boldsymbol{l}). \tag{66}$$

To obtain the equivalent of $\mathrm{curl}\boldsymbol{v}_s = 0$ for ^{3}He–A we take both derivatives in (66) to be spatial gradients to obtain

$$(\mathrm{curl}\boldsymbol{v}_s)_k = \nabla_i v_{sj} - \nabla_j v_{si} = \frac{\hbar}{2m_3}\boldsymbol{l}.(\nabla_j\boldsymbol{l} \times \nabla_i\boldsymbol{l}), \tag{67}$$

so that $\mathrm{curl}\boldsymbol{v}_s$ is non-zero in general in ^{3}He–A if there is a spatially varying $\boldsymbol{l}$ texture. To obtain the superfluid acceleration equation for ^{3}He–A we take one of the differentials in (66) to be ∇_i and the other to be $\partial/\partial t$ to obtain

$$\nabla_i\frac{\partial\alpha}{\partial t} - \frac{\partial}{\partial t}\nabla_i\alpha = \boldsymbol{l}.\left(\frac{\partial\boldsymbol{l}}{\partial t} \times \nabla_i\boldsymbol{l}\right)$$

which on using (24) ($\rightarrow \hbar\partial\alpha/\partial t = \mu$) and (64) becomes

$$2m_3\frac{\partial\boldsymbol{v}_s}{\partial t} = -\nabla_i\mu + \hbar\boldsymbol{l}.\left(\frac{\partial\boldsymbol{l}}{\partial t} \times \nabla_i\boldsymbol{l}\right). \tag{68}$$

The second term on the right hand sign indicates the possibility that superflow in ^{3}He–A can decay if there is a space- and time-dependent $\boldsymbol{l}$ texture.

To begin to explore the implications of these results we consider first ^{3}He–A in an infinitely long cylinder. Provided that the radius is not too small the $\boldsymbol{l}$ texture of lowest energy is believed to be the Mermin-Ho (1976) texture shown in Fig. 36; $\boldsymbol{l}$ has to be perpendicular to the walls to satisfy the boundary condition (53) but bends to become parallel to the axis in the centre of the cylinder. The resulting texture is singularity-free but the bending of $\boldsymbol{l}$ causes superflow as is explained in the right-hand part of the figure.

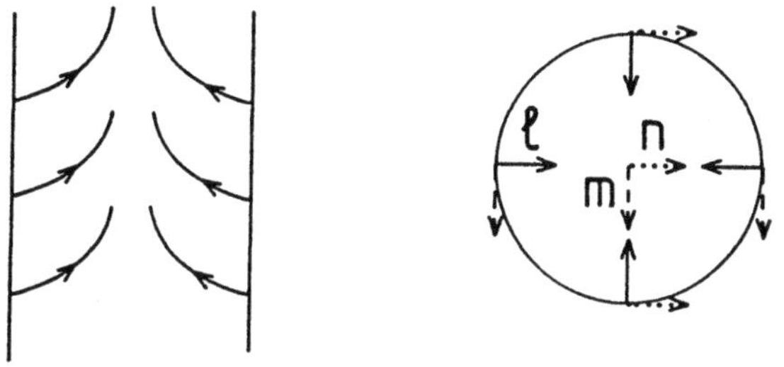

Fig. 36. The left-hand figure shows the spatial variation of $\boldsymbol{l}$ for the Mermin-Ho texture in an infinitely long cylinder. The spatial variation of $\boldsymbol{l}$, $\boldsymbol{m}$ and $\boldsymbol{n}$ for this texture is shown in the right-hand figure.

Suppose that, along the cylinder axis, $\boldsymbol{m}$ and $\boldsymbol{n}$ have the orientations indicated. On going radially outwards from the axis the bending of $\boldsymbol{l}$ causes the orientations of $\boldsymbol{m}$ and $\boldsymbol{n}$ to change in the manner indicated in the four directions shown. Inspection of the resulting texture demonstates that $\boldsymbol{m}$ and $\boldsymbol{n}$ rotate about $\boldsymbol{l}$ by 360° on encircling the *circumference* of the cylinder once. This corresponds to a circulation $\oint \boldsymbol{v}_s . d\boldsymbol{r}$ of one quantum $(h/2m_3)$ around the circumference of the cylinder. The Mermin-Ho texture is a simple example of a texture for which curl$\boldsymbol{v}_s$ is non-zero, thus allowing the circulation around a circle centred on the cylinder axis to increase continuously from zero to one quantum as the radius of the circle increases from zero to the cylinder radius.

A more elegant and straightforward way of calculating the circulation around any contour C is provided by the Ho circulation theorem (Ho 1978), obtained by integrating (67) around C,

$$\oint \boldsymbol{v}_s . d\boldsymbol{r} = \frac{\hbar}{2m_3}\left(\sigma(D) + 2\pi n\right), \tag{69}$$

where n is an integer and $\sigma(D)$ has the following geometrical interpretation. Let the direction of $\boldsymbol{l}$ at any point on C be represented by a point on a unit sphere, constructed by drawing the vector $\boldsymbol{l}$ from the centre of the sphere. Then the texture on C is mapped onto a closed curve D on the surface of the unit sphere. $\sigma(D)$ is the net area on the surface of the unit sphere to the left of D when D is traversed in the sense corresponding to the direction of integration along C. To calculate the circulation around the circumference of a cylinder for the Mermin-Ho texture therefore it is necessary only to note that the $\boldsymbol{l}$ texture along a circular contour C_1 of vanishingly small radius around the cylinder axis is represented by a point at the north or south pole on the unit sphere and the $\boldsymbol{l}$ texture around a circumference contour C_2 is represented by a curve D around the equator of the sphere. The change in area on the unit sphere swept out by the curve D on going from C_1 to C_2 is thus 2π corresponding to a change in circulation of 1 quantum according to (69).

Another useful application of the Ho circulation theorem is to calculate the circulation around a contour C_θ on the surface of a torus as shown in Fig. 37(a). Since a torus is essentially an infinitely long cylinder, the texture is again assumed to be the singularity-free Mermin-Ho texture (Fig. 37(b)). The contour D_θ on the unit sphere that corresponds to the $\boldsymbol{l}$ texture along C_θ is also shown in Fig. 37 and the area $\sigma(D)$ is shaded. It is left as an exercise to the reader to deduce that the circulation around C_θ is

$$\oint \boldsymbol{v}_s . d\boldsymbol{r} = \frac{h}{2m_3}(n + 1 + \sin\theta) . \tag{70}$$

The circulation therefore varies between n and $n+2$ quanta as θ varies. There are $n+1$ quanta around a contour along the centre of the channel if $\boldsymbol{l}$ is taken to be parallel to this contour.

Because there is in general a non-zero curl$\boldsymbol{v}_s$ associated with a spatially dependent $\boldsymbol{l}$ texture, it is possible that, within a rotating container, ^{3}He–A will rotate by the formation of a suitable $\boldsymbol{l}$ texture rather than by the formation of singular cored vortices. As an illustration of the type of $\boldsymbol{l}$ texture that allows ^{3}He–A to rotate, consider the coreless vortex of Anderson and Toulouse (1977), shown in Fig. 38. The $\boldsymbol{l}$ texture is cylindrically symmetric with respect to the dashed line; $\boldsymbol{l}$ rotates by 180° on going radially outwards from the centre of the coreless vortex in any direction. The contour D_θ on the unit sphere representing the texture moves from the north pole to the south pole with increasing radius, sweeping out the whole area of the unit sphere in the process. Application of the Ho circulation theorem (69) therefore shows that there are two quanta of circulation $(2h/2m_3)$ around a contour of large radius encircling the vortex. Since the supercurrent density does not diverge at the centre of the vortex there is no reason for the order parameter to depart from its A-phase form in this region.

Coreless vortices of the Anderson-Toulouse type cannot be stacked together to form an array. It is possible, however, to invent textures which correspond to an array of coreless vortices. This can be achieved for example by stacking together textures similar to the Mermin-Ho texture in a cylinder, which is essentially a single quantum coreless vortex; the reader should consult Vollhardt and Wölfle (1990) or Fetter (1986) for the details. The experiments that have been performed on rotating ^{3}He–A (see Hakonen et al. 1989 for a review of these) have generally been performed with an applied magnetic field, much bigger than the dipolar field H_D (see Sect. 6), parallel to the rotation axis. The $\boldsymbol{d}$ vector is everywhere anchored perpendicular to such a strong field by the anisotropy of the magnetic susceptibility. The spatially varying $\boldsymbol{l}$ texture associated with a coreless vortex

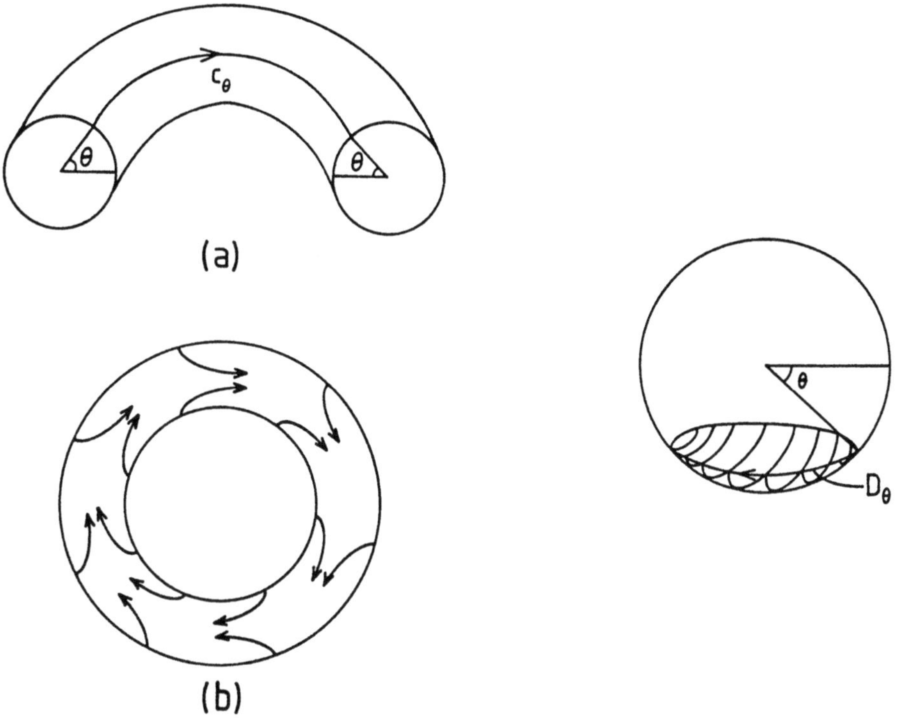

Fig. 37. (a) Definition of the surface contour C_θ on the torus. (b) Mermin-Ho texture in the torus. The mapping D_θ of the texture along C_θ onto the unit sphere is shown in the right hand figure.

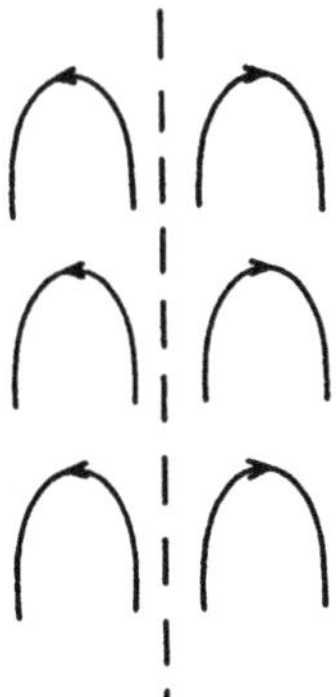

Fig. 38. The $\boldsymbol{l}$ texture of the coreless vortex of Anderson and Toulouse.

must therefore be dipole-unlocked and thus limited to a region of order the dipolar length L_D ($\sim 10\,\mu$m) in size surrounding the centre of the vortex. This dipole-unlocked region is sometimes referred to as a *soft core*. The term *hard core* is then reserved for the central region of order a coherence length ($\sim 50\,$nm) in size for vortices, such as those described for the B phase in Sect. 5, in which the order parameter is forced to vary from its bulk form. Since L_D is much less than the vortex separation at currently achievable rotation rates the soft cores are well separated. The $\boldsymbol{l}$ texture within the soft core region of one type of A phase doubly-quantized coreless vortex is shown in Fig. 39.

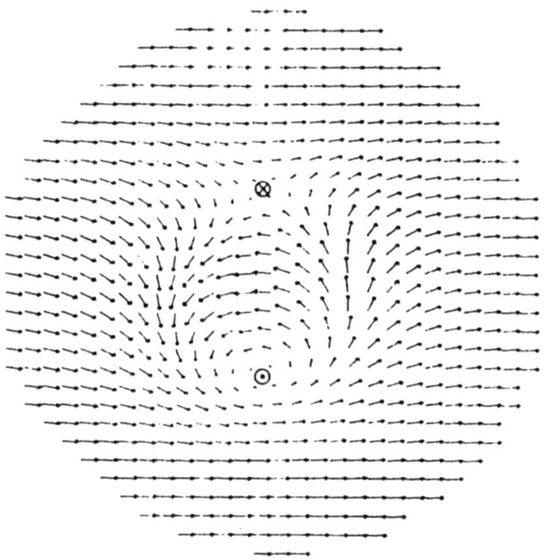

Fig. 39. The $\boldsymbol{l}$ texture associated with the doubly-quantized non-axisymmetric continuous (coreless) w-vortex (see Hakonen et al 1989 for references) of ^{3}He–A.

The $\boldsymbol{l}$ texture outside the soft core is dipole-locked along $\boldsymbol{d}$. There are two 'Mermin-Ho' type regions within the soft core each of which contributes one quantum to the total circulation; in one of the regions $\boldsymbol{l}$ bends through 90° out of the plane of the paper and in the other it bends through 90° into the plane of the paper. A vortex of this type is consistent with almost all the experimental observations made on rotating ^{3}He–A to date. Although hard-cored vortices also exist for ^{3}He–A (e.g. a uniform $\boldsymbol{l}$ texture with $\boldsymbol{m}$ and $\boldsymbol{n}$ rotating through 360° around a contour encircling the vortex) and theoretical calculations suggest that they are lower in energy, the experimental evidence suggests that doubly-quantized soft-cored vortices are preferentially nucleated. The reason for this may be that it is possible to nucleate such a vortex by distortion of a uniform $\boldsymbol{l}$ texture with only a small energy barrier.

The creation of coreless vortices also provides a mechanism for the decay of a persistent current around a torus, the change in $\boldsymbol{v}_s$ being related to the time-dependent $\boldsymbol{l}$ texture by (68). This has been discussed by Ho (1978) in some detail. Equation (70) would suggest however that the circulation around a contour on the surface is fixed by the boundary condition on $\boldsymbol{l}$. A type of surface singularity, named a boojum by Mermin (see Mermin 1981 for an interesting account of his success in getting the name accepted by the editors of Physical Review Letters), however, provides a mechanism by which even the surface circulation can decay. A boojum is essentially the texture that arises when an Anderson-Toulouse coreless vortex (Fig. 38) meets a surface as shown in Fig. 40(a). That such a texture is singular and thus must have a 'point' hard core can be seen by noting that $\boldsymbol{l}$ is directed everywhere into the surface except at the centre of the boojum where it is outwards. The boojum is also believed to be involved in the $\boldsymbol{l}$ texture of minimum energy within a spherical container as illustrated in Fig. 40(b); Mermin (1977) has shown that a non-singular texture is not possible within a simply connected container.

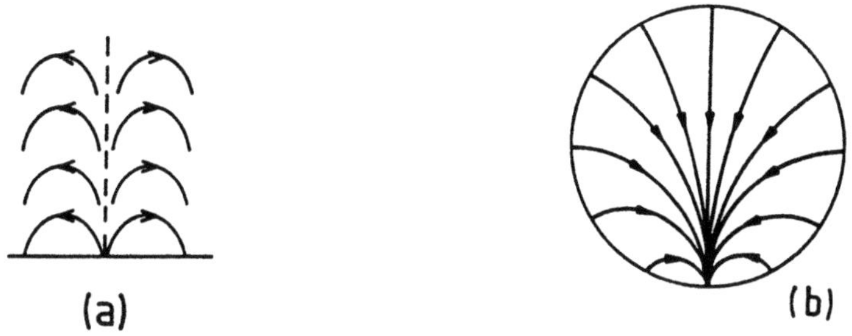

Fig. 40. (a) A point singularity of the $\boldsymbol{l}$ texture (boojum) occurs where an Anderson-Toulouse coreless vortex meets a surface. (b) Boojum texture within a spherical container.

Extrapolation of the circulation integral around the coreless vortex shows that there are two quanta of circulation along a surface contour surrounding a boojum. Passage of a boojum across the surface contour C_θ on the torus of Fig. 37 would therefore result in a change in circulation along C_θ of two quanta. A two-boojum texture which could result in the decay of a persistent current around a torus is shown in Fig. 41.

The interaction between the liquid-crystal-like and superfluid properties of ^{3}He–A that we have discussed in this section has not yet been explored experimentally to any significant extent. This is just one area in which superfluid ^{3}He is likely to provide new surprises to keep both experimentalists and theoreticians busy for several years to come.

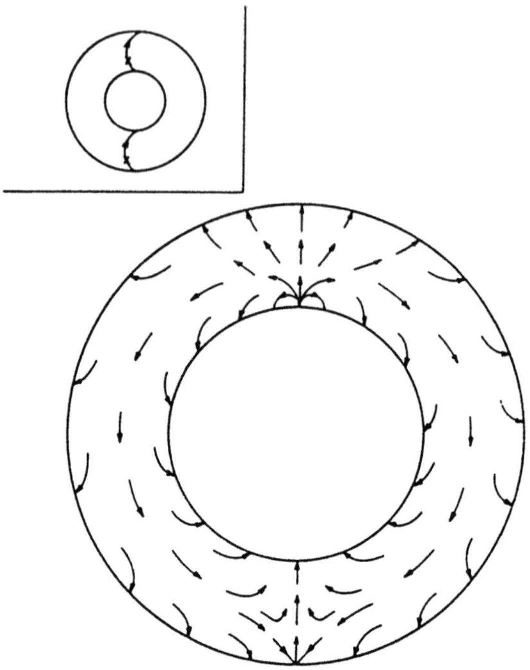

Fig. 41. Two-boojum texture in a torus. The boojum at the top is a circular one like those in Fig. 40; that at the bottom has been called a 'semi-hyperbolic' one by Mermin (1977) because it looks like a circular one in the plane perpendicular to the paper. The inset shows the motion of the boojums necessary to cause superflow decay.

References

Anderson, P.W., and Toulouse, G. (1977): Phys. Rev. Lett. **38** 508

Aslamasow, L.G., Larkin, A.I. (1969): Sov. Phys. JETP **9** 87

Avenel, O., Varoquaux, E. (1985): Phys. Rev. Lett. **55** 2704 and **57** 921(C)

Avenel, O. (1987): Jap. J. of Appl. Phys. **26** (supplement 26-3) 1798

Brinkman, W.F., Cross, M.C. (1978): "Progress in Low Temperature Physics", ed. by D.F. Brewer (North-Holland, Amsterdam) Vol VIIa, p. 105

Davis, J.C., Close, J.D., Zieve, R., Packard, R.E. (1991): Phys. Rev. Lett. **66** 329

de Gennes, P.G. (1974): "The Physics of Liquid Crystals", (Clarendon press, Oxford)

Fetter, A.L. (1986): "Progress in Low Temperature Physics", ed. by D.F. Brewer (North-Holland, Amsterdam) Vol X, p. 1

Fréedericksz, V., Repiewa, A. (1927): Z. Phys. **42** 532

Gammel, P.L., Hall, H.E., Reppy, J.D. (1984): Phys. Rev. Lett. **52** 121

Gammel, P.L., Ho, T.-L., Reppy, J.D. (1984): Phys. Rev. Lett. **52** 1701

Gay, R., Bagley, M., Hook, J.R., Sandiford, D.J., Hall, H.E. (1983): J. Low Temp. Phys. **51** 227

Hakonen, P., Lounasmaa, O.V., Simola J. (1989): Physica B **160** 1

Halperin, W.P., Pitaevskii, L.P. (1990): "Helium Three", (North-Holland, Amsterdam)
Ho, T.-L. (1978): Phys. Rev. B **18** 1144
Ho, T.-L., Mermin, N.D. (1980): Phys. Rev. B **21** 5190
Hook, J.R. (1989): J. Low Temp. Phys. **74** 19
Hook, J.R., Faraj, E., Gould, S.G., Hall, H.E. (1989): J. Low Temp. Phys. **74** 45
Hook, J.R. (1990): "Helium Three", ed. by W.P. Halperin and L.P. Pitaevskii (North-Holland, Amsterdam) p. 135
Kondo, Y., Korhonen, J.S., Krusius, M., Dmitriev, V.V., Mukharskiy, Y.M., Sonin, E.B., Volovik, G.E. (1991): preprint
Kopnin, N.B., Salomaa, M.M. (1990): Phys. Rev. B **41** 2601
Lamb, H. (1932): "Hydrodynamics", 6th edition (Cambridge University Press)
Langer, J.S., Reppy, J.D. (1970): "Progress in Low Temperature Physics", ed. by C.J. Gorter (North-Holland, Amsterdam) Vol. VI, p. 1
Leggett, A.J. (1975): Rev. Mod. Phys. **47** 331
Mermin, N.D., Ho, T.-L. (1976): Phys. Rev. Lett. **36** 594
Mermin, N.D. (1977): "Quantum Fluids and Solids", ed. by S.B. Trickey, E.D. Adams and J.W. Dufty (Plenum, New York) p. 3
Mermin, N.D. (1981): Phys. Today **34**, No. 4, 46
Monien, H., Tewordt, L. (1987): Can. J. Phys. **65** 1388
Osheroff, D.D., Richardson, R.C., Lee, D.M. (1972): Phys. Rev. Lett. **28** 885
Pekola, J.P., Simola, J.T. (1985): J. Low Temp. Phys. **58** 555
Richards, P.L., Anderson, P.W. (1965): Phys. Rev. Lett. **14** 540
Salomaa, M.M., Volovik, G.E. (1983): Phys. Rev. Lett. **51** 2040
Salomaa, M.M., Volovik, G.E. (1985): Phys. Rev. B **31** 203
Spencer, G.F., Ihas, G.G. (1982): Phys. Rev. Lett. **48** 1118
Thuneberg, E.V. (1986): Phys. Rev. Lett. **56** 359
Vinen, W.F. (1961): Proc. Roy. Soc. (London) A **260** 218
Vollhardt, D., Wölfle, P. (1990): "The Superfluid Phases of Helium 3", (Taylor and Francis, London)
Wheatley, J.C. (1975): Rev. Mod. Phys. **47** 415
Whitmore, S.C., Zimmermann Jr., W. (1968): Phys. Rev. **166** 181

NON-HYDRODYNAMIC TRANSPORT IN NORMAL AND SUPERFLUID FERMI LIQUIDS

Dietrich Einzel
Walther-Meißner-Institut für Tieftemperaturforschung
W-8046 Garching bei München, FRG

Preface

This series of lectures is devoted to a detailed study of the transport properties of Fermi fluids and superfluids. Examples of such fluids, which have been investigated experimentally quite extensively, are normal and superfluid phases of liquid ^{3}He and liquid solutions of ^{3}He in ^{4}He. The plan is to calculate all relevant transport parameters for such systems and to compare them with experimental results. The basis for the calculations will be Landau's theory of Fermi liquids and its generalization to Fermi superfluids. One particular aspect of such a theory, namely that it maps the properties of the liquid state into those of a dilute gas of ballistically moving thermal excitations, will be of central importance for these lectures. The calculations are imbedded in a discussion of general aspects like
the phenomenological hydrodynamic description of normal and superfluid Fermi liquids (lecture 1),
the transport properties of normal Fermi liquids (lecture 2),
the two fluid picture of pair-correlated Fermi superfluids (lecture 3),
the properties of the superfluid component (condensate): broken symmetries, s– vs. p–wave pairing (lectures 3 and 9),
the properties of the normal component (quasiparticles): transport and dilute gas behavior (lectures 4 and 5),
the velocity slip effect (lecture 6),
the phenomenological description of flow experiments (lecture 7),
the microscopic description of flow experiments: velocity slip, Knudsen flow and surface impedance of dilute quasiparticle gases (lectures 6 and 8), and
the anomalous spin transport properties of quasiparticle gases induced by the exchange–interaction (lecture 9).

1 Introduction

The transport properties of neutral Fermi liquids such as ^{3}He and mixtures of ^{3}He in ^{4}He have been investigated for many decades. The discovery of the superfluid phases of liquid ^{3}He at the beginning of the seventies and their subsequent theoretical interpretation as a Fermi liquid with Cooper pairs correlated in relative p–wave states has reintensified the interest of theorists and experimentalists in the transport and relaxation properties of the new low temperature phases of this Fermi liquid. Being the neutral analogue of a superconductor and the fermion analogue of superfluid ^{4}He, superfluid ^{3}He has various aspects in common with these two well known systems. It turned out soon, however, that it is the dissimilarities which are most exciting. Therefore, a few general remarks about those similarities and dissimilarities seem to be desirable:

The theoretical picture for the dissipative transport of mass and spin in a normal or a (pair–correlated) superfluid Fermi liquid is that of inelastic two–particle collisions in a dilute gas of thermal excitations. In a normal Fermi liquid thermal excitations are created in the vicinity of the Fermi surface as quasiparticles above and as quasiholes below the Fermi level, their number density vanishing proportional to the temperature T. The dominant thermal excitations in a superfluid Fermi liquid are superpositions of normal state quasiparticles and quasiholes, the so–called Bogoliubov quasiparticles (BQP or "bogolons"), which are characterized by an energy gap Δ_k in their excitation spectrum.

In these lectures I shall restrict my considerations to Fermi superfluids in which pairing correlations lead to a momentum–independent gap $\Delta_k \equiv \Delta$ in the excitation spectrum everywhere on the Fermi surface. (This applies to the superfluid B–phase of liquid ^{3}He, hypothetical s–wave superfluids and, with some caveats, to superconductors.) In this case the excitation number density has an exponential temperature variation $\propto \exp(-\Delta/k_BT)$ in the low temperature limit. As a consequence, the low temperature lifetimes and mean free paths diverge $\propto T^{-2}$ in normal Fermi liquids and $\propto \exp(\Delta/k_BT)$ in isotropic superfluids.

A class of (s– and p–wave) Fermi superfluids with momentum independent gaps is characterized by scalar (isotropic) mass transport parameters (shear viscosity η, second viscosities ζ_i, $i = 1, ...4$ and diffusive thermal conductivity κ).

Let me introduce these transport parameters in context with the linear hydrodynamic conservation laws of mass (continuity equation)

$$\frac{\partial}{\partial t}\delta\rho(\vec{r},t) + \vec{\nabla}\cdot\vec{g}(\vec{r},t) = 0 \tag{1.1}$$

momentum (Navier–Stokes equation)

$$\frac{\partial}{\partial t}g_i(\vec{r},t) + \nabla_j\Pi_{ij}(\vec{r},t) = 0 \tag{1.2}$$

and entropy

$$\frac{\partial}{\partial t}\delta\sigma(\vec{r},t) + \vec{\nabla}\vec{f}(\vec{r},t) = O\left((\vec{\nabla}\cdot\vec{v}^{\,n})^2, (\vec{\nabla}T)^2, ...\right) \tag{1.3}$$

The entropy production on the right hand side of Eq. (1.3) is quadratic in the amplitudes and hence irrelevant for the considerations to follow.

These equations apply to a situation in which both the normal component (i.e. the gas of thermal excitations) and the superfluid component (the condensate of Cooper pairs) are in *local equilibrium* with the external disturbances. In case of the normal component this can be achieved if the typical frequencies of external disturbances are less than the collision frequency of the thermal excitations (*hydrodynamic regime*). At frequencies well below the gap frequency $\Delta/\hbar$, in the so–called *macroscopic regime*, the mass dynamics of the condensate is local and governed entirely by the space and time dependent phase of the condensate wave function $\phi(\vec{r},t) = \phi_0 + \delta\phi(\vec{r},t)$. Its spatial gradients give rise to mass superflow with velocity

$$\vec{v}^{\,s}(\vec{r},t) = \frac{\hbar}{2m}\vec{\nabla}\delta\phi(\vec{r},t) \tag{1.4}$$

whereas its change in time gives rise to a chemical potential shift

$$\delta\mu(\vec{r},t) = -\frac{\hbar}{2}\frac{\partial}{\partial t}\delta\phi(\vec{r},t) \tag{1.5}$$

These two quantities are related through a quasi–conservation law (or acceleration equation) for the condensate variables

$$\frac{\partial}{\partial t}m\vec{v}^{\,s}(\vec{r},t) + \vec{\nabla}\delta\mu(\vec{r},t) = 0 \tag{1.6}$$

The system of equations (1.1), (1.2), (1.3) and (1.6) completely determines the spin-independent hydrodynamic behavior of the system. Let me now specify the currents. The mass current (momentum density) is a conserved quantity and has a well known decomposition into an normal and a superfluid contribution

$$\vec{g}(\vec{r},t) = \rho^{\mathrm{n}}(T)\vec{v}^{\,\mathrm{n}}(\vec{r},t) + \rho^{\mathrm{s}}(T)\vec{v}^{\,\mathrm{s}}(\vec{r},t) \tag{1.7}$$

where $\vec{v}^{\,\mathrm{n}}$ refers to the velocity of the gas of thermal excitations or normal component. All remaining currents are not conserved. They can therefore be decomposed into a reactive and a dissipative contribution. The momentum current can be written as

$$\Pi_{ij}(\vec{r},t) = \delta P(\vec{r},t)\delta_{ij} + \Pi'_{ij}(\vec{r},t) \tag{1.8}$$

Here

$$\delta P(\vec{r},t) = \frac{n}{m\chi_{\mathrm{n}}}\delta\rho(\vec{r},t) = \frac{1}{\kappa}\frac{\delta\rho(\vec{r},t)}{\rho}$$

is the pressure with n the particle density and $\kappa = \chi_{\mathrm{n}}/n^2$ the compressibility. The dissipative part of the momentum current tensor has the form of a hydrodynamic constitutive relation in that it is related to first order gradients in its thermodynamically conjugate (momentum) density:

$$\Pi'_{ij} = -\eta v^{\mathrm{n}}_{ij} - \delta_{ij}\left[\zeta_1\vec{\nabla}\cdot(\vec{g}-\rho\vec{v}^{\,\mathrm{n}}) + \zeta_2\vec{\nabla}\cdot\vec{v}^{\,\mathrm{n}}\right] \tag{1.9}$$

Here η, ζ_1, ζ_2 are the transport coefficients of shear and second viscosities, respectively. In (1.9) the abbreviation $v^{\mathrm{n}}_{ij} = \nabla_i v^{\mathrm{n}}_j + \nabla_j v^{\mathrm{n}}_i - \frac{2}{3}\delta_{ij}\vec{\nabla}\vec{v}^{\,\mathrm{n}}$ has been introduced. The entropy flux has the decomposition

$$\vec{f}(\vec{r},t) = \sigma\vec{v}^{\,\mathrm{n}} - \frac{\kappa}{T}\vec{\nabla}T \tag{1.10}$$

with σ the entropy density and κ the coefficient of thermal conductivity. The chemical potential shift $\delta\mu$ consists of the contributions

$$\frac{1}{m}\delta\mu(\vec{r},t) = \frac{\delta P}{\rho} - \zeta_3\vec{\nabla}\cdot(\vec{g}-\rho\vec{v}^{\,\mathrm{n}}) - \zeta_4\vec{\nabla}\cdot\vec{v}^{\,\mathrm{n}} \tag{1.11}$$

where ζ_3 and ζ_4 are further coefficients of second viscosity. The hydrodynamic equations presented above are formally identical with Khalatnikov's two–fluid hydrodynamic equations for superfluid He–II [1]. One important physical difference between Fermi and Bose liquids, however, originates from the fact that the existence of a Fermi level $\mu = k_{\mathrm{B}}T_{\mathrm{F}}$

is the cause of a strong reduction of the entropy $\sigma = O(T/T_F)$ (and hence of the thermomechanical coupling) as well as of the viscosity coefficients $\zeta_1 = \zeta_4 = O(T^2/T_F^2)$ and $\zeta_2 = O(T^4/T_F^4)$.

The spin-dependent transport and relaxation parameters have no analogue in superfluid ^{4}He. They are isotropic in systems with s-wave pairing correlations but anisotropic in superfluid ^{3}He-B even though the gap is momentum independent. In superfluid ^{3}He-B there exists, besides the ordinary spin diffusion, a further transport parameter, namely the spin analogue Ξ of the second viscosity coefficient ζ_3. It has become well known in the literature under the name Leggett-Takagi or intrinsic spin relaxation of the quasiparticle magnetization and is absent in systems with s-wave pairing. The differences between systems with s- and p-wave pairing can be traced back to the broken symmetries characteristic of the condensate of Cooper pairs. In addition to the broken gauge symmetry in superconductors, superfluid ^{3}He displays what is referred to as spontaneously broken spin orbit symmetry (SBSOS), which gives rise to a complex intrinsic spin structure of the p-wave order parameter and leads to macroscopic anisotropies even for momentum independent gaps. As a consequence, the spin diffusion of superfluid ^{3}He-B is a tensor quantity with (at least) two nonvanishing eigenvalues.

A comprehensive discussion of a phenomenological description of the spin dynamics in p-wave superfluids can be found in the article by Leggett and Takagi [2]. Analogue to the mass transport case, the spin transport parameters of a neutral isotropic superfluid can be introduced by considering the conservation laws for the magnetization densities and currents. The magnetization conservation law is of the form (Bloch-Leggett-Takagi equation [2]):

$$\frac{\partial}{\partial t}\vec{M}(\vec{r},t) + \nabla_i J_i^M(\vec{r},t) = \gamma \vec{M} \times \vec{H}^{\rm ext}(\vec{r},t) + \gamma \vec{R}_{\rm D} \tag{1.12}$$

Here γ denotes the gyromagnetic ratio of the spin system under consideration and $\vec{H}^{\rm ext}$ is the applied external magnetic field. The broken spin-orbit-symmetry manifests itself in a spin analogue of the pair of relations (1.4) and (1.5). I shall demonstrate in lecture 3 that the spin and orbital degrees of freedom of the order parameter for superfluid ^{3}He-B are related through a three dimensional rotation matrix $R_{\mu\nu}(\hat{n},\theta) = R_{\mu\nu}(\hat{n}_0,\theta_0) - \epsilon_{\mu\nu\alpha}\delta\theta_\alpha$ with $\delta\vec{\theta} = \hat{n}\delta\theta$, which is specified by an axis $\hat{n}$ and an angle θ. The degeneracy of the p-wave order parameter with respect to the rotation angle θ is removed by the dipole interaction resulting in a finite equilibrium value $\theta_0 = \arccos(-\frac{1}{4})$, the so-called Leggett-angle. The equilibrium orientation $\hat{n}_0$ turns out to be preferably along the direction of the magnetic field. The dipole interaction — tiny in normal Fermi liquids — is amplified by the presence of p-wave pairing correlations to a macroscopic restoring force $\vec{R}_D \propto \delta\vec{\theta}$ in the Bloch equation (1.12), the so-called dipole torque. Spatial gradients in the spin-orbit phase variable $\delta\vec{\theta}$ give rise to spin superflow:

$$\vec{v}_i^{\,s}(\vec{r},t) = \frac{\hbar}{2m}\nabla_i\delta\vec{\theta}(\vec{r},t) \tag{1.13}$$

Its time variation causes a change $\delta\vec{\mu}$ in the magnetic energy

$$\delta\vec{\mu}(\vec{r},t) = -\frac{\hbar}{2}\frac{\partial}{\partial t}\delta\vec{\theta}(\vec{r},t) \tag{1.14}$$

The variables $\vec{v}_i^{\,s}$ and $\delta\vec{\mu}$ govern exclusively the spin dynamics of the condensate in the macroscopic limit and are related by a spin analogue of the quasi-conservation law (1.6):

$$\frac{\partial}{\partial t}m\vec{v}_i^{\,s}(\vec{r},t) + \nabla_i\delta\vec{\mu}(\vec{r},t) = 0 \tag{1.15}$$

The magnetization current has a decomposition into a reactive (condensate) and a dissipative (quasiparticle) contribution

$$\vec{J}_i^M(\vec{r},t) = \frac{\gamma\hbar}{2}\overset{\leftrightarrow s}{n}_{ij}(T)\vec{v}_j^{\,s}(\vec{r},t) - \overset{\leftrightarrow}{D}_{ij}(T)\nabla_j\delta\vec{M}(\vec{r},t) \tag{1.16}$$

where $\overset{\leftrightarrow s}{n}_{ij}$ is a spin superfluid density tensor which characterizes the reactive spin current response, $\overset{\leftrightarrow}{D}_{ij}$ is the spin diffusion tensor, and $\delta\vec{M} = \vec{M} - \chi_M\vec{H}^{ext}$ with χ_M the magnetic (spin) susceptibility. Similar to Eq. (1.11) the quantity $\delta\vec{\mu}$ has a decomposition

$$\delta\vec{\mu}(\vec{r},t) = \frac{\gamma\hbar}{2}\frac{1}{\chi_M}\left[1 + \Xi\frac{\partial}{\partial t}\right]\delta\vec{M}(\vec{r},t) \tag{1.17}$$

into a reactive and a dissipative part, where the first term is the spin analogue of the pressure and Ξ is the parameter of intrinsic spin relaxation which represents the dissipative part. Note that all terms involving the vectorial phase $\delta\vec{\theta}$ are absent in systems with s-wave pairing correlations.

Let me now briefly summarize what is known about the temperature dependence of the transport parameters, introduced above. For such considerations it is useful to keep in mind that Landau's theory of Fermi liquids, which establishes a one-to-one correspondence between real particles and quasiparticles, is de facto no longer a theory of *liquids* but rather of *dilute gases* of thermal excitations (Landau and Bogoliubov quasiparticles). A simple gas–kinetic picture applied to the thermal excitations of a normal Fermi liquid leads to a temperature variation $\propto T^{-2}$ for the viscosity η, the diffusive thermal conductivity κ/T and the spin diffusion constant D. A corresponding bogolon gas–kinetic picture, in which the exponential decrease of the normal fluid density is compensated by an exponential increase in the lifetime leads to transport parameters which have a constant low temperature limit or at most some power law variation $\propto T^{\nu}$. In the B–phase of superfluid ^{3}He, ν turns out to be 0 in case of both the shear viscosity η and the second viscosity ζ_3 [3]. The exponent ν equals -1 for the diffusive thermal conductivity κ [3], the components of the spin diffusion tensor $\overleftrightarrow{D}$ [4] and the relaxation parameter Ξ of the Bogoliubov quasiparticle magnetization [4].

A satisfactory understanding of experimental data for the transport parameters has been possible only in the last few years for the normal Fermi liquid phase and even more recently for the superfluid phases. This is due to the experimental conditions necessary for

(i) entraining the Fermi liquid into measuring cells of finite size and

(ii) applying finite magnetic fields which force the Fermi liquid out of its hydrodynamic regime.

One is therefore faced with the problem of incorporating these experimental conditions into the description of transport.

Two kinds of nonhydrodynamic effects can drastically modify the transport properties of a gas of thermal excitations:

The first (spatial confinement or finite size) effect is connected with quasiparticle mean free paths becoming comparable to or larger than the typical size of the measuring cell at low temperatures. As a consequence for fluid flow, macroscopic (normal) velocity field profiles no longer match the velocities of the confining walls but are seen to stay finite and extrapolate to zero only a length ζ_0 (slip length) behind the walls (velocity slip effect). The slip length generally scales with the mean free path of the thermal excitations and enters the transport theory as an additional temperature and pressure dependent parameter [5]. Experiments particularly on the shear viscosity, performed in both the slip and the wall scattering dominated (or Knudsen) regime, reveal a dramatic nonhydrodynamic "droop" by several orders of magnitude below the expected plateau value for infinitely extended normal and superfluid Fermi liquids. [6], [7] and [8].

The second (temporal) nonhydrodynamic effect occurs whenever the frequency of the external field (e. g. an acoustic or a magnetic field) becomes comparable with or larger than the quasiparticle collision rate. In this so-called *collisionless regime* the influence of the Landau quasiparticle interaction is known to dominate the transport of the system. A typical example for the latter kind of nonhydrodynamic behavior of a transport parameter is the Leggett–Rice effect on the spin diffusion of a normal Fermi liquid in a finite magnetic gradient field [9]. The spin diffusion coefficient

(i) becomes a tensor quantity with components parallel and perpendicular to the polarization direction.

(ii) Its perpendicular component vanishes at low temperature as T^2 instead of diverging as T^{-2} ("diffusion blocking") and

(iii) it acquires a reactive component which gives rise to spin waves [10] and multiple spin echoes [11].

This effect has its origin in the exchange contribution to the Landau Fermi liquid interaction between quasiparticles, which acts so as to rotate the macroscopic magnetization current about the direction of spin polarization. (Therefore, the effect is sometimes referred to as spin current rotation effect in the literature.) At the superfluid transition this nonhydrodynamic effect carries over into the superfluid phases where it remains as a property of the normal component (i.e. the gas of Bogoliubov quasiparticles).

It is the purpose of this series of lectures to establish first a two fluid description for pair correlated superfluid Fermi liquids on the basis of which the set of hydrodynamic equations, discussed above, can be derived. The mathematical framework will be the quasiclassical kinetic equations of the Landau–Boltzmann type. Second, I wish to calculate *all* relevant transport parameters for such systems, namely the shear viscosity η, the second viscosity ζ_3, the thermal conductivity κ, the components of the spin diffusion tensor $\overleftrightarrow{D}$ and the intrinsic spin relaxation parameter Ξ. Third, I would like to discuss the necessary extensions of such a theory to include realistic experimental conditions such as flow channels of finite (narrow) size and pulsed NMR in finite (large) magnetic fields.

2 Transport in normal Fermi liquids

The properties of normal Fermi liquids sufficiently below their degeneracy temperature $T_F = E_F/k_B$ are well accounted for by Landau's semiphenomenological quasiparticle theory [12]. In this theory the state of a normal Fermi liquid at a temperature T is described by a spin density matrix distribution function

$$\mathbf{n}_k(\vec{r},t) = \begin{pmatrix} n_{k\uparrow\uparrow}(\vec{r},t) & n_{k\uparrow\downarrow}(\vec{r},t) \\ n_{k\downarrow\uparrow}(\vec{r},t) & n_{k\downarrow\downarrow}(\vec{r},t) \end{pmatrix} = n_k(\vec{r},t)\tau^0 + \vec{n}_k(\vec{r},t)\cdot\vec{\tau} \tag{2.1}$$

for a quasiparticle with momentum $\hbar\vec{k}$ and spin $(\hbar/2)\vec{\tau}$ at a space point $\vec{r}$ and time t. Here τ^0 is the 2×2 unit matrix and $\vec{\tau} = \{\tau^x, \tau^y, \tau^z\}$ is the vector of the Pauli spin matrices. In *global* equilibrium $\mathbf{n}_k$ is given by the equilibrium Fermi distribution

$$n_k^0(\epsilon_k) = \frac{1}{\exp(\epsilon_k - \mu)/k_B T + 1} \tag{2.2}$$

where $\epsilon_k = \hbar^2\vec{k}^2/2m^* = \xi_k + \mu$ is the equilibrium quasiparticle energy, $\mu \equiv E_F$ the chemical potential (= Fermi energy) and m^* the quasiparticle effective mass. The energy ξ_k measures the quasiparticle energy from the Fermi level and allows the distinction between *particle*-like ($\xi_k > 0$) and *hole*-like ($\xi_k < 0$) thermal excitations. The total number density $n_q(T)$ of thermal excitations in equilibrium can be calculated from the equilibrium Fermi distribution (2.2) by associating a positive excitation energy $|\xi_k|$ with *both* thc quasiparticles and the quasiholes:

$$n_q(T) = 2\sum_{\vec{k}} n_k^0(|\xi_k|) = 2\ln 2\; N_F k_B T \tag{2.3}$$

Here $N_F = m^* p_F/\pi^2\hbar^3 = (3/2)n/E_F$ is the quasiparticle density of states for both spin projections. As a consequence, the thermal excitations have a number density $\propto k_B T/E_F$ and form a dilute gas at low temperature.

A quantity intimitely connected with the excitation density n_q is the entropy density

$\sigma(T)$, which is exclusively determined by the properties of the dilute quasiparticle gas:

$$\begin{aligned}\sigma(T) &= k_B\, 2\sum_{\vec{k}}\left[n_k^0 \ln\, n_k^0 + (1-n_k^0)\ln(1-n_k^0)\right] \\ &= 2\sum_{\vec{k}} \frac{\xi_k^2}{T}\varphi_k = \frac{\pi^2}{3} N_F k_B^2 T = \frac{\pi^2 k_B}{6\ln 2}\, n_q(T) \\ \varphi_k &= -\frac{\partial n_k^0}{\partial \xi_k}\end{aligned} \tag{2.4}$$

One further important property specifying the dilute gas behavior of the thermal excitations is the T^{-2} temperature variation of the quasiparticle lifetime

$$\tau_N(\xi_k, T) = \frac{\tau_N(0,T)}{I(\xi_k,T)} \quad ; \quad I(\xi_k,T) = 1 + \left(\frac{\xi_k}{\pi k_B T}\right)^2 \tag{2.5}$$

and its thermal average

$$\bar{\tau}_N(T) = \frac{\tau_N(0,T)}{\int_{-\infty}^{+\infty} d\xi_k \varphi_k\; I(\xi_k,T)} = \frac{3}{4}\tau_N(0,T) \tag{2.6}$$

with $\tau_N(0,T) = 32 E_F \hbar/(\pi k_B T)^2 < W >_a$. Here the quantity $W = W(\theta,\phi)$ is the effective quasiparticle scattering cross section which depends on the two so-called Abrikosov–Khalatnikov angles [13] θ and ϕ only and can be expressed by the singlet and triplet quasiparticle scattering amplitudes $A_0(\theta,\phi)$ and $A_1(\theta,\phi)$, [14] respectively, as $W = \pi[A_0^2 + 3A_1^2]/4$. The angular average is defined as $< \cdots >_a = 2\int_0^1 d\cos(\theta/2) \int_0^{2\pi} d\phi/2\pi\; \cdots$.

Let me now consider slight deviations from global equilibrium caused by the presence of external fields, which I assume, for the time being, to vary only weakly in space and time. They induce changes in the quasiparticle energy which are of the general form:

$$\delta\epsilon_k^{ext}(\vec{r},t) = -\left\{[eU^{ext}(\vec{r},t) + \vec{p}\cdot\vec{v}^{ext}(\vec{r},t)]\tau^0 + [\frac{\gamma\hbar}{2}\vec{H}^{ext}(\vec{r},t) + \cdots]\cdot\vec{\tau}\right\} \tag{2.7}$$

The first term in (2.7) is relevant for charged fermions (electrons) with U^{ext} a scalar potential and e the (electronic) charge. The second term describes the coupling to an external velocity field $\vec{v}^{ext}$ for neutral fermions and to a vector potential $\vec{v}^{ext} \equiv e\vec{A}^{ext}/mc$ for electrons. The spin part of the energy change is characterized by an external magnetic field $\vec{H}^{ext}$ and γ denotes the gyromagnetic ratio of the fermions under consideration. The external fields induce a change $\delta n_k(\vec{r},t) = [n_k^0 + \delta n_k]\tau^0 + \delta\vec{n}_k \cdot \vec{\tau}$ in the quasiparticle distribution which, in turn, give rise to a Landau molecular field contribution $\delta\epsilon_k^{FL}$ to the change in the quasiparticle energy:

$$\begin{aligned}\delta\epsilon_k^{\rm tot}(\vec r,t) &= \delta\epsilon_k^{\rm ext}(\vec r,t)+\delta\epsilon_k^{\rm FL}(\vec r,t)\\ \delta\epsilon_k^{\rm FL}(\vec r,t) &= \frac{2}{N_{\rm F}}\sum_\ell\sum_{\vec k'} P_\ell(\hat k\cdot\hat k')\left[F_\ell^{\rm s}\delta n_k\tau^0+F_\ell^{\rm a}\delta\vec n_k\cdot\vec\tau\right](\vec r,t)\end{aligned} \tag{2.8}$$

The Fermi liquid interaction between quasiparticles in quantum states $|\hbar\vec k|$ and $|\hbar\vec k'|$ is essentially restricted to momenta $|\hbar\vec k|, |\hbar\vec k'| \approx p_{\rm F}$ close to the Fermi momentum. The Landau molecular field contribution can therefore be written as a Legendre expansion, which is characterized by a set of dimensionless interaction parameters, the so–called Landau parameters $F_\ell^{\rm s}$ (spin–symmetric case) and $F_\ell^{\rm a}$ (spin–antisymmetric case), the first few of which ($\ell = 0,1,2$) can be determined from experiment (see below). I should note that for Galilean invariant Fermi systems the Landau parameter $F_1^{\rm s}$ determines the effective mass m^* via the so–called effective mass relation $m^*/m = 1 + F_1^{\rm s}/3$. In addition to $\delta\epsilon_k^{\rm FL}$ let us now also allow for local changes in the temperature T and the chemical potential μ:

$$\begin{aligned}T(\vec r,t) &= T\ +\delta T(\vec r,t)\\ \mu(\vec r,t) &= E_{\rm F}+\delta\mu(\vec r,t)\end{aligned}$$

If the variation in space and time of these field variables is weak, one may introduce the very useful concept of *local equilibrium* by writing the deviation of the distribution function in the presence of the fields from that representing global equilibrium as:

$$\begin{aligned}\delta {\rm n}_k^{\rm loc}(\vec r,t) &= n_k^0\left[\xi_k\tau^0+\delta\epsilon_k^{\rm tot}-\left(\delta\mu+\frac{\xi_k}{T}\delta T\right)\tau^0\right]-n_k^0(\xi_k)\\ &= -\varphi_k\left[\delta\epsilon_k^{\rm tot}-(\delta\mu+\frac{\xi_k}{T}\delta T)\tau^0\right]\end{aligned} \tag{2.9}$$

where the derivative of the Fermi function φ_k has been defined through Eq. (2.4). In Eq. (2.9) the implicit structure of Fermi liquid theory becomes transparent with the dependence of the distribution function on the quasiparticle energy and vice versa. With the aid of (2.9) one may now compute macroscopic observable densities in local equilibrium according to

$$\delta A_\mu^{\rm loc}(\vec r,t)=2\sum_{\vec k} a_k\,\frac{1}{2}{\rm tr}[\tau^\mu\delta{\rm n}_k^{\rm loc}(\vec r,t)] \tag{2.10}$$

Examples of such observable densities are summarized in table 1. We are now able to study the local response properties of the Fermi liquid by inserting the local equilibrium

OBSERVABLE DENSITY	$\delta A_\mu(\vec{r},t)$	a_k	μ
PARTICLE	$\delta n(\vec{r},t)$	1	0
MASS	$\delta\rho(\vec{r},t)$	m	0
ENERGY	$\delta\epsilon(\vec{r},t)$	$\xi_k+\mu$	0
ENTROPY	$\delta\sigma(\vec{r},t)$	ξ_k/T	0
SPIN	$S_\mu(\vec{r},t)$	$\hbar/2$	x,y,z
MAGNETIZATION	$M_\mu(\vec{r},t)$	$\gamma\hbar/2$	x,y,z

Table 1: Observable densities in a normal Fermi liquid

distribution function (2.9) into the expression for the macroscopic observable densities (2.10). From such a procedure one gets

(i) the local density response

$$\delta n(\vec{r},t) = \chi_n[eU^{\rm ext}(\vec{r},t) + \delta\mu(\vec{r},t)] \;\; ; \;\; \chi_n = \frac{N_{\rm F}}{1+F_0^s} \tag{2.11}$$

characterized by a renormalized temperature independent density response function χ_n or compressibility $\kappa = \chi_n/n^2$,

(ii) the local entropy response

$$\delta\sigma(\vec{r},t) = \frac{C_{\rm V}(T)}{T}\delta T(\vec{r},t) \;\; ; \;\; C_{\rm V}(T) = \frac{\pi^2}{3}N_{\rm F}k_{\rm B}^2T \equiv \sigma(T) \tag{2.12}$$

with the specific heat $C_{\rm V}(T)$ and

(iii) the local magnetization response

$$\vec{M}(\vec{r},t) = \chi_M\vec{H}^{\rm ext}(\vec{r},t) \;\; ; \;\; \chi_M = \left(\frac{\gamma\hbar}{2}\right)^2 \frac{N_{\rm F}}{1+F_0^a} \tag{2.13}$$

with an exchange–enhanced temperature independent (Pauli) spin susceptibility χ_M.

As a next step let us now consider perturbations which drive the system away from local equilibrium and give rise to transport processes. In such a situation, the mass and spin dynamics of the Fermi liquid is governed by the quasiclassical Landau–Boltzmann–Silin equation [15]

$$\frac{\partial}{\partial t}\delta \mathrm{n}_k(\vec{r},t) + \vec{v}_k\cdot\vec{\nabla}\,\mathrm{h}_k(\vec{r},t) = \delta\mathrm{I}_k(\vec{r},t) - \frac{2}{\hbar}\left[\delta\vec{n}_k(\vec{r},t)\times\delta\vec{\epsilon}_k^{\,\rm tot}(\vec{r},t)\right]\cdot\vec{\tau} \tag{2.14}$$

In (2.14) $\vec{v}_k = \hbar\vec{k}/m^*$ is the quasiparticle group velocity and

$$\mathrm{h}_k(\vec{r},t) = \delta \mathrm{n}_k(\vec{r},t) + \varphi_k\, \delta\epsilon_k^{\mathrm{tot}}(\vec{r},t)$$

is the deviation of the distribution from its local equilibrium form. Note that the (nonlinear) precession term involves the *total* quasiparticle energy change including the exchange parts of the Landau molecular field.

The collision integral $\delta \mathrm{I}_k$ describes inelastic two–quasiparticle collision processes, which dominate at low temperature. It is a functional of the deviation from local equilibrium, h_k, only:

$$\delta \mathrm{I}_k = -\frac{1}{\tau_{\mathrm{N}}(0,T)} \frac{2}{N_{\mathrm{F}}} \sum_{\vec{k}'} J(\xi_k, \xi_{k'})$$
$$\left\{ \mathrm{h}_k - \frac{\varphi_k}{\varphi_{k'}} \sum_\ell (2\ell+1)\, P_\ell(\hat{k}\cdot\hat{k}')\, [\lambda_\ell^t\, h_{k'}^{(t)} \tau^0 + \lambda_\ell^a\, \vec{h}_{k'} \cdot \vec{\tau}] \right\} \tag{2.15}$$

Here $t = \pm 1$ denotes the parity of the spin–independent part of the distribution function h_k with respect to the operation $\xi_k \to -\xi_k$. An inspection of Eq. (2.10) shows that t is negative only in the case of the entropy (energy) density. The collision operator J in Eq. (2.15) reads ($x = \xi/k_{\mathrm{B}}T$):

$$J(x,x') = \frac{1}{\pi^2 k_{\mathrm{B}} T} \frac{(x-x')}{\sinh([x-x']/2)} \frac{\cosh(x/2)}{\cosh(x'/2)}$$

It has the properties

$$\begin{aligned} \frac{\varphi_k}{\varphi_{k'}} J(\xi,\xi') &= J(\xi',\xi)) \\ \int_{-\infty}^{\infty} d\xi' J(\xi,\xi') &= I(\xi,T) \end{aligned}$$

In (2.15) the quantities λ_ℓ are pressure–dependent scattering parameters

$$\begin{aligned} \lambda_\ell^t &= \frac{1}{\langle W \rangle_a} \left\langle W[-tP_\ell(\hat{k}_1\cdot\hat{k}_2) + P_\ell(\hat{k}_1\cdot\hat{k}_3) + P_\ell(\hat{k}_1\cdot\hat{k}_4)] \right\rangle_a \\ \lambda_\ell^a &= \frac{1}{\langle W \rangle_a} \left\langle -W_2 P_\ell(\hat{k}_1\cdot\hat{k}_2) + W_3 P_\ell(\hat{k}_1\cdot\hat{k}_3) + W_4 P_\ell(\hat{k}_1\cdot\hat{k}_4) \right\rangle_a \end{aligned} \tag{2.16}$$

which are weighted angular averages over the quasiparticle scattering cross sections $W = -W_2 + W_3 + W_4$ and $W_2 = \pi[A_1^2 - A_0^2]/4$, $W_3 = \pi[A_1 + A_0]A_1/2$ and $W_4 = \pi[A_1 - A_0]A_1/2$. The unit vectors $\hat{k}_i, i = 1,...4$ point into the direction of the momenta (on the Fermi surface) of incoming and outgoing quasiparticles, respectively.

μ	ℓ	m	$a^{\mu}_{\ell m}(\xi)$	CONSERVATION LAW FOR
0	0	0	1	MASS
0	1	-1,0,1	1	MOMENTUM
0	0	0	$\mathrm{sgn}(\xi_k)$	ENERGY
x,y,z	0	0	1	SPIN

Table 2: Collisional invariants of a normal Fermi liquid

An inspection of the definition (2.16) shows that there are a few special cases of pressure–independent scattering parameters λ_ℓ, namely $\lambda_0^+ = \lambda_1^+ = \lambda_0^-/3 = \lambda_0^a \equiv 1$. I shall demonstrate below that these are related to the conservation laws for mass, momentum, energy and spin. For a discussion of the conservation properties of the collision integral it is convenient to evaluate the moments

$$\begin{aligned} 2\sum_k \frac{1}{2}\mathrm{tr}[\mathbf{a}_k \delta \mathbf{I}_k] &= -\frac{1}{\tau_N(0,T)}\frac{2}{N_F}\sum_{\vec{k}} 2\sum_{\vec{k}'} J(\xi_k, \xi_{k'})\delta a^{\mu}_{kk'} h_{k'\mu} \\ \delta a^{\mu}_{kk'} &= \sum_{\ell,m}\left[a^{\mu}_{\ell m}(\xi) - \lambda_\ell a^{\mu}_{\ell m}(\xi')\right] Y_{\ell m}(\hat{k}) \end{aligned}$$

Clearly, these moments vanish for the set of $a^{\mu}_{\ell m}$ shown in table 2 which is therefore referred to as collisional invariants of the system (8 in a normal Fermi liquid). These collisional invariants are particularly important for the choice of approximations for the collision integral. In what follows I would like to list two possibilities for approximating the collision operator $J(\xi,\xi')$ which is currently applied in the literature:

(i) the separate kernel approximation

$$\begin{aligned} J(\xi,\xi') &= \varphi_{k'} I(\xi_k) I(\xi_{k'}) \left[\frac{1}{< I(\xi) >_\varphi}\frac{1+t}{2} + \frac{1}{3}\frac{\xi_k \xi_{k'}}{< \xi^2 I(\xi) >_\varphi}\frac{1-t}{2}\right] \\ < A(\xi) >_\varphi &= \int_{-\infty}^{\infty} d\xi_k \varphi_k A(\xi_k) \end{aligned} \tag{2.17}$$

(ii) the conserving relaxation time approximation

$$J(\xi,\xi') = \varphi_{k'} < I(\xi) >_\varphi \left[\frac{1+t}{2} + \frac{1}{3}\frac{\xi_k \xi_{k'}}{< \xi^2 >_\varphi}\frac{1-t}{2}\right] \tag{2.18}$$

One can easily convince onself that the approximations (2.17) and (2.18) preserve the collisional invariants of the exact collision operator. The separate kernel approximation

(2.17), which keeps essential parts of the energy dependence of the collision operator, is known to lead to results for the transport parameters, which are off available exact results [16] by typically one percent [17]. The conserving relaxation time approximation (2.18) is the simplest approximation compatible with the conservation laws, it nevertheless has an accuracy of a few percent compared with the exact solution of the transport equation. In what follows I restrict all further discussions of collision effects to the use of the conserving relaxation time approximation. To illustrate its use, let me discuss the conservation laws for macroscopic observables as obtained from the Landau–Boltzmann–Silin equation (2.14). Generalizing the definition for the moments (2.10) to the case of the full nonequilibrium distribution δn_k, one obtains linearized conservation laws

(i) for the mass density (continuity equation):

$$\frac{\partial}{\partial t}\delta\rho(\vec{r},t) + \vec{\nabla}\cdot\vec{g}(\vec{r},t) = -\frac{1}{\bar{\tau}_{\mathrm{N}}(T)}(1-\lambda_0^+)\;\delta\rho'(\vec{r},t) \quad ; \quad \lambda_0^+ \equiv 1 \tag{2.19}$$

with momentum density $\vec{g}$ (or fluid velocity $\vec{v}$)

$$\vec{g}(\vec{r},t) = m\;2\sum_{\vec{k}} \vec{v}_k\; h_k(\vec{r},t) = \rho(\vec{v}-\vec{v}^{\mathrm{ext}})$$

(ii) for the momentum density (Navier–Stokes equation):

$$\frac{\partial}{\partial t}g_i(\vec{r},t) + \nabla_j\Pi_{ij}(\vec{r},t) = -\frac{1}{\bar{\tau}_{\mathrm{N}}(T)}(1-\lambda_1^+)\;g_i'(\vec{r},t) \quad ; \quad \lambda_1^+ \equiv 1 \tag{2.20}$$

with momentum current

$$\Pi_{ij}(\vec{r},t) = p_{\mathrm{F}}v_{\mathrm{F}}\;2\sum_{\vec{k}}\{\frac{1}{3}\delta_{ij} + (\hat{k}_i\hat{k}_j - \frac{1}{3}\delta_{ij})\}\;h_k(\vec{r},t) = \delta P(\vec{r},t)\delta_{ij} + \Pi'_{ij}(\vec{r},t)$$

and $\delta P(\vec{r},t) = (1+F_0^s)\delta n(\vec{r},t)/N_{\mathrm{F}}$ the pressure,

(iii) for the entropy density

$$\frac{\partial}{\partial t}\delta\sigma(\vec{r},t) + \vec{\nabla}\cdot\vec{f}(\vec{r},t) = -\frac{1}{\bar{\tau}_{\mathrm{N}}(T)}(1-\frac{\lambda_0^-}{3})\;\delta\sigma(\vec{r},t) \quad ; \quad \lambda_0^- \equiv 3 \tag{2.21}$$

with entropy flux

$$\vec{f}(\vec{r},t) = 2\sum_{\vec{k}} \vec{v}_k\frac{\xi_k}{T}\;h_k(\vec{r},t)$$

and

(iv) the magnetization density (Bloch equation)

$$\frac{\partial}{\partial t}\vec{M}(\vec{r},t) + \nabla_i\vec{J}_i^M(\vec{r},t) = \vec{M}(\vec{r},t)\times\vec{\omega}_{\mathrm{L}}(\vec{r}) - \frac{1}{\bar{\tau}_{\mathrm{N}}(T)}(1-\lambda_0^a)\;\vec{M}' \quad ; \quad \lambda_0^a \equiv 1 \tag{2.22}$$

with $\vec{\omega}_{\rm L}(\vec{r}) = \gamma \vec{H}^{\rm ext}(\vec{r})$ the local Larmor frequency and with a magnetization current

$$\vec{J}_i^M(\vec{r},t) = \frac{\gamma\hbar}{2}\, 2\sum_{\vec{k}} v_{ki}\, \vec{h}_k(\vec{r},t)$$

The primed quantities on the right hand sides of the conservation laws denote the deviation of the density under consideration from its local equilibrium value. The conserving relaxation time approximation is in fact not only capable of reproducing correctly the conservation laws, it provides also a rather good approximate treatment of the relaxation properties of the Fermi liquid, i.e. the transport parameters, as I shall demonstrate now. In the following I list the current densities which obey relaxation rather than conservation laws. The hydrodynamic transport parameters of the normal Fermi liquid are then obtained by a first order gradient expansion performed in the kinetic equation, which leads in general to (so-called constitutive) relations between the decaying currents and gradients in the thermodynamically conjugate densities:

(v) momentum current

$$\begin{aligned} \frac{\partial}{\partial t}\Pi'_{ij}(\vec{r},t) + \ldots &= -\frac{1}{\tau_\eta(T)}\Pi'_{ij}(\vec{r},t) \quad ; \quad \tau_\eta(T) = \bar{\tau}_{\rm N}(T)/(1-\lambda_2^+) \\ \Pi'_{ij}(\vec{r},t) &= -\eta(T) v_{ij}(\vec{r},t) \quad ; \quad \eta(T) = \frac{1}{5} n p_{\rm F} v_{\rm F} \tau_\eta(T) \end{aligned} \tag{2.23}$$

with a viscous transport (or momentum current decay) time $\tau_\eta(T)$, the coefficient of shear viscosity $\eta(T)$ and the definition $v_{ij} = \nabla_i v_j + \nabla_j v_i - \frac{2}{3}\delta_{ij}(\vec{\nabla}\cdot\vec{v})$,

(vi) entropy flux

$$\begin{aligned} \frac{\partial}{\partial t}\vec{f}(\vec{r},t) + \ldots &= -\frac{1}{\tau_\kappa(T)}\vec{f}(\vec{r},t) \quad ; \quad \tau_\kappa(T) = \bar{\tau}_{\rm N}(T)/(1-\frac{\lambda_1^-}{3}) \\ \vec{f}(\vec{r},t) &= -\frac{\kappa(T)}{T}\,\vec{\nabla}\delta T(\vec{r},t) \quad ; \quad \kappa(T) = \frac{1}{3} C_{\rm V}(T) v_{\rm F}^2 \tau_\kappa(T) \end{aligned} \tag{2.24}$$

with $\tau_\kappa(T)$ the thermal conductivity transport (or entropy flux decay) time and $\kappa(T)$ the coefficient of diffusive thermal conductivity.

(vii) hydrodynamic magnetization current ($\vec{\omega}_{\rm L} \to 0$)

$$\begin{aligned} \frac{\partial}{\partial t}\vec{J}_i^M(\vec{r},t) + \ldots &= -\frac{1}{\tau_{\rm D}(T)}\vec{J}_i^M(\vec{r},t) \quad ; \quad \tau_{\rm D}(T) = \bar{\tau}_{\rm N}(T)/(1-\lambda_1^a) \\ \vec{J}_i^M(\vec{r},t) &= -D_0(T)\nabla_i \delta\vec{M}(\vec{r},t) \quad ; \quad D_0(T) = \frac{v_{\rm F}^2}{3}\,(1+F_0^a)\,\tau_{\rm D}(T) \end{aligned} \tag{2.25}$$

with spin diffusion transport (or magnetization current decay) time $\tau_{\rm D}(T)$. In (2.25) I have defined $\delta\vec{M}(\vec{r},t) = \vec{M}(\vec{r},t) - \chi_M \vec{H}^{\rm ext}(\vec{r},t)$.

This concludes the list of hydrodynamic transport parameters of a normal Fermi liquid and we can now turn to a generalization of these concepts to such Fermi liquids in which (s– and p–wave) pairing correlations lead to the occurrence of superfluidity or superconductivity.

3 Kinetic theory of pair–correlated Fermi liquids

In this lecture, I would like to discuss the necessary extensions of Fermi liquid transport theory to the case of pair–correlated superfluid Fermi liquids. Being primarily interested in the dissipative transport of such systems, which is a property of the gas of thermal excitations alone, I shall concentrate on the dynamics of the relevant thermal excitations (the Bogoliubov quasiparticles) which form the normal component in a two fluid description. In doing so, I shall restrict myself to cases where the pairing correlations lead to a momentum–independent gap $\Delta(T)$ in the spectrum of thermal excitations. This happens in the B–phase of superfluid ^{3}He (neutral fermions experience p–wave pairing, mediated by the exchange of virtual "paramagnons" [18]), in ordinary superconductors (charged fermions experience s–wave pairing, mediated by the exchange of virtual phonons) and in its neutral analogue, a ficticious Fermi liquid with s–wave pairing correlations. It is illustrative to discuss such a system together with superfluid ^{3}He–B because the exotic features of the latter (new broken symmetries, anisotropies, etc.) can be understood more clearly by this comparison. The assumption of isotropic gaps is mainly motivated by the fact that in this case the dilute quasiparticle gas behavior, which leads to nonhydrodynamic effects, particularly mean free path effects and the superfluid analogue of the Leggett–Rice effect, is most pronounced. To keep things on a simple level I shall limit the discussion of the quasiparticle dynamics to the small amplitude (linear response) regime of external fields. Let me start with a description of the order parameter representing the superfluid component of pair–correlated superfluids. In general, the order parameter has a 2 × 2 matrix structure in spin space which originates from the coupling of the pair spins (in units of $\hbar$) of magnitude 1/2 and projections σ_1 and σ_2 to a total spin s and the corresponding total projection m_s via the Clebsch–Gordon coefficient

$$C^{s\ m_s}_{\sigma_1\sigma_2} = \left(\begin{array}{cc|c} \frac{1}{2} & \frac{1}{2} & s \\ \sigma_1 & \sigma_2 & m_s \end{array} \right) = \begin{pmatrix} \delta_{s,1}\delta_{m_s,1} & \delta_{m_s,0}/\sqrt{2} \\ (-1)^{s+1}\delta_{m_s,0}/\sqrt{2} & \delta_{s,1}\delta_{m_s,-1} \end{pmatrix}$$

For singlet pairing ($s = 0$) this matrix is proportional to the Pauli matrix $i\tau^y$. In the case of triplet pairing ($s = 1$), however, this matrix is characterized by all three components of the vector Pauli matrix $\vec{\tau} i\tau^y$, which represent the spin triplet $m_s = 0, \pm 1$. Accordingly, a general and convenient way of writing the order parameter is by the following vector representation:

$$\Delta_k(T) = \left[d_0(T)\tau^0 + \vec{d}_k(T) \cdot \vec{\tau}\right] i\tau^y \tag{3.1}$$

The quantity $d_0(T)$ represents the order parameter for singlet pairing

$$d_0(T) = \Delta(T)e^{i\phi}$$

characteristic of ordinary superconductors. Here $\Delta(T)$ is the temperature dependent magnitude of the gap and ϕ, the phase of the order parameter, is related to the broken gauge symmetry of the superconducting ground state. The vector $\vec{d}_k(T)$ is a representation of the components $m_s = 0, \pm 1$ of the triplet order parameter, characteristic of the superfluid phases of liquid ^{3}He:

$$\vec{d}_k(T) = \Delta(T)e^{i\phi}\hat{d}(\hat{k})$$

with $\Delta(T)$ and ϕ specifying again the magnitude and phase of the order parameter. A total Cooper pair spin $s = 1$ implies (via the Pauli principle) a pair orbital angular momentum $\ell = 1$ and the components of the unit vector $\hat{d}(\hat{k})$ describe therefore the spin–orbit correlations present in the condensate. They are linear functions of $\hat{k}$ in the case of p–wave pairing. Their explicit form reads for the so–called Balian–Werthamer state, believed to represent the B–phase of superfluid ^{3}He [19]:

$$\hat{d}(\hat{k}) = \overleftrightarrow{R}(\hat{n}, \theta) \cdot \hat{k}$$

The rotation matrix $\overleftrightarrow{R}(\hat{n}, \theta)$ correlates spin and orbital degrees of freedom; the rotation axis $\hat{n}$ and the rotation angle θ are manifestations of an additional broken symmetry, not present in ordinary superconductors, namely the spontaneously broken spin–orbit symmetry (SBSOS). The concept of SBSOS was first introduced by Tony Leggett [20] to explain the exotic NMR features of superfluid ^{3}He soon after their discovery. The spin–orbit degeneracy described by the rotation matrix $\overleftrightarrow{R}$ is lifted by the nuclear dipole interaction. In contrast to the case of normal paramagnetic Fermi liquids this interaction is not tiny but amplified as a conseqence of the phase coherence of the Cooper pairs.

The dipole energy is minimized for a "magic" rotation angle $\theta_0 = \arccos(-1/4)$ (which is sometimes also referred to as "Leggett-angle"). The remaining degeneracy with respect to the direction of the $\hat{n}$-vector is removed in finite magnetic fields which tend to align $\hat{n}$ parallel to it ($\hat{n}_0 \parallel \vec{H}^{ext}$). In the case of p-wave pairing the equilibrium order parameter reads therefore:

$$\vec{d}_k^{\,0}(T) = \Delta(T)\, e^{i\phi_0}\, \hat{d}_k^{\,0} \;\; ; \;\; \hat{d}_k^{\,0} = \overleftrightarrow{R}(\hat{n}_0, \theta_0) \cdot \hat{k} \tag{3.2}$$

For further details the reader is recommended to see the review by Leggett [20] or the book by Vollhardt and Wölfle [21].

The gap (3.1) determines the spectrum of quasiparticle excitations from the superconducting or superfluid ground state:

$$E_k = \sqrt{\xi_k^2 + \Delta_k \Delta_k^{\dagger}} \tag{3.3}$$

Let me now proceed with a description of the pair-correlated Fermi liquid in terms of momentum distribution functions in phase space. A normal Fermi liquid may be described by the "diagonal" distribution function of particle-like excitations

$$(\mathrm{n}_{kk})_{\sigma\sigma'} = \left\langle c_{k\sigma}^{\dagger} c_{k\sigma'} \right\rangle$$

or, alternatively, its hole counterpart

$$(\mathrm{n}_{-k-k})_{\sigma'\sigma} = \left\langle c_{-k\sigma'} c_{-k\sigma}^{\dagger} \right\rangle$$

Here $< ... >$ denotes the statistical average over some nonequilibrium ensemble. The pair correlations in a Fermi superfluid lead to the existence of a new "off-diagonal" thermal average (pair amplitude):

$$(\mathrm{g}_{kk})_{\sigma\sigma'} = \langle c_{-k\sigma} c_{k\sigma'} \rangle$$

The pair amplitude g_{kk} describes the structure and the dynamics of the off-diagonal long range order present in the system. Although external fields do not directly couple to g_{kk}, the dynamics of the order parameter can be observed indirectly through its coupling to n_{kk}. It has turned out that the resulting coupled dynamics of particle-like, hole-like and off-diagonal averages can be conveniently described by combining these quantities to 2×2 matrices in particle-hole space (i.e. 4×4 matrices in particle-hole and spin space), the so-called Nambu-matrices [22], which I shall denote, in what follows, with an underbar:

$$\underline{n}_{kk} = \begin{pmatrix} \mathrm{n}_{kk} & \mathrm{g}_{kk} \\ \mathrm{g}^{\dagger}_{kk} & -\mathrm{n}^{T}_{-k,-k} \end{pmatrix} \tag{3.4}$$

Correspondingly, the normal state equilibrium quasiparticle energy ξ_k, measured from the Fermi energy μ, together with its hole counterpart $-\xi_k$ has to be amended by the off-diagonal mean fields Δ_k to combine to an equilibrium energy matrix $\underline{\epsilon}^0_k$:

$$\underline{\epsilon}^0_k = \begin{pmatrix} \xi_k & \Delta_k \\ \Delta^{\dagger}_k & -\xi_k \end{pmatrix} \tag{3.5}$$

The equilibrium form of the distribution (3.4) is a Fermi function taken at the matrix energy $\underline{\epsilon}^0_k$:

$$\underline{n}^0_k = \frac{1}{2}\left[1 - \frac{\underline{\epsilon}^0_k}{E_k}\tanh\frac{E_k}{2k_{\mathrm{B}}T}\right] \tag{3.6}$$

The equilibrium form of the pairing amplitude is obtained from (3.6) as

$$\mathrm{g}^0_k = \mathrm{g}^0_{kk} = -\Theta_k\,\Delta_k \quad ; \quad \Theta_k = \frac{1}{2E_k}\tanh\frac{E_k}{2k_{\mathrm{B}}T}$$

If the pairing interaction is denoted by $\Gamma_{kk'}$, the self-consistency (or gap-) equation, by which the magnitude of the gap (3.1), $\Delta(T)$, can be determined, is of the form

$$\Delta_k = \sum_{\vec{k}'} \Gamma_{kk'}\,\mathrm{g}^0_{k'} \tag{3.7}$$

Its solution can be interpolated between its limiting forms at zero temperature $\Delta(0) = \delta_{sc}k_{\mathrm{B}}T_{\mathrm{c}}$ and at T_{c} $\Delta(T \to T_{\mathrm{c}}) = \pi k_{\mathrm{B}}T_{\mathrm{c}}\sqrt{(2\Delta C/3C_{\mathrm{N}})x}$ with $x = 1 - T/T_{\mathrm{c}}$ as

$$\Delta(T) = \delta_{sc}k_{\mathrm{B}}T_{\mathrm{c}}\tanh\left\{\frac{\pi}{\delta_{sc}}\sqrt{\frac{2}{3}\frac{\Delta C}{C_{\mathrm{N}}}\frac{x}{1-x}[1 + c_1x + c_2x^2]}\right\} \tag{3.8}$$

where the quantities δ_{sc} and $\Delta C/C_{\mathrm{N}}$ parametrize the zero temperature gap and the specific heat discontinuity. They may either be taken to assume their weak coupling values of 1.764... and 1.426..., respectively (certainly true for ^{3}He–B at 0 bar), or treated as unknown strong coupling parameters which have to be taken from experiment. The coefficients $c_1 = 0.1916...$ and $c_2 = 0.2065...$ originate from an expansion of the gap equation in $(\Delta/\pi k_{\mathrm{B}}T_{\mathrm{c}})^2 \propto x$ near the transition. In the weak coupling limit the gap interpolation formula (3.8) reproduces a numerical solution of the gap equation [23] within a few tenths of a percent.

It is convenient, at this stage, to introduce the analogue in particle–hole space of the 2×2 unit matrix and the set of Pauli–matrices τ^μ; $\mu = x, y, z$:

$$\underline{\alpha}^\mu_{(s)} = \begin{pmatrix} \tau^\mu & 0 \\ 0 & -s\tau^{\mu\,T} \end{pmatrix} \quad ; \; \mu = 0,\; x, y, z; \; s = \pm 1 \tag{3.9}$$

where T denotes the matrix transpose, together with the equilibrium gap matrix

$$\underline{\hat{\Delta}}_k = \frac{1}{|\Delta_k|}\begin{pmatrix} 0 & \Delta_k \\ \Delta_k^\dagger & 0 \end{pmatrix} \tag{3.10}$$

Let me now turn to the case that a pair correlated Fermi liquid is exposed to the same external fields that I have already discussed for the normal Fermi liquid. (cf. (2.7)). With the aid of (3.9) this kind of external perturbation aquires the following particle–hole matrix structure:

$$\delta\underline{\epsilon}_k^{\mathrm{ext}}(\vec{r}, t) = -\left\{[eU^{\mathrm{ext}}(\vec{r}, t)\, \underline{\alpha}^0_{(+)} + \vec{p}\cdot\vec{v}^{\mathrm{ext}}(\vec{r}, t)\, \underline{\alpha}^0_{(-)}] + \frac{\gamma\hbar}{2}\vec{H}^{\mathrm{ext}}(\vec{r}, t)\cdot\underline{\vec{\alpha}}_{(+)}\right\} \tag{3.11}$$

If the variation in space of the external perturbations is of the plane wave type $\propto \exp(i\vec{q}\cdot\vec{r})$ the nonequilibrium distribution function becomes a momentum matrix $\underline{n}_{p,p'}(t) \equiv \underline{n}_k(\vec{q}, t)$, where $\vec{p} = \hbar(\vec{k} + \vec{q}/2)$ and $\vec{p}\,' = \hbar(\vec{k} - \vec{q}/2)$:

$$\underline{n}_{p,p'}(t) = \underline{n}_k(\vec{q}, t) = \underline{n}_k^0\, \delta_{\vec{q},0} + \delta\underline{n}_k(\vec{q}, t)$$

The quasiparticle energy changes accordingly:

$$\underline{\epsilon}_{p,p'}(t) = \underline{\epsilon}_k(\vec{q}, t) = \underline{\epsilon}_k^0\, \delta_{\vec{q},0} + \delta\underline{\epsilon}_k(\vec{q}, t)$$

The evolution of the nonequilibrium matrix distribution function in time and space is governed by a matrix–kinetic (von Neumann's) equation [24]:

$$i\frac{\partial}{\partial t}\underline{n}_{p,p'}(t) = \frac{1}{\hbar}\sum_{p''}\left[\underline{n}_{p,p''}, \underline{\epsilon}_{p'',p'}\right]_- + i\underline{I}_{p,p'} \tag{3.12}$$

in which the full quasiparticle energy $\underline{\epsilon}_{p,p'}$ plays the role of the Hamiltonian of the system. If $\underline{n}_{p,p'}$ is interpreted as a quantum–mechanical Wigner distribution function (the diagonal element of which describes the probability amplitude for the excitatiuon of a particle–hole pair with momenta $\vec{p} = \hbar(\vec{k}+\vec{q}/2)$ (particle) and $\vec{p}\,' = \hbar(\vec{k}-\vec{q}/2)$ (hole)), the applicability of (3.12) extends to external perturbations with frequencies $\omega \ll E_{\mathrm{F}}/\hbar$ and wavevectors

$\hbar|\vec{q}| \ll p_{\rm F}$. My aim is to derive the matrix analogue of the Landau–Boltzmann–Silin equation (cf. Eq. (2.14)). For this purpose let me as a first step perform a long-wavelength or small-$\hbar\vec{v}_k \cdot \vec{q}$-expansion of the (quantum) matrix kinetic equation (3.12) in order to eleminate Panck's constant to leading order. The expansions for the equilibrium energy and distribution read:

$$\begin{aligned} \underline{\epsilon}^0_{k\pm q/2} &= \underline{\epsilon}^0_k \pm \frac{\hbar}{2}\vec{v}_k \cdot \vec{q}\; \underline{\alpha}^0_{(+)} \\ \underline{n}^0_{k\pm q/2} &= \underline{n}^0_k \mp \frac{\hbar}{2}\vec{v}_k \cdot \vec{q}\left(-\frac{\partial \underline{n}^0_k}{\partial \xi_k}\right) \\ \left(-\frac{\partial \underline{n}^0_k}{\partial \xi_k}\right) &= \Phi_k \underline{\alpha}^0_{(+)} + \Xi_k\, \hat{\underline{\Delta}}_k \end{aligned} \tag{3.13}$$

where

$$\begin{aligned} \Phi_k &= \frac{\Delta^2}{E_k^2}\Theta_k + \frac{\xi_k^2}{E_k^2}\varphi_k \\ \Xi_k &= \Delta\frac{\partial \Theta_k}{\partial \xi_k} = \frac{\xi_k \Delta}{E_k^2}[\varphi_k - \Theta_k] \\ \varphi_k &= -\frac{\partial f(E_k)}{\partial E_k} \end{aligned} \tag{3.14}$$

The linearized form of the matrix–kinetic equation reads in the long wavelength limit after transforming back into real space ($\vec{q} \to -i\vec{\nabla}$)

$$\begin{aligned} \frac{\partial}{\partial t}\underline{n}_k(\vec{r},t) &+ \vec{v}_k \cdot \vec{\nabla}\frac{1}{2}\left\{\underline{\alpha}^0_{(+)}, \delta\underline{n}_k(\vec{r},t) + \Phi_k \delta\underline{\epsilon}_k(\vec{r},t)\right\}_+ \\ &+ \vec{v}_k \cdot \vec{\nabla}\frac{1}{2}\left\{\hat{\underline{\Delta}}_k, \Xi_k \delta\underline{\epsilon}_k(\vec{r},t)\right\}_+ \\ &= \delta\underline{I}_k(\vec{r},t) - \frac{i}{\hbar}\left[\delta\underline{n}_k, \delta\underline{\epsilon}_k\right]_- + \frac{i}{\hbar}\left[\underline{\epsilon}^0_k, \delta\underline{n}_k + \Theta_k\delta\underline{\epsilon}_k\right]_- \end{aligned} \tag{3.15}$$

The term involving Ξ_k is odd in ξ_k and will therefore play a role only in context with the derivation of a transport equation for the thermal conductivity. These coupled equations of motion for the diagonal and off–diagonal averages describe the dynamics of the superfluid in the whole quasiclassical frequency regime $0 < \omega \ll E_{\rm F}/\hbar$ which includes particularly precessional motion (described by the second term on the r.h.s. of (3.15)) and order parameter collective modes (described by the third term on the r.h.s. of (3.15)). They have to be amended by self–consistency relations for the diagonal part (Fermi liquid forces, cf. (2.8))

$$\{\delta\underline{\epsilon}_k(\vec{r},t)\}_{++} = \delta\epsilon_k^{\rm ext}(\vec{r},t) + \frac{2}{N_{\rm F}}\sum_\ell\sum_{\vec{k}'} P_\ell(\hat{k}\cdot\hat{k}')\left[F_\ell^{\rm s}\delta n_k\tau^0 + F_\ell^{\rm a}\delta\vec{n}_k\cdot\vec{\tau}\right](\vec{r},t) \tag{3.16}$$

and the off–diagonal part (gap equation)

$$\{\delta \underline{\epsilon}_k(\vec{r},t)\}_{+-} = \{\delta \underline{\Delta}_k(\vec{r},t)\}_{+-} = \delta\Delta_k(\vec{r},t) = \sum_{k'} \Gamma_{kk'}\ \delta g_{k'}(\vec{r},t) \tag{3.17}$$

of the matrix quasiparticle energy change. The latter equation can be obtained by straightforward variation of the gap equation (3.7) at constant pairing interaction (weak coupling assumption).

As my main concern is the description of transport in quasiparticle gases, I may confine myself to a somewhat more restricted range of frequencies, namely $0 < \omega \ll \Delta(T)/\hbar$, in which the *magnitude* of the order parameter is *not* affected by the external fields (i.e. no excitation of order parameter collective modes) but in which ω may still become comparable with or larger than the inverse of the characteristic lifetime of quasiparticle excitations. In such a situation only the normal component displays nonhydrodynamic behavior. The scale for typical zero temperature gap frequencies $\Delta(0)/\hbar$ and quasiparticle collision frequencies at the transition $\tau_N^{-1}(0,T_c)$ in superfluid ^{3}He are a few hundred MHz and a few MHz, respectively. The order parameter dynamics is, in such a restricted frequency regime, comletely describable in terms of the phase variables which correspond to the broken symmetries of the superconducting or superfluid ground state. In the case of s–wave singlet pairing correlations, the dynamics of the order parameter is simply characterized by the dynamics of the phase ϕ which represents the broken gauge symmetry:

$$\delta d_0(\vec{r},t) = d_0(T)\ i\delta\phi(\vec{r},t) \tag{3.18}$$

In the case of the superfluid B–phase of liquid ^{3}He one encounters triplet p–wave pairing and the dynamics of the order parameter is governed by both the phase ϕ and the variables $\hat{n}$ and θ describing the spin–orbit correlations:

$$\delta\vec{d}_k(\vec{r},t) = \vec{d}_k^{\,0}(T)\ i\delta\phi(\vec{r},t) - \vec{d}_k^{\,0}(T) \times \delta\vec{\theta}(\vec{r},t) \tag{3.19}$$

where I have defined $\delta\vec{\theta} = \hat{n}\delta\theta$. The regime of frequencies where (3.18) and (3.19) hold is referred to as *macroscopic* because the dynamics is entirely determined by the variables describing macroscopic quantum coherence of the Cooper pair wave function. It will turn out that it is in this regime, that the treatment of the matrix kinetic equation (3.15) greatly simplifies and a two fluid description in the spirit of Khalatnikov's treatment of superfluid ^{4}He becomes possible. The simple form of the order parameter dynamics

(3.18) and (3.19) suggests the application of a gauge transformation of the matrix–kinetic equation, by which the phase variables $\delta\phi$ and $\delta\vec{\theta}$ can be eliminated from the off–diagonal matrix elements of the nonequilibrium energy matrix:

$$\underline{U}(\vec{r},t) = \exp\left\{-\frac{i}{2}\left[\delta\phi(\vec{r},t)\ \underline{\alpha}^{0}_{(+)} + \delta\vec{\theta}(\vec{r},t)\cdot\underline{\vec{\alpha}}_{(+)}\right]\right\} \tag{3.20}$$

An application of the transformation (3.20) to matrices $\underline{A}_k$ of the general type

$$\underline{A}_k = \begin{pmatrix} \mathbf{a}_k & \mathbf{\Delta}_k \\ \mathbf{\Delta}_k^{\dagger} & -\mathbf{a}_{-k}^{T} \end{pmatrix}$$

with $\mathbf{a}_k = a_{k0}\tau^0 + \vec{a}_k\cdot\vec{\tau}$ shows that it has the property

$$\left\{\underline{U}\ \underline{A}_k\ \underline{U}^{\dagger}\right\} = \begin{pmatrix} a_{k0}\tau^0 + \vec{\tau}\cdot\overleftrightarrow{R}(\hat{n},-\theta)\cdot\vec{a}_k & e^{-i\phi}\mathbf{\Delta}_k + [\overleftrightarrow{R}(\hat{n},-\theta)\cdot\vec{d}^{0}_{k}]\vec{\tau}i\tau^{y} \\ e^{i\phi}\mathbf{\Delta}_k^{\dagger} + [\overleftrightarrow{R}(\hat{n},-\theta)\cdot\vec{d}^{0*}_{k}]\vec{\tau}i\tau^{y} & -a_{-k0}\tau^0 - \vec{\tau}^{T}\cdot\overleftrightarrow{R}(\hat{n},-\theta)\cdot\vec{a}_{-k} \end{pmatrix}$$

of removing completely all phase variables from its off–diagonal matrix elements and hence the dynamics of the order parameter. When applied to the matrix kinetic equation for the Nambu–distribution function in the form (3.12) the gauge transformation generates time derivatives and gradients of the phase variables describing the condensate. After linearization these read:

$$\delta\mu_0(\vec{r},t) = -\frac{\hbar}{2}\frac{\partial}{\partial t}\delta\phi(\vec{r},t) \quad ; \quad \delta\vec{\mu}(\vec{r},t) = -\frac{\hbar}{2}\frac{\partial}{\partial t}\delta\vec{\theta}(\vec{r},t) \tag{3.21}$$

$$v^{s}_{i0}(\vec{r},t) = \frac{\hbar}{2m}\nabla_i\delta\phi(\vec{r},t) \quad ; \quad \vec{v}_i^{\,s}(\vec{r},t) = \frac{\hbar}{2m}\nabla_i\delta\vec{\theta}(\vec{r},t) \tag{3.22}$$

Here the quantities $v^{s}_{i\nu}$ (ν=0, x,y,z,) can be interpreted, respectively, as the mass superfluid velocity ($\nu = 0$) and its spin analogue (ν=x,y,z) which exists for the p–wave pairing case only. Correspondingly the quantities $\delta\mu_\nu$ (ν=0, x,y,z,), have the meaning of chemical potential shifts. Acceleration equations for the condensate can be derived from Eqs. (3.22) by differentiating $v^{s}_{i\nu}(\vec{r},t)$ with respect to the time t:

$$\frac{\partial}{\partial t}mv^{s}_{i\nu}(\vec{r},t) + \nabla_i\delta\mu_\nu(\vec{r},t) = 0 \quad ; \quad \nu = 0,x,y,z \tag{3.23}$$

Again, the spin analogue of the condensate acceleration equation is only present in systems with triplet p–wave pairing correlations. Let me write down a few further steps towards a gauge–transformed matrix kinetic equation. The linearized distribution reads after the transformation:

$$\delta\underline{\tilde{n}}_k = \left\{\underline{U}\ \delta\underline{n}_k\ \underline{U}^{\dagger}\right\} = \delta\underline{n}_k + \Theta_k\delta\underline{\Delta}_k - \Phi_k\sum_{\nu=0,x,y,z} v_{ki}mv^{s}_{i\nu}\ \underline{\alpha}^{\nu}_{(-)} \tag{3.24}$$

The corresponding result for the transformed quasiparticle energy matrix is:

$$\begin{aligned}\delta\tilde{\underline{\epsilon}}_k &= \left\{\underline{U}\,\delta\underline{\epsilon}_k\,\underline{U}^\dagger\right\} = \delta\underline{\epsilon}_k - \delta\underline{\Delta}_k - \sum_{\nu=0,x,y,z}\left[\delta\mu_\nu\,\underline{\alpha}^\nu_{(+)} - v_{ki} m v^s_{i\nu}\,\underline{\alpha}^\nu_{(-)}\right] \\ &= \delta\tilde{\underline{\epsilon}}^{\text{ext}}_k + \delta\tilde{\underline{\epsilon}}^{\text{FL}}_k \end{aligned} \tag{3.25}$$

In (3.24) and (3.25) I have made use of the effective mass relation $m^*/m = 1 + F_1^s/3$. The result (3.25) for the transformed matrix quasiparticle energy implies that the dynamics of the condensate induces additional contributions to the diagonal elements which appear now in a gauge invariant form as effective forces on the quasiparticle system:

$$\begin{aligned}\delta\tilde{\underline{\epsilon}}^{\text{ext}}_k = &- \left[eU^{\text{ext}} + \delta\mu_0\right]\underline{\alpha}^0_{(+)} + p_i\left[v^s_{i0} - v^{\text{ext}}_i\right]\underline{\alpha}^0_{(-)} \\ &- \left[\frac{\gamma\hbar}{2}\vec{H}^{\text{ext}} + \delta\vec{\mu}\right]\cdot\vec{\underline{\alpha}}_{(+)} + \left(1 + \frac{F_1^a}{3}\right) v_{ki}\, m\vec{v}_i^{\,s}\cdot\vec{\underline{\alpha}}_{(-)}\end{aligned} \tag{3.26}$$

The molecular field contributions $\delta\tilde{\underline{\epsilon}}^{\text{FL}}_k$ to the transformed quasiparticle energy are of the usual form (3.16) but with $\delta n_k(\vec{r},t)$ replaced by the diagonal element of (3.24), $\delta\tilde{n}(\vec{r},t)$. The linearized matrix kinetic equation which governs the dynamics of the superfluid is, after the gauge transformation, still of the form (3.15) but with $\delta\underline{n}_k$ replaced by $\delta\tilde{\underline{n}}_k$ and $\delta\underline{\epsilon}_k$ replaced by $\delta\tilde{\underline{\epsilon}}_k$. It is greatly simplified in that $\delta\tilde{\underline{\epsilon}}_k$ is now a diagonal matrix in particle–hole space. The dynamics of the condensate has now the simple effect of introducing new force terms which add to the external forces acting on the quasiparticles. In what follows I shall deal only with the gauge transformed version of (3.15) and therefore the tilde on the matrix distribution $\delta\underline{n}$ and the matrix energy $\delta\underline{\epsilon}$ will be omitted.

The structure of the collision integral $\delta\underline{I}_k$ in particle hole space is rather complex and cannot be discussed here for lack of space. The interested reader may see ref. [25] for details. At this stage I should mention that the matrix collision integral preserves the conservation laws of the normal Fermi liquid, i.e. it is characterized by a set of collisional invariants $\underline{A}_k$

$$2\sum_{\vec{k}}\frac{1}{4}\text{tr}[\underline{A}_k\,\delta\underline{I}_k] = 0$$

which correspond to conservation laws for mass ($\underline{A}_k = m\underline{\alpha}^0_{(+)}$), momentum ($\underline{A}_k = \vec{p}\,\underline{\alpha}^0_{(-)}$), energy ($\underline{A}_k = \underline{\epsilon}^0_k$) and spin ($\underline{A}_k = (\hbar/2)\,\vec{\underline{\alpha}}_{(+)}$).

We are now sufficiently prepared to attack the problem of deriving conservation laws for macroscopic observable densities from the matrix kinetic equation. As a first step let me note that, as in the normal state, the diagonal part δn_k of the distribution function can

OBSERVABLE	CONSERVATION LAW	DENSITY	CURRENT
MASS	$\frac{\partial\rho}{\partial t}+\vec{\nabla}\cdot\vec{g}=0$	$\delta\rho = m2\sum_{\vec{k}}\delta n_k^{(+)}$	$g_i = 2\sum_{\vec{k}} v_{ki}[\delta n_k^{(-)}$ $+\Phi_k\delta\epsilon_k^{(-)}]$
MOMENTUM	$\frac{\partial g_i}{\partial t}+\nabla_j\Pi_{ij}=0$	$g_i = 2\sum_{\vec{k}} p_i\delta n_k^{(-)}$ $+\rho[v_{i0}^s - v_i^{ext}]$	$\Pi_{ij} = \delta P\delta_{ij} + \Pi'_{ij}$ $\delta P = n\delta\mu_0$ $+\frac{n}{N_F}2\sum_{\vec{k}}[\delta n_k^{(+)}$ $+\Phi_k\delta\epsilon_k^{(+)}]$ $\Pi'_{ij} = 2\sum_{\vec{k}} p_i v_{kj}[\delta n_k^{(+)}$ $+\Phi_k\delta\epsilon_k^{(+)}]$
ENTROPY	$\frac{\partial\sigma}{\partial t}+\vec{\nabla}\cdot\vec{f}=0$	$\sigma = 2\sum_{\vec{k}}\delta\sigma_k$	$T\vec{f} = 2\sum_{\vec{k}}\vec{v}_k\xi_k[\delta n_k^{(-)}$ $+\varphi_k\delta\epsilon_k^{(-)}]$
MAGNETIZATION	$\frac{\partial\vec{M}}{\partial t}+\nabla_i\vec{J}_i^M =$ $\gamma\vec{M}\times\vec{H}^{ext}+\gamma\vec{R}_D$	$\vec{M} = \frac{\gamma\hbar}{2}2\sum_{\vec{k}}\delta\vec{n}_k^{(+)}$	$\vec{J}_i^M = \frac{\gamma\hbar}{2}2\sum_{\vec{k}} v_{ki}[\delta\vec{n}_k^{(-)}$ $+\Phi_k\delta\vec{\epsilon}_k^{(-)}]$

Table 3: Observable densities and currents in a pair–correlated superfluid

be classified according to its parities s and t with respect to the operations $\hat{k}\to-\hat{k}$ and $\xi_k\to-\xi_k$, respectively. The diagonal elements of the matrix distribution function $\delta\underline{n}_k$ can be combined to scalar projections

$$\delta n_{k\mu}^{(s)} = \frac{1}{4}\mathrm{tr}_4\left[\underline{\alpha}^{\mu}_{(s)}\delta\underline{n}_k\right] = \frac{1}{2}\left[\delta n_{k\mu} + s\delta n_{-k\mu}\right]$$

in the case of positive parity $t=1$ and

$$T\delta\sigma_k = \frac{1}{4}\mathrm{tr}_4\left[\underline{\epsilon}_k^0\delta\underline{n}_k\right]$$

in the case of negative parity $t=-1$. Here tr_4 denotes the trace over 4×4 matrices in particle–hole space. The parity projected scalar distribution functions $\delta n_{k\mu}^{(s)}$ and $\delta\sigma_k$ obey a pair of scalar kinetic equations which are strongly reminiscent the normal state Landau–Boltzmann–Silin equation:

$$\frac{\partial}{\partial t}\delta n_{k\mu}^{(s)} + \vec{v}_k\cdot\vec{\nabla}\left[\delta n_{k\mu}^{(-s)} + \Phi_k\delta\epsilon_k^{(-s)}\right] = \frac{1}{4}\mathrm{tr}_4\left[\underline{\alpha}^{\mu}_{(s)}\delta\underline{I}_k\right] - \frac{2}{\hbar}\epsilon_{\mu\alpha\beta}\left[\delta n_{k\alpha}\delta\epsilon_{k\beta}\right]^{(s)} \tag{3.27}$$

$$T\frac{\partial}{\partial t}\delta\sigma_k + \xi_k\,\vec{v}_k\cdot\vec{\nabla}\left[\delta n_k + \varphi_k\delta\epsilon_k\right] = \frac{1}{4}\mathrm{tr}_4\left[\underline{\epsilon}_k^0\delta\underline{I}_k\right] \tag{3.28}$$

Here terms which vanish upon summation over $\vec{k}$ as a consequence of the self–consistency relation (3.17) have been omitted. The parity–projected kinetic equations (3.27) and (3.28) can now be used to determine the macroscopic observable densities and currents and their conservation properties, now generalized to include condensate contributions $\propto \delta\mu_\nu$ and $v^s_{i\nu}$. The results are summarized in table 3.

4 The two fluid description

As an important step towards the derivation of a two fluid model we have now to decompose the distributions $\delta n_{k\mu}$ and $\delta\sigma_k$ which determine the physical observables (cf. table 3) into their quasiparticle and condensate contributions. This can be achieved by performing a Bogoliubov–Valatin transformation which is defined as the unitary transformation that diagonalizes the equilibrium matrix energy:

$$\underline{U}_k\ \underline{\epsilon}_k^0\ \underline{U}_k^\dagger = E_k\ \underline{\alpha}_{(+)}^0 \tag{4.1}$$

where

$$\underline{U}_k = \begin{pmatrix} u_k & \mathrm{v}_k \\ -\mathrm{v}_k^\dagger & u_k \end{pmatrix}$$

with $u_k = \sqrt{(1+\xi_k/E_k)/2}$ and $\mathrm{v}_k = \hat{\Delta}_k\sqrt{(1-\xi_k/E_k)/2}$. The Bogoliubov transformation of the equilibrium matrix distribution $\underline{n}_k^0$ consists of the equilibrium Fermi distribution $\nu^0(\pm E_k)$ for Bogoliubov quasiparticles (E_k) and -holes ($-E_k$):

$$\underline{\nu}_k^0 = \underline{U}_k\ \underline{n}_k^0\ \underline{U}_k^\dagger = \begin{pmatrix} \nu^0(E_k) & 0 \\ 0 & \nu^0(-E_k) \end{pmatrix}$$

The presence of external perturbations changes $\underline{\nu}_k^0$ into a nonequilibrium quasiparticle distribution matrix:

$$\delta\underline{\nu}_{p,p'} = \underline{U}_p\ \delta\underline{n}_{p,p'}\ \underline{U}_{p'}^\dagger = \begin{pmatrix} \delta\nu_{p,p'} & \delta\gamma_{p,p'} \\ \delta\gamma_{p,p'}^\dagger & -\delta\nu_{-p',-p}^T \end{pmatrix}$$

The Bogoliubov transformation of the nonequilibrium quasiparticle energy matrix is defined to have the matrix elements

$$\delta\underline{E}_{p,p'} = \underline{U}_p\ \delta\underline{\epsilon}_{p,p'}\ \underline{U}_{p'}^\dagger = \begin{pmatrix} \delta\mathbf{E}_{p,p'} & \delta\mathbf{D}_{p,p'} \\ \delta\mathbf{D}_{p,p'}^\dagger & -\delta\mathbf{E}_{-p',-p}^T \end{pmatrix}$$

One may now perform a Bogoliubov transformation to the matrix kinetic equation (3.12) by multiplying from the left with $\underline{U}_p$ and from the right with $\underline{U}_{p'}$. After a subsequent expansion in small $\hbar\vec{v}_k \cdot \vec{q}$ one obtains the following set of coupled kinetic equations for the new distribution functions $\delta\nu_k$ and $\delta\gamma_k$:

$$\frac{\partial}{\partial t}\delta\nu_k(\vec{r},t) + \vec{V}_k \cdot \vec{\nabla}\left[\delta\nu_k + \varphi_k\delta\mathbf{E}_k\right](\vec{r},t) = \delta\mathbf{I}_{k++} - \frac{2}{\hbar}\left[\vec{h}_k \times \delta\vec{E}_k\right](\vec{r},t)\cdot\vec{\tau} \tag{4.2}$$

$$\frac{\partial}{\partial t}\delta\gamma_k(\vec{r},t) - \frac{2i}{\hbar}\left[\delta\gamma_k + \Theta_k\delta\mathbf{D}_k\right](\vec{r},t) \;=\; \delta\mathbf{I}_{k+-} - \ldots \tag{4.3}$$

In (4.2) $\vec{V}_k = \partial E_k/\partial\hbar\vec{k} = (\xi_k/E_k)\,\vec{v}_k$ denotes a quasiparticle group velocity, which vanishes at the Fermi surface. We are interested in solutions of these coupled kinetic equations (4.2) and (4.3) in the macroscopic limit, $\omega \ll \Delta(T)/\hbar$ which I have discussed already in the preceding lecture. An inspection of Eq. (4.3) immediately shows that in this limit the distribution function $\delta\gamma_k$, which describes the Cooper pair response, has the local equilibrium form

$$\delta\gamma_k(\vec{r},t) = -\Theta_k\delta\mathbf{D}_k(\vec{r},t) + O\left(\frac{\hbar\omega}{\Delta(T)}, \frac{\hbar v_{\mathrm{F}}|\vec{q}|}{\Delta(T)}, \frac{\hbar}{\tau_{+-}\Delta(T)}, \ldots\right)$$

where τ_{+-} is a relaxation time characterizing the off-diagonal component $\delta\mathbf{I}_{k+-}$ of the transformed collision integral. Splitting off contributions of the order $O(\hbar\omega/\Delta)$ also in the quasiparticle kinetic equation (4.2) leaves us with

$$\begin{aligned}\delta\nu_k(\vec{r},t) &= -\varphi_k\delta\mathbf{E}_k(\vec{r},t) + \mathbf{h}_k(\vec{r},t) + O\left(\frac{\hbar\omega}{\Delta(T)}, \frac{\hbar v_{\mathrm{F}}|\vec{q}|}{\Delta(T)}, \frac{\hbar}{\tau_{++}\Delta(T)}, \ldots\right)\\ \mathbf{h}_k &= \delta\nu_k + \varphi_k\delta\mathbf{E}_k\end{aligned} \tag{4.4}$$

Here τ_{++} is a time scale which specifies the diagonal component of the collision integral; it will be discussed in detail later. In the macroscopic regime $\omega \ll \Delta(T)/\hbar$ the quasiparticle distribution function $\delta\nu_k$ has a local equilibrium contribution $\delta\nu_k^{\mathrm{loc}} = -\varphi_k\delta\mathbf{E}_k$, but, unlike the Cooper pair part $\delta\gamma_k$, also a contribution $\mathbf{h}_k$, which describes the deviation from local equilibrium. We shall see below that it is the distribution function $\mathbf{h}_k$ which describes exclusively the dynamics of the *normal component* in the macroscopic limit.

We are now able to decompose the distribution functions $\delta n_{k\mu}^{(s)}$ and $\delta\sigma_k$ into their quasiparticle and condensate part by writing

$$\begin{aligned}\delta n_{k\mu}^{(s)} &= \frac{1}{4}\mathrm{tr}_4\left[\underline{U}_k\;\underline{\alpha}_{(s)}^{\mu}\;\underline{U}_k^{\dagger}\;\delta\underline{\nu}_k\right]\\ T\delta\sigma_k &= \frac{1}{4}\mathrm{tr}_4\left[\underline{U}_k\;\underline{\epsilon}_k^{0}\;\underline{U}_k^{\dagger}\;\delta\underline{\nu}_k\right] = E_k\delta\nu_k\end{aligned} \tag{4.5}$$

A two fluid description of superfluid Fermi liquids emerges from (3.27) and (3.28) together with (4.5) if one associates the local contributions $\delta\nu_k^{\rm loc}$ and $\delta\gamma_k$ with the local condensate response whereas the normal component is represented by the distribution function $\mathbf{h}_k$. The normal–superfluid decomposition can then be summarized as follows:

(i) spin–independent case:

$$\begin{aligned} \delta n_k^{(s)} &= s_k^{(s)} h_k^{(s)} - [P_k^{(s)} + Q_k^{(s)}]\delta\epsilon_k^{(s)} \\ P_k^{(s)} &= P_k \frac{1+s}{2} \quad ; \quad P_k = \frac{\Delta^2}{E_k^2}\Theta_k \\ Q_k^{(s)} &= \varphi_k s_k^{(s)2} \quad ; \quad s_k^{(s)} = \frac{\xi_k}{E_k}\frac{1+s}{2} + \frac{1-s}{2} \end{aligned} \tag{4.6}$$

(ii) spin–dependent case

$$\begin{aligned} \delta\vec{n}_k^{(s)} &= \overleftrightarrow{S}_k^{(s)} \vec{h}_k^{(s)} - [\overleftrightarrow{P}_k^{(s)} + \overleftrightarrow{Q}_k^{(s)}]\delta\vec{\epsilon}_k^{(s)} \\ \overleftrightarrow{P}_k^{(s)} &= P_k\left[(\overleftrightarrow{1} - \overleftrightarrow{\Lambda}_k)\frac{1+s}{2} + \overleftrightarrow{\Lambda}_k \frac{1-s}{2}\right] \quad ; \quad \overleftrightarrow{Q}_k^{(s)} = \varphi_k \overleftrightarrow{S}_k^{(s)2} \\ \overleftrightarrow{S}_k^{(s)} &= \left[\frac{\xi_k}{E_k}(\overleftrightarrow{1} - \overleftrightarrow{\Lambda}_k) + \overleftrightarrow{\Lambda}_k\right]\frac{1+s}{2} + \left[\frac{\xi_k}{E_k}\overleftrightarrow{\Lambda}_k + \overleftrightarrow{1} - \overleftrightarrow{\Lambda}_k\right]\frac{1-s}{2} \end{aligned} \tag{4.7}$$

Here $\overleftrightarrow{\Lambda}_k = \hat{d}_k^{\,0} : \hat{d}_k^{\,0}$ is a tensor which projects on the direction parallel to the equilibrium $\vec{d}$–vector. The difference between s– and p–wave pairing correlations has consequences only for the spin–dependent part of the quasiparticle distribution. The s–wave case is obtained from the general result (4.7) by simply replacing the projector $\overleftrightarrow{\Lambda}_k \rightarrow \overleftrightarrow{1}$. The local response functions Q_k (quasiparticles) and P_k (pairs) have the important property

$$P_k^{(+)} + Q_k^{(+)} = \Phi_k \quad ; \quad \int_{-\infty}^{\infty} d\xi_k \Phi_k = 1$$

As a final step we now have to insert the decompositions for the distribution functions $\delta n_k^{(s)}$ into the expressions for the densities and currents (cf. table 3). In doing so one encounters energy and angular integrals of the kernels P_k and Q_k which have the physical meaning of local response functions (susceptibilities) of normal and superfluid component, respectively. They read in the spin–independent case:

$$\begin{aligned} \chi_0^{q(s)}(T) &= \int_{-\infty}^{\infty} d\xi_k Q_k^{(s)} = Y_2(T)\frac{1+s}{2} + Y_0(T)\frac{1-s}{2} \\ \chi_0^{p(s)}(T) &= \int_{-\infty}^{\infty} d\xi_k P_k^{(s)} = [1 - Y_2(T)]\frac{1+s}{2} \\ \chi_0^{(s)}(T) &= \chi_0^{q(s)}(T) + \chi_0^{p(s)}(T) = \frac{1+s}{2} + Y_0(T)\frac{1-s}{2} \end{aligned} \tag{4.8}$$

Here $Y_n(T)$, $n = 0, 1, 2, \ldots$ is a set of functions which describe the temperature dependence of the local quasiparticle response. They are defined by:

$$Y_n(T) = \int_{-\infty}^{\infty} d\xi_k \varphi_k \left|\frac{\xi_k}{E_k}\right|^n \tag{4.9}$$

In the spin–dependent case the local response functions are in general tensor quantities of the form:

$$\begin{aligned}
\overset{\leftrightarrow}{\chi}{}^{q(s_\mu)}_{\mu}(T) &= (2\mu+1)\int \frac{d\Omega_k}{4\pi} P^2_\mu(\hat{k}\cdot\hat{q}) \int_{-\infty}^{\infty} d\xi_k \, \overset{\leftrightarrow}{Q}{}^{(s_\mu)}_k \quad ; \quad s_\mu = (-1)^\mu \\
\overset{\leftrightarrow}{\chi}{}^{p(s_\mu)}_{\mu}(T) &= (2\mu+1)\int \frac{d\Omega_k}{4\pi} P^2_\mu(\hat{k}\cdot\hat{q}) \int_{-\infty}^{\infty} d\xi_k \, \overset{\leftrightarrow}{P}{}^{(s_\mu)}_k \\
\overset{\leftrightarrow}{\chi}{}^{(s_\mu)}_{\mu}(T) &= \overset{\leftrightarrow}{\chi}{}^{q(s_\mu)}_{\mu}(T) + \overset{\leftrightarrow}{\chi}{}^{p(s_\mu)}_{\mu}(T)
\end{aligned} \tag{4.10}$$

The unit vector $\hat{q}$ in the definitions (4.10) defines the direction of spatial variation of macroscopic observables. In later applications, however, $\hat{q}$ will be assumed to be oriented along the direction of the magnetic field $\vec{H}^{ext} \parallel \hat{z}$. It turns out that in the spin–dependent case the following temperature dependent susceptibilities are important:

(i) s–wave pairing ($\overset{\leftrightarrow}{\Lambda}_k \equiv \overset{\leftrightarrow}{1}$):

$$\begin{aligned}
\overset{\leftrightarrow}{\chi}{}^{q(+)}_{0}(T) &= Y_0(T)\,\overset{\leftrightarrow}{1} \\
\overset{\leftrightarrow}{\chi}{}^{p(+)}_{0}(T) &= 0 \\
\overset{\leftrightarrow}{\chi}{}^{(+)}_{0}(T) &= Y_0(T)\,\overset{\leftrightarrow}{1} \equiv \chi_0(T)\,\overset{\leftrightarrow}{1} \\
\overset{\leftrightarrow}{\chi}{}^{q(-)}_{1}(T) &= Y_2(T)\,\overset{\leftrightarrow}{1} \\
\overset{\leftrightarrow}{\chi}{}^{p(-)}_{1}(T) &= [1 - Y_2(T)]\,\overset{\leftrightarrow}{1} \\
\overset{\leftrightarrow}{\chi}{}^{(-)}_{1}(T) &= \overset{\leftrightarrow}{1}
\end{aligned} \tag{4.11}$$

The function $\chi_0(T) = Y_0(T)$ has first been introduced by Yosida [26] in context with a discussion of the temperature dependent spin susceptibility in superconductors. I shall therefore refer to the whole set of functions $Y_n(T)$ as Yoshida functions in what follows.

(ii) p–wave pairing

$$\begin{aligned}
\overset{\leftrightarrow}{\chi}{}^{q(+)}_{0}(T) &= \left[\frac{1}{3}Y_0(T) + \frac{2}{3}Y_2(T)\right]\overset{\leftrightarrow}{1} \equiv \chi^q_0(T)\,\overset{\leftrightarrow}{1} \\
\overset{\leftrightarrow}{\chi}{}^{p(+)}_{0}(T) &= \frac{2}{3}[1 - Y_2(T)]\,\overset{\leftrightarrow}{1} \equiv \chi^p_0(T)\,\overset{\leftrightarrow}{1}
\end{aligned}$$

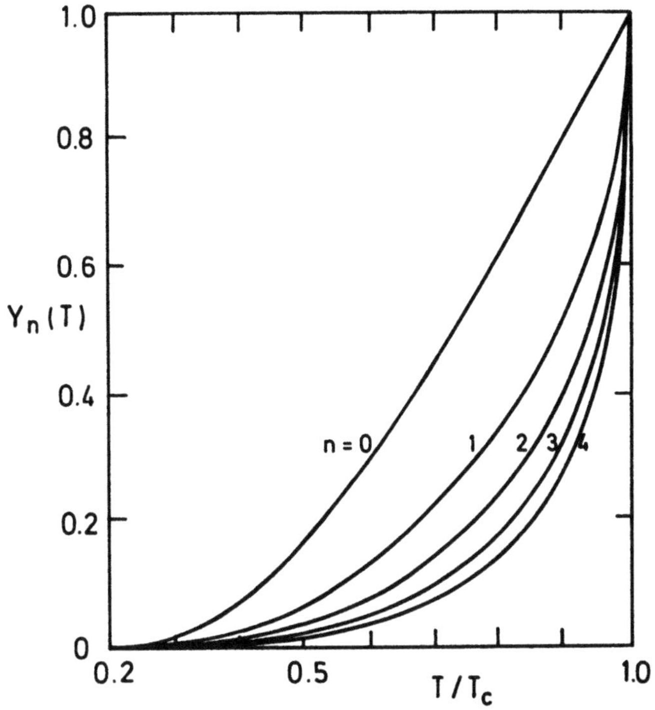

Fig. 1 The generalized Yosida functions $Y_n(T)$ for isotropic pair–correlated superfluids in the weak coupling limit as a function of reduced temperature T/T_c for $n = 0, 1, 2, 3, 4$.

$$
\begin{aligned}
\overleftrightarrow{\chi}_0^{(+)}(T) &= \left[\frac{2}{3} + \frac{1}{3}Y_0(T)\right]\overleftrightarrow{1} \;\equiv\; \chi_0(T)\,\overleftrightarrow{1} \\
\overleftrightarrow{\chi}_1^{q(-)}(T) &= \overleftrightarrow{\Lambda}_q\left[\frac{2}{5}Y_0(T) + \frac{3}{5}Y_2(T)\right] + (\overleftrightarrow{1} - \overleftrightarrow{\Lambda}_q)\left[\frac{4}{5}Y_0(T) + \frac{1}{5}Y_2(T)\right] \\
\overleftrightarrow{\chi}_1^{p(-)}(T) &= \left[\frac{3}{5}\overleftrightarrow{\Lambda}_q + \frac{1}{5}(\overleftrightarrow{1} - \overleftrightarrow{\Lambda}_q)\right][1 - Y_2(T)] \\
\overleftrightarrow{\chi}_1^{(-)}(T) &= \overleftrightarrow{\Lambda}_q\left[\frac{3}{5} + \frac{2}{5}Y_0(T)\right] + (\overleftrightarrow{1} - \overleftrightarrow{\Lambda}_q)\left[\frac{1}{5} + \frac{4}{5}Y_0(T)\right]
\end{aligned}
\tag{4.12}
$$

Here $\overleftrightarrow{\Lambda}_q = \hat{d}_q^{\,0} : \hat{d}_q^{\,0}$, $\hat{d}_q^{\,0} = \overleftrightarrow{R}(\hat{n}_0, \theta_0) \cdot \hat{q}$ is a projection operator, which describes the correlation between spin ($\hat{d}$) and the direction of spatial variation $\hat{q}$. It is hence the spin orbit correlations in p–wave superfluids which are the cause of anisotropies in the susceptibilities and, as we shall see later, also in the transport coefficients even though the gap $\Delta(T)$ in the spectrum of excitations is isotropic (momentum–independent).

The dependence of the first few Yosida functions $Y_n(T)$, which determine the temperature dependence of the local response functions χ, on reduced temperature T/T_c is visualized

OBSERVABLE DENSITY	SUPERFLUID COMPONENT	NORMAL COMPONENT
MASS $\delta\rho$	$\rho n \kappa [eU^{\text{ext}} + \delta\mu_0]$	$[1+F_0^{s}]^{-1} 2\sum_{\vec{k}} m(\xi_k/E_k) h_k^{(+)}$
ENTROPY $T\delta\sigma$		$C_V \delta T + 2\sum_{\vec{k}} E_k h_k^{(+)}$
MAGNETIZATION $\vec{M}$		
s–wave	$\chi_M \vec{H}^{\text{ext}}$	$[1+F_0^{a}\chi_0]^{-1} \frac{\gamma\hbar}{2} 2\sum_{\vec{k}} \vec{h}_k^{(+)}$
p–wave	$\chi_M [\vec{H}^{\text{ext}} + \frac{2}{\gamma\hbar}\delta\vec{\mu}]$	$[1+F_0^{a}\chi_0]^{-1} \frac{\gamma\hbar}{2} 2\sum_{\vec{k}} \overleftrightarrow{S}_k^{(+)}\, \vec{h}_k^{(+)}$
MOMENTUM g_i	$\rho^{s}(v_{i0}^{s} - v_i^{\text{ext}})$	$[1+(F_1^{s}/3)Y_0]^{-1} 2\sum_{\vec{k}} p_i h_k^{(-)}$
PRESSURE δP	$n\delta\mu_0$	$\sigma\delta T + (n/N_F) 2\sum_{\vec{k}} (\xi_k/E_k) h_k^{(+)}$
STRESS–T. Π'_{ij}		$2\sum_{\vec{k}} p_i V_{kj} h_k^{(+)}$
ENTROPY CURRENT $T\vec{f}$		$T\sigma\vec{v}^{\,\text{n}} + 2\sum_{\vec{k}} \vec{v}_k \xi_k h_k^{(-)}$
MAGNETIZATION CURRENT $\vec{J}_z^M$		
s–wave		$\frac{\gamma\hbar}{2} 2\sum_{\vec{k}} V_{kz} \vec{h}_k^{(-)}$
p–wave	$\frac{\gamma\hbar}{2} \overleftrightarrow{n}_{zz}^{s}\, \vec{v}_z^{\,s}$	$\frac{\gamma\hbar}{2} \overleftrightarrow{H}_1^{a}\, 2\sum_{\vec{k}} \overleftrightarrow{V}_{kz}\, \vec{h}_k^{(-)}$

Table 4: Two fluid decomposition of densities and currents in pair–correlated superfluids

in Fig. 1. With these susceptibilities I may now summarize the final result for the two–fluid decomposition of densities and currents in table 4. In this table I have introduced the generalized group velocities

$$\vec{V}_k = \vec{v}_k \frac{\xi_k}{E_k} \quad ; \quad \overleftrightarrow{V}_{kz} = v_{kz}\, \overleftrightarrow{S}_k^{(-)} \tag{4.13}$$

Furthermore I have used the abbreviation

$$\overleftrightarrow{H}_1^{a} = \left(1 + \frac{F_1^{a}}{3}\right)\left\{1 + \frac{F_1^{a}}{3}\overleftrightarrow{X}_1^{(-)}\right\}^{-1} \tag{4.14}$$

From the normal contribution to the mass current (or momentum density) one can identify the normal velocity field as

$$\rho^{\text{n}}(T)\vec{v}^{\,\text{n}}(\vec{r},t) = \{1 + \frac{F_1^{s}}{3} Y_0(T)\}^{-1}\; 2\sum_{\vec{k}} p_i h_k^{(-)}(\vec{r},t) \tag{4.15}$$

The temperature dependent susceptibilities which characterize the local response of the macroscopic observables are

(i) the compressibility (cf. (2.11))

$$\kappa = \frac{1}{n^2}\frac{N_F}{1+F_0^s} \tag{4.16}$$

(ii) the specific heat (cf. (2.12))

$$C_V(T) = 2\sum_{\vec{k}} E_k \frac{\partial \nu_k^0}{\partial T} = 2\sum_{\vec{k}} \varphi_k \frac{E_k}{T}\left[E_k - T\frac{\partial E_k}{\partial T}\right] \tag{4.17}$$

(iii) the magnetic spin susceptibility (cf. (2.13))

$$\chi_M(T) = \left(\frac{\gamma\hbar}{2}\right)^2 N_F \chi(T) \quad ; \quad \chi(T) = \frac{\chi_0(T)}{1+F_0^a \chi_0(T)} \tag{4.18}$$

(iv) the normal and superfluid density

$$\begin{aligned} \rho^n(T) &= \rho\frac{m^*}{m}\frac{Y_0(T)}{1+(F_1^s/3)Y_0(T)} \\ \rho^s(T) &= \frac{\rho[1-Y_0(T)]}{1+(F_1^s/3)Y_0(T)} = \rho - \rho^n(T) \end{aligned} \tag{4.19}$$

and

(v) the spin superfluid density

$$\overleftrightarrow{n}^s_{zz} = \frac{\overleftrightarrow{H}^a_1}{1+F_1^s/3}\,\overleftrightarrow{n}^s_{0zz} \tag{4.20}$$

$$\overleftrightarrow{n}^s_{0zz} = n[1-\overleftrightarrow{X}_1^{(-)}] = n[1-Y_0(T)]\; 3\int \frac{d\Omega_k}{4\pi}\hat{k}_z^2(\overleftrightarrow{1} - \overleftrightarrow{\Lambda}_k) \tag{4.21}$$

The last equality in the second of the equations (4.19) holds only for Galilean invariant systems.

Our next task will now be to treat the case that the response of the system is dissipative, i.e. the transport properties of the quasiparticle gas. For this purpose it is necessary to specify the collision integral $\delta \mathrm{I}_{k++}$ in (4.2). The details of the Bogoliubov transformation of the matrix collision integral cannot be presented here for lack of space. The interested reader is referred to Refs. [3] and [25]. It turns out that the (++)-component of transformed matrix collision integral $\underline{U}_k\, \delta\underline{I}_k\, \underline{U}_k^\dagger$ has a form quite reminiscent of the result for the normal Fermi liquid (cf. (2.15)):

$$\begin{aligned} \delta \mathrm{I}_{k++} = &-\frac{1}{\tau_N(0,T)}\frac{2}{N_F}\sum_{\vec{k}'} J(\xi_k,\xi_{k'}) \\ &\left\{\mathrm{h}_k - \frac{\varphi_k}{\varphi_{k'}}\sum_{\ell}\,[\lambda_\ell^t\, h_\ell^{(t)}(\vec{k})\tau^0 + \lambda_\ell^a\, \vec{h}_\ell(\vec{k})\cdot\vec{\tau} + \ldots]\right\} \end{aligned} \tag{4.22}$$

where a few terms involving new superfluid scattering parameters, which have no consequences for the following discussion, have been left out to avoid confusion. Here the quantities $\mathbf{h}_\ell$ are moments of the quasiparticle distribution function $\mathbf{h}_k$ which are defined through

$$\mathbf{h}_\ell^{(t)}(\vec{k}) = (2\ell+1)2\sum_{\vec{k}'} P_\ell(\hat{k}\cdot\hat{k}')\mathbf{C}_{kk'}^{ts}\mathbf{h}_{k'} \tag{4.23}$$

The Bogoliubov transformation of the collision integral generates the coherence tensor $\mathbf{C}_{kk'}^{ts}$, which is of the form ($t=\pm 1, s=\pm 1$):

$$\mathbf{C}_{kk'}^{ts} = \tau^0\left[s_k^{(s)}s_{k'}^{(s)}\frac{1+t}{2}+\frac{Y_0(T)}{3}\frac{E_kE_{k'}}{<E^2>_\varphi}s_k^{(-s)}s_{k'}^{(-s)}\frac{1-t}{2}\right]\tau^0+\vec{\tau}\cdot\overleftrightarrow{S}_k^{(s)}\overleftrightarrow{S}_{k'}^{(s)}\cdot\vec{\tau} \tag{4.24}$$

Here the scalar ($s_k^{(s)}$) and tensorial ($\overleftrightarrow{S}_k^{(s)}$) coherence factors have been introduced via Eqs. (4.6) and (4.7). The case of s–wave pairing correlations is contained in the spin part of (4.24) as the limit $\overleftrightarrow{\Lambda}_k \to \overleftrightarrow{1}$. The integral operator $J(E,E')$ in the scalar quasiparticle collision integral has the integral property

$$\int_{-\infty}^{\infty} d\xi' J(E_k,E_{k'}) = I(E_k,T)$$

yielding, like in the normal Fermi liquid, an energy and temperature dependent function $I(E_k,T)$, which describes the energy dependence of the lifetime of Bogoliubov quasiparticle excitations:

$$\tau_S(E_k,T) = \frac{\tau_N(0,T)}{I(E_k,T)} = \frac{\tau_N(0,T_c)}{I(E_k,T)}\left(\frac{T_c}{T}\right)^2 \tag{4.25}$$

In contrast to the Landau quasiparticle excitations from the ground state of a normal Fermi liquid, the Bogoliubov quasiparticles are characterized by a gap in the excitation spectrum, which is reflected in the energy dependence of the function $I(E_k,T)$. At arbitrary temperatures this energy dependence cannot be given in a closed form. There are analytical expressions for this quantity, however, near T_c [27]:

$$\lim_{T\to T_c} I(E_k,T) = 1+\left(\frac{E_k}{\pi k_B T}\right)^2 + O\left(\frac{\Delta^2}{(k_BT)^2}\right)$$

and at low temperature [3]

$$\lim_{T\to 0} I(E_k,T) = I_0(T)\left\{w_0+\frac{k_BT}{\Delta}\left[\frac{3}{4}(1+x_k^2)w_0-(1+2x_k^2)(\frac{\gamma_0}{3}+\delta_0)\right]\right\} \tag{4.26}$$

with new superfluid scattering parameters γ_0, δ_0 and $w_0 = 1-2\gamma_0/3+\delta_0$, which, in the case of ^{3}He, turn out to be only weakly pressure dependent and may be taken to be 0.1 and 0.3, respectively [25]. The function $I_0(T) = 3\Delta Y_0(T)/2\pi k_B T$ describes the asymptotic behavior with temperature of the quasiparticle lifetime in the limit as $T \rightarrow 0$; the dependence on the dimensionless energy variable $x_k = \xi_k/\sqrt{2k_B T\Delta}$ characterizes the first order finite temperature correction terms $\propto k_B T/\Delta$.

At intermediate temperatures an approximate form for the function $I(E_k, T)$ has been found, which interpolates between the two limits and has an accuracy of a few percent compared with the full numerical evaluation [50].

Let me now discuss the conservation properties of the quasiparticle collision integral (4.22). This can be done by studying the time evolution of the quasiparticle moments

$$\delta\nu_a = 2\sum_{\vec{k}} \frac{1}{2}\mathrm{tr}\left[\mathrm{a}_k \delta\nu_k\right]$$

with the aid of the scalar kinetic equation (4.2):

$$\frac{\partial}{\partial t}\delta\nu_a + \vec{\nabla}\cdot\vec{\mathrm{j}}_a = \delta\mathrm{I}_a + \vec{P}_a\cdot\vec{\tau} \tag{4.27}$$

Here $\vec{\mathrm{j}}_a$ is the associated quasiparticle current

$$\vec{\mathrm{j}}_a = 2\sum_{\vec{k}} \vec{V}_k \frac{1}{2}\mathrm{tr}\left[\mathrm{a}_k \mathrm{h}_k\right]$$

and $\vec{P}_a$ the precession term

$$\vec{P}_a = 2\sum_{\vec{k}} \frac{1}{2}\mathrm{tr}\left[\mathrm{a}_k(\vec{h}_k \times \delta\vec{E}_k)\cdot\vec{\tau}\right]$$

It turns out that the corresponding moments of the collision integral

$$\delta\mathrm{I}_a = 2\sum_{\vec{k}} \frac{1}{2}\mathrm{tr}\left[\mathrm{a}_k \delta\mathrm{I}_k\right]$$

vanish for the following set of collisional invariants a_k:

$$\mathrm{a}_k = \begin{cases} \hbar\vec{k}\tau^0 & \text{MOMENTUM CONSERVATION} \\ E_k\tau^0 & \text{ENERGY CONSERVATION} \\ \frac{\hbar}{2}\vec{\tau} & \text{SPIN CONSERVATION (ONLY S} - \text{WAVE)} \end{cases} \tag{4.28}$$

This important result requires some comments: Due to the lack of number and spin conservation, Bogoliubov quasiparticles are characterized by less collision invariants than the thermal excitations from a normal Fermi liquid. Only momentum, energy and spin (s–wave case) of Bogoliubov quasiparticles are conserved in two particle collision processes.

As we shall see below, this opens the possibility of a mutual disequilibrium between the normal and superfluid number and spin density contributions to the total densities, respectively. A mutual disequilibrium of this kind is possible only in superfluids, in which *both* normal *and* superfluid component contribute to the number and spin density. This is generally the case for the number density (characterized by a superfluid contribution $\propto \chi_0^{p(+)}(T)$) and, only in the case of triplet pairing, for the spin density ($\propto \chi_0^{p}(T)$). As the magnetization in systems with singlet pairing correlations is entirely due to quasiparticles (cf. table 4) a corresponding intrinsic spin relaxation mechanism does not exist in these systems.

Exact solutions of the transport integral equations resulting from (4.2) with the full expression (4.22) for the BQP collision integral are possible only in asymptotic form near T_c [29] and in the low temperature limit [3],[30]. As I am interested in the full temperature dependence of the transport parameters, I proceed therefore with an approximate treatment of the collision integral. As in the normal Fermi liquid I shall from now on use the "conserving" relaxation time approximation for the collision operator J, which is a straightforward generalization of (2.18):

$$J(E_k, E_{k'}) = \varphi_{k'} < I(E) >_\varphi \left[\frac{1}{Y_0(T)}\frac{1+t}{2} + \frac{1}{3}\frac{E_k E_{k'}}{< E^2 >_\varphi}\frac{1-t}{2}\right] \tag{4.29}$$

A major advantage of such an approximate treatment is that all results for the transport properties will emerge in a very simple and physically transparent form. I should like to note, however, that instead of using approximations for the collision operator, one may as well approximate the solution of the transport equation. The resulting variational treatment has been applied by Ono et al. [31] to the calulation of the shear viscosity and leads to results which are identical (at least within a few percent) with that of the conserving relaxation time approximation. It is immediately clear that the application of the conserving relaxation time approximation amounts to replacing the energy dependent quasiparticle lifetime by its thermal average:

$$\bar{\tau}_S(T) = \frac{Y_0(T)}{\int\limits_{-\infty}^{\infty} d\xi_k \varphi_k / \tau_S(E_k, T)} \tag{4.30}$$

In the low temperature limit the averaged lifetime $\bar{\tau}_S$ becomes identical with the limit $k_BT/\Delta \to 0$ of the energy dependent lifetime (4.26)

$$\lim_{T\to 0} \bar{\tau}_S(T) = \lim_{T\to 0} \tau_S(E_k, T) = \frac{\tau_N(0,T)}{I_0(T)w_0}$$

It will turn out to be important later that the product of the averaged lifetime $\bar{\tau}_S(T)$ and the Yosida function $Y_0(T)$ reduces to a simple temperature power law

$$\lim_{T\to 0} Y_0(T)\bar{\tau}_S(T) = \frac{2\pi}{3w_0\delta_{sc}} \frac{T_c}{T} \tau_N(0, T_c) \tag{4.31}$$

in the low temperature limit due to a cancellation of the exponential factors. As we shall see below, this is the cause of the low temperature power law behavior of almost all transport parameters in isotropic superfluid Fermi liquids.

It is illustrative to investigate the conservation and relaxation properties of the approximated collision operator. For this purpose we study again the time evolution of the quasiparticle observable densities $\delta\nu_a$. As a result of such an analysis one finds that the corresponding moments $\delta \mathrm{I}_a$ of the approximated scalar collision integral can be written in the simple form

$$\delta \mathrm{I}_a = -\frac{1}{\tau_a}\,\mathrm{h}_a \quad ; \quad \mathrm{h}_a = 2\sum_{\vec{k}} \frac{1}{2}\mathrm{tr}[\mathrm{a}_k\,\mathrm{h}_k] \tag{4.32}$$

The moments (4.32) are characterized by a set of transport times $\tau_a(T)$ which are displayed in table 5. It is clear from table 5 that the conserving relaxation time approximation reproduces in fact the conservation laws for momentum and energy, relevant for the Fermi superfluid under consideration. It furthermore yields approximate expressions for the decay times of observables, which are not conserved. The case of quasiparticle mass and spin density (the latter, however, only for p–wave pairing!) is exceptional in that these quantities relax in the superfluid phase whereas they are conserved in the normal Fermi liquid. This fact is reflected also in the approximate treatment of the collision operator through the divergence of the associated decay times $\tau_0(T)$ and $\tau_{CE}(T) \propto 1/\Delta(T)$ in the limit $T \to T_c^-$. A few further comments on the intrinsic spin relaxation seem to be useful. The intrinsic spin relaxation time $\tau_{CE}(T)$ has been introduced by Combescot and Ebisawa [32] into the literature and describes the relaxation of the quasiparticle magnetization $\vec{M}^q$ to a local equilibrium value $\vec{M}^{q\ \mathrm{loc}} = (\gamma\hbar/2)\chi_0^q N_F \vec{X}$ with an effective field $\vec{X} = (\gamma\hbar/2)\vec{H}^{\mathrm{ext}} + \delta\vec{\mu} - F_0^a \vec{M}/(N_F\gamma\hbar/2)$:

$$\frac{\partial}{\partial t}\vec{M}^q + \ldots = -\frac{1}{\tau_{CE}(T)}\vec{M}^{q\prime} \quad ; \quad \vec{M}^{q\prime} = \vec{M}^q - \vec{M}^{q\ \mathrm{loc}} \tag{4.33}$$

QUASIPARTICLE DENSITY	a_k	RELAXATION RATE τ_a^{-1}	$\lim_{T\to T_c}\tau_a^{-1}$
MASS	$m(\xi_k/E_k)$	$\bar{\tau}_S^{-1}(1-\lambda_0^+ Y_2/Y_0) \equiv \tau_0^{-1}$	0
ENTROPY	E_k/T	$\bar{\tau}_S^{-1}(1-\lambda_0^-/3) = 0$	0
MAGNETIZATION	$\gamma\hbar/2\ \overleftrightarrow{S}_k^{(+)}$	$\bar{\tau}_S^{-1}(1-\lambda_0^a\chi_0^q/Y_0) \equiv \tau_{CE}^{-1}$	0
MOMENTUM	p_i	$\bar{\tau}_S^{-1}(1-\lambda_1^+) = 0$	0
STRESS TENSOR	$p_i V_{kj}$	$\bar{\tau}_S^{-1}(1-\lambda_2^+ Y_2/Y_0) \equiv \tau_\eta^{-1}$	τ_η^{-1}
ENTROPY CURR.	$v_{ki}\xi_k$	$\bar{\tau}_S^{-1}(1-\frac{\lambda_1^-}{3}\frac{<\xi^2>_\varphi}{<E^2>_\varphi}) \equiv \tau_\kappa^{-1}$	τ_κ^{-1}
MAGN. CURR.	$(\gamma\hbar/2)\ \overleftrightarrow{V}_{kz}$	$\bar{\tau}_S^{-1}(1-\lambda_1^a\ \overleftrightarrow{X}_1^{q-}/Y_0) \equiv \overleftrightarrow{\tau}_D^{-1}$	τ_D^{-1}

Table 5: Conservation and relaxation properties of the Bogoliubov quasiparticle collision integral in the conserving relaxation time approximation

The quantity $\vec{M}^{q'}$ can be physically interpreted as the excess quasiparticle magnetization that describes the *intrinsic* disequilibrium between quasiparticle ($\vec{M}^q$) and pair ($\vec{M}^p$) contributions to the constant total magnetization:

$$\vec{M}^q = \frac{\chi_0^q}{\chi_0}\vec{M} + \frac{\chi_0^p}{\chi_0}\vec{M}^{q'} \quad ; \quad \vec{M}^p = \frac{\chi_0^p}{\chi_0}\vec{M} - \frac{\chi_0^p}{\chi_0}\vec{M}^{q'} \tag{4.34}$$

As it is the fraction χ_0^q/χ_0 of the total magnetization to which $\vec{M}^q$ relaxes in a real NMR situation, one has to rewrite (4.33) accordingly, by inserting $\vec{M}^{q'}$ from (4.34):

$$\frac{\partial}{\partial t}\vec{M}^q + ... = -\frac{1}{\tau_{LT}(T)}[\vec{M}^q - \frac{\chi_0^q}{\chi_0}\vec{M}] \tag{4.35}$$

Here $\tau_{LT}(T) = \tau_{CE}(T)\chi_0^p/\chi_0$ is the decay time characteristic of a relaxation process at constant total magnetization. It has been introduced by Leggett and Takagi [2] in the context of their phenomenological description of spin dynamics of the Helium–3 superfluids.

Let me conclude this lecture with the remark that in the case of p–wave pairing the magnetization current decay time $\overleftrightarrow{\tau}_D$ is a tensor quantity. As in the case of the susceptibilities $\overleftrightarrow{X}_1^{q(-)}$, $\overleftrightarrow{X}_1^{p(-)}$ and $\overleftrightarrow{X}_1^{(-)}$ this can be traced back to the presence of spin–orbit correlations which give rise to macroscopic anisotropies.

5 Transport in infinitely extended Fermi superfluids

After having discussed the local response properties of the normal and superfluid component of a Fermi superfluid in the last lecture, I shall now turn to the transport parameters, characteristic for the dissipative response of a quasiparticle gas. For this purpose we have to calculate the dissipative part of the currents, introduced in the last lecture, which, as an inspection of table 4 shows, requires the knowledge of the distribution function h_k. Let me therefore first rewrite the scalar kinetic equation for Bogoliubov quasiparticles in terms of h_k by splitting off the local contribution to the quasiparticle distribution function $\delta\nu_k$:

$$\delta\nu_k(\vec{r},t) = \mathrm{h}_k(\vec{r},t) - \varphi_k \delta \mathrm{E}_k(\vec{r},t)$$

The local contribution $\propto \delta\partial \mathrm{E}_k/\partial t$ enters the right hand side of the kinetic equation (4.2) as a collection of force terms which originate from external, condensate (gauge–) and Fermi liquid potentials:

$$\left[\frac{\partial}{\partial t} + \vec{V}_k \cdot \vec{\nabla}\right] \mathrm{h}_k(\vec{r},t) = \varphi_k \frac{\partial}{\partial t}\delta \mathrm{E}_k + \delta \mathrm{I}_{k++} - \frac{2}{\hbar}\left[\vec{h}_k \times \vec{H}^{\mathrm{ext}}\right]\cdot \vec{\tau} \tag{5.1}$$

The quasiparticle energy change can, with the aid of the normal–superfluid decompositions (4.6) and (4.7) be expressed by the distribution function h_k, or more precisely, by a Legendre expansion in the moments $\mathrm{h}_\ell(\vec{k})$:

(i) spin–independent case ($s_\ell = (-1)^\ell$):

$$\begin{aligned} \delta E_k^{(s)} &= \left[-\frac{\xi_k}{E_k}\frac{eU^{\mathrm{ext}} + \delta\mu_0}{1+F_0^s}\frac{1+s}{2} + \frac{p_i[v_i^s - v_i^{\mathrm{ext}}]}{1+F_1^s Y_0/3}\frac{1-s}{2}\right] \\ &+ \frac{1}{N_F}\sum_\ell s_k^{(s_\ell)} A_\ell^s(T)\, h_\ell(\vec{k}) \\ A_\ell^s(T) &= \frac{F_\ell^s/(2\ell+1)}{1+F_\ell^s \chi_\ell^{(s_\ell)}/(2\ell+1)} \end{aligned} \tag{5.2}$$

(ii) spin–dependent case

$$\delta \vec{E}_k^{(s)} = \left[-\overset{\leftrightarrow}{S}{}_k^{(+)} \cdot \frac{(\gamma\hbar/2)\vec{H}^{\mathrm{ext}} + \delta\vec{\mu}}{1+F_0^a\chi_0}\frac{1+s}{2} + \overset{\leftrightarrow}{H}{}_1^a \overset{\leftrightarrow}{V}_{kz}\, m\vec{v}_z^{\,s}\frac{1-s}{2}\right]$$

$$+ \quad \frac{1}{N_F}\sum_\ell \overleftrightarrow{S}_k^{(s_\ell)} \overleftrightarrow{A}_\ell^{a}(T)\, \vec{h}_\ell(\vec{k}) \tag{5.3}$$

$$\overleftrightarrow{A}_\ell^{a}(T) = \frac{F_\ell^a/(2\ell+1)}{1+F_\ell^a \overleftrightarrow{\chi}_\ell^{(s_\ell)}/(2\ell+1)}$$

Here the parity dependent coherence factors $s_k^{(s)}$ and tensors $\overleftrightarrow{S}_k^{(s)}$ have been introduced via Eqs. (4.6) and (4.7), respectively. Assuming a harmonic time dependence $\partial/\partial t \to -i\omega$ in what follows, one observes that the Legendre expansion of the Fermi liquid contributions to the time dependent quasiparticle energy change $\delta \mathbf{E}_k$ simply adds to the Legendre expansion of the collision integral and leads to a replacement of the collision parameters λ_ℓ with complex quantities Λ_ℓ:

$$\begin{aligned} \lambda_\ell^s \to \Lambda_\ell^s(\omega) &= \lambda_\ell^s - i\omega\bar{\tau}_S Y_0(T)\, A_\ell^s(T) \\ \lambda_\ell^a \to \overleftrightarrow{\Lambda}_\ell^{a}(\omega) &= \lambda_\ell^a - i\omega\bar{\tau}_S Y_0(T)\, \overleftrightarrow{A}_\ell^{a}(T) \end{aligned} \tag{5.4}$$

In the odd–parity case with respect to the variable ξ_k these Fermi liquid effects are seen to contribute at most to the order T_c/T_F and will therefore not be discussed further. Inserting the Legendre expansions (5.2) and (5.3) into (5.1) on arrives at very general forms of the transport equations for the distribution function $\mathbf{h}_k$. They read in the

(i) spin–independent case ($t = +1$)

$$\begin{aligned} \left[1+\tilde{\tau}_S \vec{V}_k\cdot\vec{\nabla}\right] h_k &= \frac{1}{1-i\omega\bar{\tau}_S}\frac{\varphi_k}{N_F Y_0}\sum_\ell \Lambda_\ell^+(\omega)\, h_\ell(\vec{k}) \\ &+ \varphi_k(i\omega\tilde{\tau}_S)\left[\frac{\xi_k}{E_k}\frac{eU^{\text{ext}}+\delta\mu_0}{1+F_0^s} - \frac{p_i[v_i^s - v_i^{\text{ext}}]}{1+F_1^s Y_0/3}\right] \end{aligned} \tag{5.5}$$

where $\tilde{\tau}_S = \bar{\tau}_S/(1-i\omega\bar{\tau}_S)$. Similarly, one obtains in the

(ii) spin–independent case ($t = -1$)

$$\left[1+\tilde{\tau}_S \vec{V}_k\cdot\vec{\nabla}\right] h_k = \frac{1}{1-i\omega\bar{\tau}_S}\frac{\varphi_k}{N_F Y_0}\sum_\ell \lambda_\ell^- h_\ell(\vec{k}) \tag{5.6}$$

Finally we have a result in the

(iii) spin–dependent case (spatial variation assumed to be in the z–direction!)

$$\begin{aligned} \left[1+\tilde{\tau}_S \vec{V}_k\cdot\vec{\nabla}\right] \vec{h}_k &= \frac{1}{1-i\omega\bar{\tau}_S}\frac{\varphi_k}{N_F Y_0}\sum_\ell \overleftrightarrow{\Lambda}_\ell^{a}(\omega)\, \vec{h}_\ell(\vec{k}) - \frac{2}{\hbar}[\vec{h}_k\times\delta\vec{E}_k]\tilde{\tau}_S \\ &+ \varphi_k(i\omega\tilde{\tau}_S)\left[\overleftrightarrow{S}_k^{(+)}\cdot\frac{(\gamma\hbar/2)\vec{H}^{\text{ext}}+\delta\vec{\mu}}{1+F_0^a\chi_0} - \overleftrightarrow{H}_1^{a}\overleftrightarrow{V}_{kz}\, m\vec{v}_z^{\,s}\right] \end{aligned} \tag{5.7}$$

The general procedure for solving the transport equations for Fermi superfluids consists essentially of properly inverting the gradient operator. This can be achieved by two methods. The first is the simple gradient expansion

$$\begin{aligned}\left\{1+\tilde{\tau}_S\vec{V}_k\cdot\vec{\nabla}\right\}^{-1} &= \left[1-\tilde{\tau}_S\vec{V}_k\cdot\vec{\nabla}\right]\hat{D}_k \\ \hat{D}_k &= 1+(\tilde{\tau}_S\vec{V}_k\cdot\vec{\nabla})^2+(\tilde{\tau}_S\vec{V}_k\cdot\vec{\nabla})^4+\dots\end{aligned} \tag{5.8}$$

which explicitly separates even and odd powers in $V_k \propto \xi_k/E_k$. It is clear from (5.8) that higher order terms in the gradient operator $\hat{D}_k$ become important with increasing values for the k-dependent quasiparticle mean free path $\lambda_k \equiv \tilde{\tau}_S V_k$. In the ideal situation, where the typical wall separation distance d of the container which entrains the Fermi liquid is large compared to λ_k, one may neglect higher order terms in the gradient expansion and put $\hat{D}_k \equiv 1$. We know, on the other hand, that the mean free path λ_k can become very large at low temperature due to the increase of the quasiparticle lifetime $\propto T^{-2}$ in the normal Fermi liquid and, even more dramatically, $\propto \exp(-\Delta/k_BT)$, in an isotropic Fermi superfluid. In the low temperature limit, there will therefore eventually arise the situation that λ_k becomes comparable with d and mean free path (or finite size) effects play an important role. It is in this case where the second method, which consists of integrating the transport equation with respect to the spatial variables, is preferable. We leave the discussion of this method to the next lecture where it will be used in context with the treatment of mean free path (or finite size) effects.

We are now able to solve the transport equations for $\mathbf{h}_k$ in the various cases of interest. All numerical calculations which will have to be performed in this context apply to liquid ^{3}He. They are based on one set of pressure dependent parameters, which is displayed in table 6. Let me begin with the **shear viscosity** of superfluid Fermi liquids.

The transport equation applied to this case has the solution

$$h_k^{(+)} = \frac{\varphi_k}{1-i\omega\bar{\tau}_S}\hat{D}_k\left\{\Lambda_2^+\frac{h_2(\vec{k})}{N_F Y_0}-\frac{1}{2}\tilde{\tau}_S\Lambda_1^+(\omega)p_i V_{kj}v_{ij}^{n}\right\} \tag{5.9}$$

Here the traceless normal velocity derivative tensor is given by $v_{ij}^{n} = \nabla_i v_j^{n} + \nabla_j v_i^{n} - \frac{2}{3}\delta_{ij}\vec{\nabla}\vec{v}^{\,n}$. The solution (5.9) generates the following expression for the dissipative part of the stress tensor:

$$\Pi'_{ij} = p_F v_F\, 2\sum_{\vec{k}}\left(\hat{k}_i\hat{k}_j-\frac{1}{3}\delta_{ij}\right)h_k^{(+)} = -\eta(\omega)\frac{\Lambda_1^+(\omega)}{1-i\omega\bar{\tau}_S}v_{ij}^{n} \tag{5.10}$$

P	v_F	F_0^a	λ_2^+	$\tau_N(0,T_c)$	$\frac{\Delta(0)}{k_B T_c}$	$\frac{\Delta C}{C_N}$
[bar]	[cm/sec]			[μsec]		
0	5910	-0.70	0.68	0.500	1.76	1.43
10	4567	-0.74	0.74	0.120	1.85	1.70
20	3868	-0.76	0.75	0.054	2.00	1.83
30	3463	-0.75	0.74	0.040	2.10	1.92
ref.	[33]	[33]	[34]	[35]	[36]	[33]

Table 6: Pressure dependence of the theoretical input parameters that enter the numerical calculations for liquid ^{3}He

where $\eta(\omega) = \eta(T)/(1 - i\omega\tau_\eta(T)/\phi_2(T))$ and $\phi_2(T) = [1 + F_2^s/5]/[1 + F_2^s(1 - Y_2)/5]$, from which, in the stationary limit ($\omega\bar{\tau}_S \to 0$), a coefficient of shear viscosity can be identified as:

$$\eta(T) = \frac{1}{5} n p_F v_F Y_2(T) \tau_\eta(T) \tag{5.11}$$

It is illustrative to rewrite the result of the shear viscosity coefficient in terms of thermal averages characteristic of the dilute quasiparticle gas. For this purpose it is convenient to introduce the averaged quasiparticle group (r.m.s.–) velocity, relevant for this case, as

$$V_{\rm rms}^2 = \frac{2\sum_{\underline{k}} \varphi_k \vec{V}_k^2}{2\sum_{\underline{k}} \varphi_k} = v_F^2 \frac{\chi_0^{q(+)}(T)}{Y_0(T)} = v_F^2 \frac{Y_2(T)}{Y_0(T)} \overset{T\to 0}{=} v_F^2 \frac{k_B T}{\Delta} \tag{5.12}$$

and write the result (5.11) for the shear viscosity in a form reminiscent of the one known from the kinetic theory of classical gases:

$$\eta(T) = \frac{1}{5} \rho_0^n(T)\, V_{\rm rms}^2(T) \tau_\eta(T) \tag{5.13}$$

Here $\rho_0^n = \rho(m^*/m)Y_0$ is the normal fluid density in the absence of Fermi liquid corrections. Now we remember that the product $Y_0 \cdot \bar{\tau}_S$ of the quasiparticle lifetime and the Yosida function varies as T^{-1} at low temperatures (cf. (4.31)) so that our result (5.13) implies a constant zero temperature limit of the shear viscosity of an isotropic Fermi liquid because of the $k_B T/\Delta$-variation of the averaged group velocity. To illustrate this plateau-like behavior, I have plotted in Fig. 2 the dependence of the normalized shear viscosity of superfluid ^{3}He–B at 20 bar pressure on reduced temperature T/T_c. Also shown in this

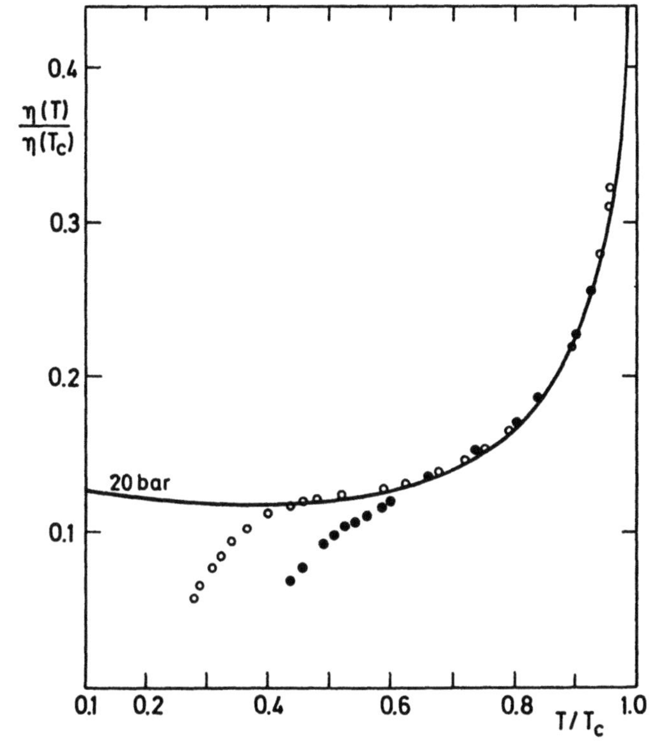

Fig. 2 Normalized shear viscosity of ^{3}He–B at 20 bar pressure vs. reduced temperature T/T_c. Full line: Theory; circles: torsional oscillator data by Archie et al. [6] (full circles), vibrating wire data by Carless et al. [7] (open circles).

figure are two sets of data by Archie et al. (94 μm torsional oscillator, closed circles) [6] and Carless et al. (58 μm vibrating wire, open circles) [7]. The agreement between theory and experiment is very good for higher temperatures. At the low temperature end the experimental points are seen to fall below the theoretical prediction (viscosity "droop"). This discrepancy is, however, dramatic only in a temperature regime, where the quasiparticle mean free path can be expected to match the height of the torsional oscillator or the diameter of the vibrating wire and the applicability of the bulk theory breaks down. In the next lecture I shall deal in detail with this problem.

Let me continue with the **second viscosity** of superfluid Fermi liquids. The $\ell = 0$

projection of the transport equation (5.5) yields the isotropic solution

$$h_k^{(+)} = \frac{\varphi_k}{N_F(1-i\omega\bar{\tau}_S)}\hat{D}_k\left\{\left[\frac{\lambda_0^+}{Y_0} - i\omega\bar{\tau}_S\right]h_0(\vec{k}) + \frac{1}{m}\frac{\xi_k}{E_k}\vec{\nabla}\cdot\left[\vec{g} - \frac{\Lambda_1^+(\omega)}{1-i\omega\bar{\tau}_S}\rho\vec{v}^{\,\mathrm{n}}\right]\right\} \tag{5.14}$$

which determines the dissipative contribution to the pressure (cf. table 4):

$$\frac{1}{N_F}2\sum_{\vec{k}}\frac{\xi_k}{E_k}h_k^{(+)} = \frac{1}{mN_F}\frac{\tau_0(T)Y_2}{1-i\omega\tau_0(T)(1-Y_2)}\vec{\nabla}\cdot\left\{\vec{g} - \frac{\Lambda_1^+(\omega)}{1-i\omega\bar{\tau}_S}\rho\vec{v}^{\,\mathrm{n}}\right\} \tag{5.15}$$

This, in turn, gives rise to a dissipative term in the acceleration equation for the superfluid, which, in the stationary limit ($\omega\bar{\tau}_S \to 0$), is of the form:

$$\frac{\partial}{\partial t}v_{i0}^{\mathrm{s}}(\vec{r},t) + \nabla_i\left\{\frac{\delta P}{\rho} - \zeta_3(T)\vec{\nabla}\cdot[\vec{g} - \rho\vec{v}^{\,\mathrm{n}}]\right\} = 0 \tag{5.16}$$

The coefficient ζ_3 of second viscosity, which characterizes this dissipation, is given by

$$\zeta_3(T) = \frac{1}{m^2 N_F}\tau_0(T)Y_2(T) \tag{5.17}$$

Due to the existence of a conservation law for the number of quasiparticles in the normal Fermi liquid, the relaxation time τ_0 for the quasiparticle density, which specifies ζ_3, diverges near the transition:

$$\lim_{T\to T_c^-}\tau_0(T) = \frac{4k_B T}{\pi\Delta(T)}\bar{\tau}_S \propto \frac{1}{\sqrt{1-T/T_c}} \tag{5.18}$$

One may furthermore show that, like the shear viscosity, the second viscosity coefficient ζ_3 has also a finite low temperature limit. The dependence on reduced temperature T/T_c of the second viscosity ζ_3 for superfluid ^{3}He–B is shown in Fig. 3 for two pressures of 1.28 and 21 bar. Also shown is a set of experimental data obtained by the Manchester group [7]. The agreement of the experimental points is good in general. The second viscosity at 21 bar is predicted to be roughly an order of magnitude smaller than that at 1.28 bar and is therefore probably more difficult to measure.

Let me now proceed with the discussion of the **diffusive thermal conductivity** of superfluid Fermi liquids. The distribution function h_k reads in this case:

$$h_k = \varphi_k\hat{D}_k\left\{\frac{\lambda_1^-}{N_F Y_0}h_1(\vec{k}) - \frac{1}{T}\bar{\tau}_S\xi_k\frac{\lambda_0^-}{3}\vec{v}_k\cdot\vec{\nabla}\delta T\right\} \tag{5.19}$$

It determines the entropy current (cf. table 4) which reads for $\hat{D}_k \to 1$:

$$T\vec{j} = 2\sum_{\vec{k}}\vec{v}_k\xi_k\,h_k^{(-)} = -\frac{n}{m^*T} < \xi^2 >_\varphi \tau_\kappa(T)\vec{\nabla}\delta T \tag{5.20}$$

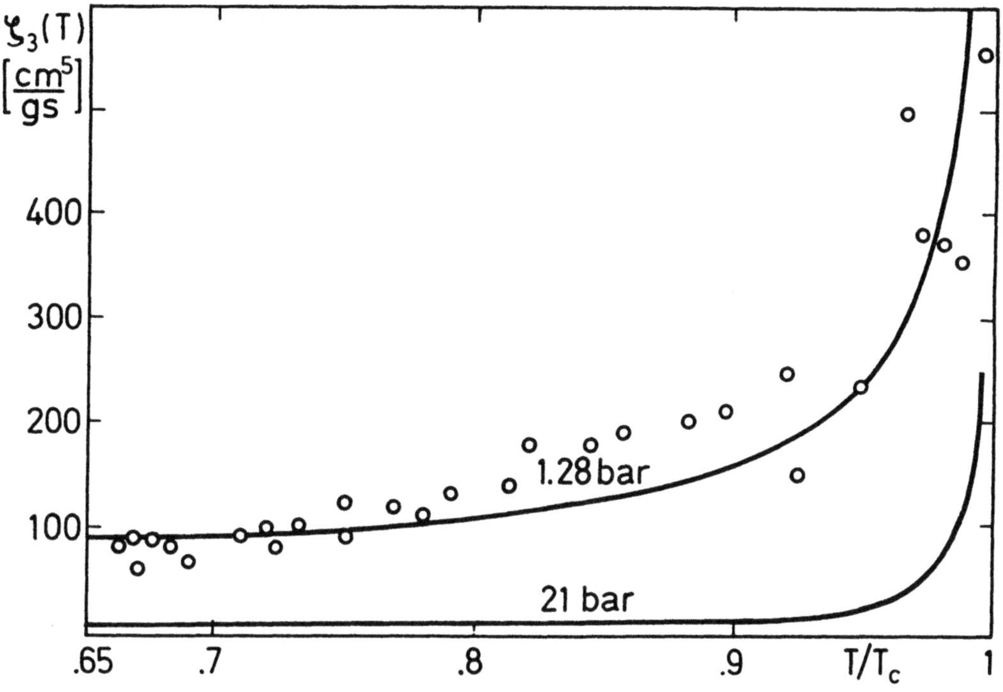

Fig. 3: Second viscosity ζ_3 of ^{3}He–B at two different pressures vs. reduced temperature. The circles refer to data by Carless, Hall and Hook [7].

From this result one may identify the coefficient $\kappa(T)$ of diffusive thermal conductivity

$$\kappa(T) = \frac{n}{m^* T} < \xi^2 >_\varphi \tau_\kappa(T) \tag{5.21}$$

In this case the thermal average of the relevant group velocity has to be defined in a slightly different way:

$$V_{\rm rms}^2 \stackrel{t=-1}{=} \frac{2\sum_{\vec{k}} \varphi_k \vec{v}_k^2 \xi_k^2}{2\sum_{\vec{k}} \varphi_k E_k^2} = v_{\rm F}^2 \frac{< \xi^2 >_\varphi}{< E^2 >_\varphi} \tag{5.22}$$

With (5.22) one obtains a simple BQP gas–kinetic form, at least at low temperatures, also for the thermal conductivity:

$$\kappa(T) \stackrel{T \lessapprox T_c}{=} \frac{1}{3} C_{\rm V}(T) V_{\rm rms}^2(T) \tau_\kappa(T) \tag{5.23}$$

The dependence of the normalized diffusive thermal conductivity of Bogoliubov quasiparticles $\kappa(T)T/\kappa(T_c)T_c$ on reduced temperature is shown in Fig. 4 for various values of the scattering parameter λ_1^-. Also shown is a set of experimental data points obtained by the

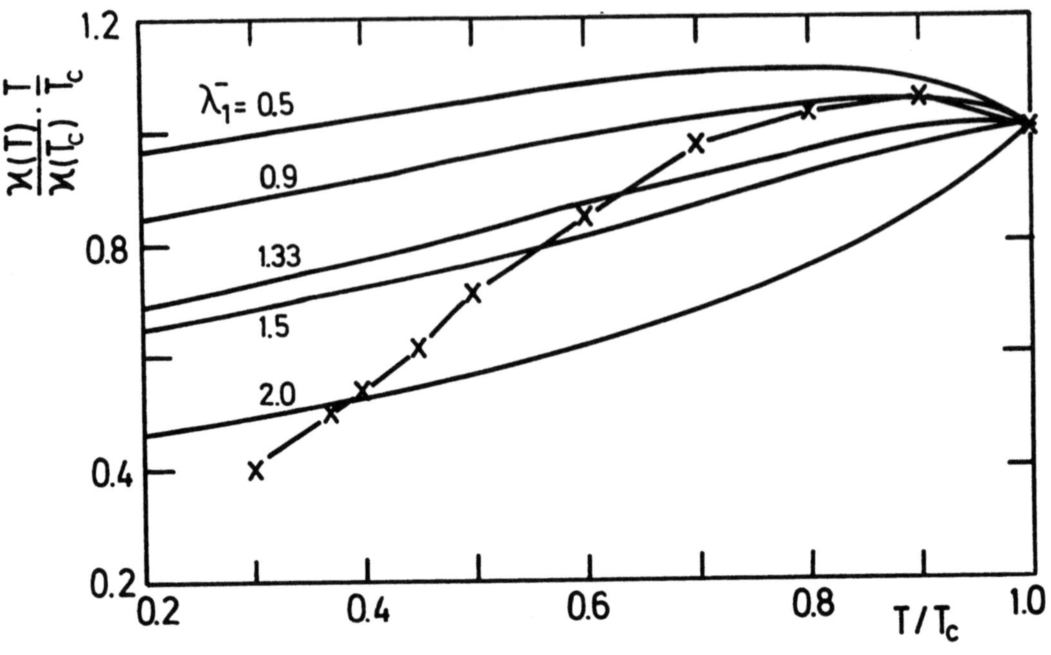

Fig. 4: Normalized thermal conductivity of ^{3}He–B vs. reduced temperature. Solid lines: theory for parameter values λ_1^- as indicated. Crosses: data by the Manchester group [37].

Manchester group [37]. Although theory and experiment agree with respect to the initial increase of the thermal conductivity below T_c for a certain range of scattering parameters λ_1^-, the data points show a much stronger decrease at the low temperature end. The origin of this deviation is still unclear, it might, however, be due to finite size effects.

Let me now turn to the spin–dependent transport properties and start with the simple case where the precession terms in the kinetic equations can be ignored (zero field limit). In such a situation the intrinsic spin relaxation of superfluid Fermi liquids can be derived from the spin part of h_k for the case of positive parity (s=1):

$$\vec{h}_k^{(+)} = \frac{\varphi_k}{N_F(1 - i\omega\bar{\tau}_S)} \hat{D}_k \left\{ \left[\frac{\lambda_0^a}{Y_0} - \frac{i\omega\bar{\tau}_S}{\chi_0} \right] \vec{h}_0(\vec{k}) + \frac{i\omega\bar{\tau}_S}{\chi_0} \frac{\vec{M}}{\gamma\hbar/2} \right\} \tag{5.24}$$

Being the spin analogue of the second viscosity in the case of triplet p–wave pairing, it gives rise to a dissipative contribution in the acceleration equation for the spin superfluid velocity ($\hat{D}_k \to 1$)

$$- i\omega m\vec{v}_z^{\,s}(\vec{r},t) + \frac{\gamma\hbar}{2}\nabla_z \left[1 - \frac{i\omega\tau_{\rm CE}(T)}{1 - i\omega\tau_{\rm CE}(T)\chi_0^{\rm p}/\chi_0}\,\frac{\chi_0^{\rm q}}{\chi_0}\,\frac{1}{1+F_0^{\rm a}\chi_0}\right]\frac{\delta\vec{M}}{\chi_M} = 0 \qquad (5.25)$$

from which we can identify the parameter Ξ of intrinsic spin relaxation in the hydrodynamic limit:

$$\Xi(T) = \tau_{\rm CE}(T)\frac{\chi_0^{\rm q}(T)}{\chi_0(T)}\frac{1}{1+F_0^{\rm a}\chi_0(T)} \qquad (5.26)$$

in the stationary limit $\omega\bar{\tau}_{\rm S} \to 0$. Due to the existence of a conservation law for the quasiparticle magnetization on the normal Fermi liquid side, the relaxation parameter Ξ diverges near T_c as

$$\lim_{T\to T_c^-}\Xi(T) = \frac{3}{2}\frac{4k_{\rm B}T}{\pi\Delta(T)}\bar{\tau}_{\rm S} \propto \frac{1}{\sqrt{1-T/T_c}} \qquad (5.27)$$

Fig. 5 shows the temperature dependence of the relaxation parameter Ξ for superfluid ^{3}He–B evaluated at a pressure of 20 bar. Also shown are three sets of experimental data points by Webb et al. [38], Eska et al. [39] and Ishikawa et al. [40]. The agreement of theory and experiment at higher temperatures is quite good. The expected plateau in the parameter of intrinsic spin relaxation is qualitatively confirmed by the low temperature data points, leaving, however, a quantitative discrepancy with respect to its absolute value.

Let me now come to the last relevant spin transport parameter, namely the **spin diffusion** of superfluid Fermi liquids. It is calculated from the spin part of the distribution function h_k which reads for the case of negative parity ($s = -1$):

$$\vec{h}_k^{(-)} = \frac{\varphi_k}{1-i\omega\bar{\tau}_{\rm S}}\hat{D}_k\left\{\overset{\leftrightarrow}{\Lambda}{}_1^{\rm a}\,\frac{\vec{h}_1(\vec{k})}{N_{\rm F}Y_0} - i\omega\bar{\tau}_{\rm S}\;\overset{\leftrightarrow}{H}{}_1^{\rm a}\overset{\leftrightarrow}{V}_{kz}\,m\vec{v}_z^{\,s}\right\} \qquad (5.28)$$

It determines the dissipative part of the magnetization current $\vec{J}_z^M$ (cf. table 4):

$$\vec{J}_z^{M\prime} = \frac{\gamma\hbar}{2}\;\overset{\leftrightarrow}{H}{}_1^{\rm a}\;2\sum_{\vec{k}}\overset{\leftrightarrow}{V}_{kz}\;\vec{h}_k^{(-)} = -\;\overset{\leftrightarrow}{D}_{zz}\,(T)\nabla_z\delta\vec{M} \qquad (5.29)$$

from which one may identify a tensorial coefficient $\overset{\leftrightarrow}{D}_{zz}$ of spin diffusion

$$\overset{\leftrightarrow}{D}_{zz} = \frac{v_{\rm F}^2}{3\chi(T)}(\overset{\leftrightarrow}{H}{}_1^{\rm a})^2\;\overset{\leftrightarrow}{\tau}_{\rm D}\,(T)\;\overset{\leftrightarrow}{\chi}{}_1^{{\rm q}(-)}\,(T) \qquad (5.30)$$

Here one may define a spin analogue of the thermally averaged quasiparticle group velocity:

$$\overset{\leftrightarrow}{V}{}^2_{\rm rms} = \frac{2\sum_{\vec{k}}\varphi_k\,\overset{\leftrightarrow}{V}{}^2_{kz}}{2\sum_{\vec{k}}\varphi_k\hat{k}_z^2} = v_{\rm F}^2\frac{\overset{\leftrightarrow}{\chi}{}_1^{{\rm q}(-)}\,(T)}{Y_0(T)} \overset{T\to 0}{=} v_{\rm F}^2\left[1+O(k_{\rm B}T/\Delta)\right] \qquad (5.31)$$

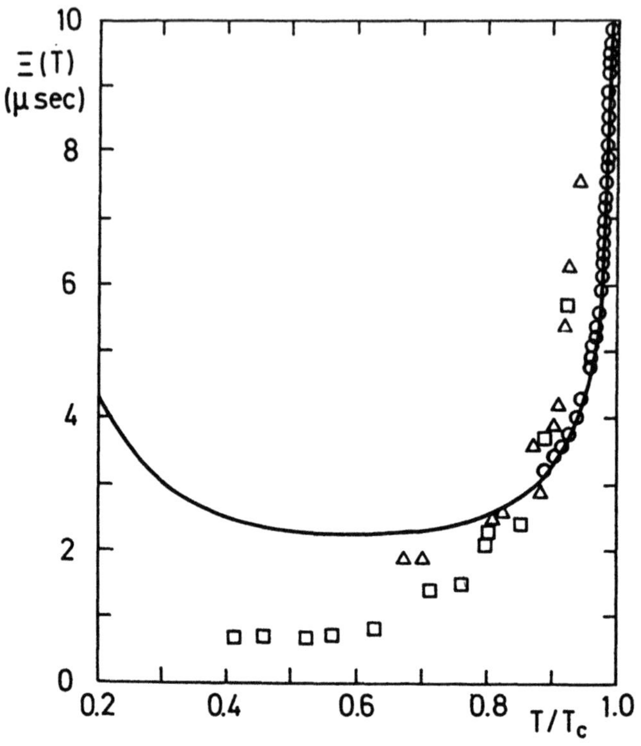

Fig. 5: Intrinsic spin relaxation parameter $\Xi(T)$ at 20 bar pressure vs. reduced temperature. Full line: theory. Data points are taken from Webb et al. [38] (circles), Eska et al. [39] (triangles) and Ishikawa et al. [40] (squares).

With the aid of (5.31) we obtain a gas–kinetic version of the result for the spin diffusion tensor in the form

$$\overleftrightarrow{D}_{zz} = \frac{\overleftrightarrow{V}^2_{\text{rms}}(T)}{3\chi(T)} (\overleftrightarrow{H}^a_1)^2 \, Y_0(T) \, \overleftrightarrow{\tau}_D(T) \tag{5.32}$$

If specialized to the case of s–wave pairing, the hydrodynamic spin diffusion tensor is of the simple form

$$\overleftrightarrow{D}_{zz}(T) = D^{\text{hyd}}(T) \, \overleftrightarrow{1} \; ; \; D^{\text{hyd}}(T) = \frac{V^2_{\text{rms}}(T)}{3} [1 + F^a_0 Y_0(T)] \tau_D(T) \tag{5.33}$$

In the p–wave case, on the other hand, the hydrodynamic spin diffusion tensor has (depending on the approximation for the collision integral at least) two nontrivial eigenvalues:

$$\begin{aligned} \overleftrightarrow{D}_{zz}(T) &= D^{\| \text{ hyd}}_{zz}(T) \, \hat{z} : \hat{z} + D^{\perp \text{ hyd}}_{zz}(T) [\overleftrightarrow{1} - \hat{z} : \hat{z}] \\ D^{\|,\perp \text{ hyd}}_{zz}(T) &= \frac{V^{\|,\perp 2}_{\text{rms}}(T)}{3\chi(T)} \, (H^{a\|,\perp}_1(T))^2 \, Y_0(T) \, \tau^{\|,\perp}_D(T) \end{aligned} \tag{5.34}$$

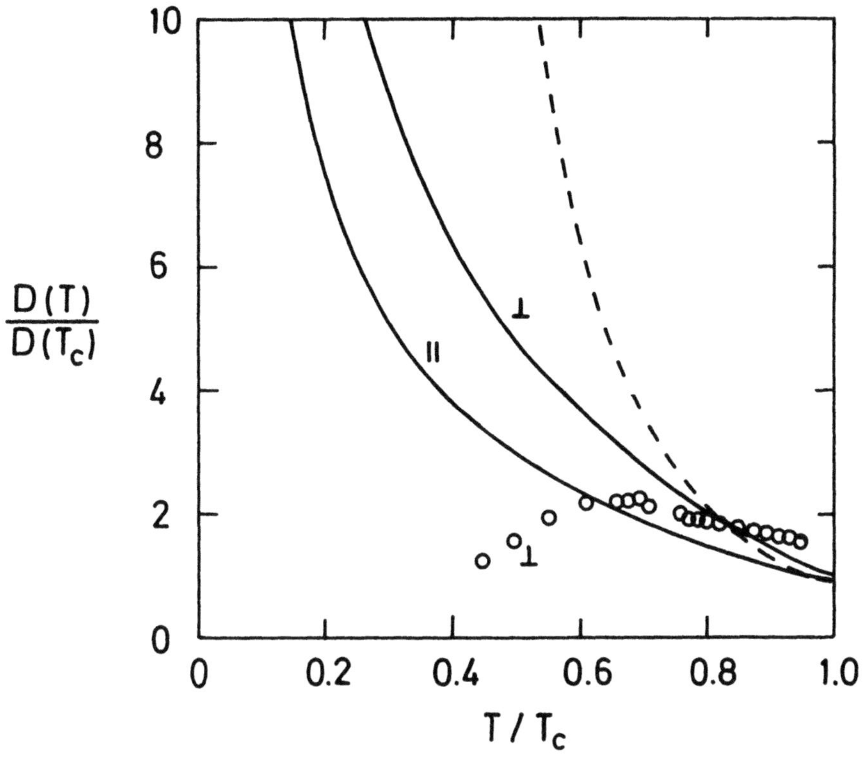

Fig. 6: Normalized components of the hydrodynamic spin diffusion tensor of superfluid ^{3}He–B at 20 bar vs. reduced temperature (full lines). The corresponding result for an s–wave superfluid is shown as dashed line. Circles are NMR data by Bunkov et al. [41].

According to (4.31) the product $Y_0(T)\tau_D(T)$ varies $\propto T^{-1}$ at low temperature. From the low temperature behavior of the averaged quasiparticle group velocity $\overleftrightarrow{V}_{rms}$ (5.31) one may immediately verify that the hydrodynamic spin diffusion diverges like $Y_0(T)^{-1}$ in the case of s–wave pairing and like T^{-1} in superfluid ^{3}He–B.
The dependence of the two components of the hydrodynamic spin diffusion tensor $\overleftrightarrow{D}_{zz}$ for superfluid ^{3}He–B at 20 bar pressure on reduced temperature is illustrated in Fig. 6 (full lines). Also shown (dashed line) is the result expected for the case of a ficticious s–wave superfluid. The circles are recent experimental data points for the transverse spin diffusion by Bunkov et al. [41]. In sharp contrast to the predicted monotonic low temperature *increase* of the spin diffusion, the temperature dependence of the experimental data is characterized by a *maximum* and a *falloff* at the low temperature end. This feature will be reconsidered in detail in lecture 9.

Let me summarize this lecture by stating that the transport parameters of superfluid ^{3}He–B seem to compare well with the experimental data with two dramatic exceptions: the shear viscosity and the spin diffusion. The explanation of the discrepancies between theory and experiment in these cases will be the topic of the remaining lectures.

6 The slip effect

In the preceding lecture it has become clear that there exists a serious qualitative discrepancy between experimental viscosity data and the theoretical prediction for infinitely extended Fermi superfluids. We shall therefore have to take a somewhat closer look at some experimental details of viscosity measurement. Most of the experimental arrangements study the purely transverse (Couette–, Poiseuille–) flow of the quantum liquid past the confining walls of the measuring cell. In such a situation the condensate remains (at least to a good approximation) unaffected and the dynamics of the system is due to that of the normal component alone, described by the quasiparticle Boltzmann equation (4.2), the condensate being completely at rest with respect to the laboratory frame. As the quasiparticle gas becomes very dilute at low temperature, one may expect consequences from the fact that their motion becomes more and more ballistic. This can be conveniently described by rewriting the gas–kinetic expression (5.13) of the shear viscosity in a form, in which the thermal average $\lambda_\eta(T)$ of the quasiparticle mean free path

$$\lambda_\eta(T) = V_{\mathrm{rms}}\tau_\eta \tag{6.1}$$

appears explicitly:

$$\eta(T) = \frac{1}{5}\rho_0^{\mathrm{n}}(T)V_{\mathrm{rms}}(T)\lambda_\eta(T) \tag{6.2}$$

In Fig. 7 I show a plot of the viscous transport mean free path λ_η as a function of reduced temperature for ^{3}He-B at four different pressures. The temperature dependence of the viscous mean free path is characterized first by an initial nonanalytical drop below the transition, a minimum and eventually an exponential increase at the low temperature end. The drop at T_c is due to the change in the character of quasiparticle excitations, different on both sides of the superfluid transition.

The experimental cell for the viscosity measurement is a flow channel with a typical size d of the order of a few 100 μm and does therefore not allow for an unlimited increase in the

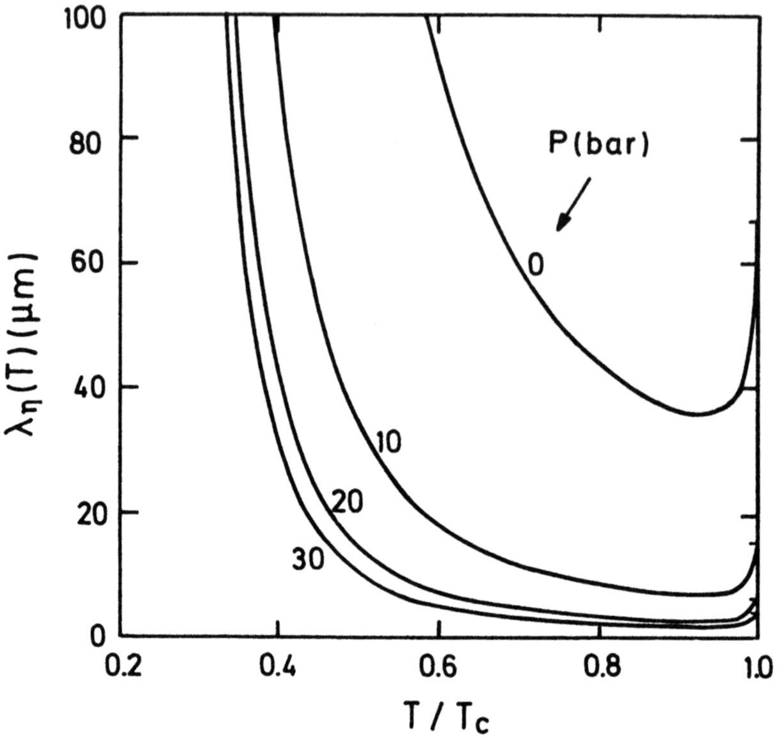

Fig. 7: Viscous mean free path of Bogoliubov quasiparticles in ^{3}He–B at four different pressures vs. reduced temperature.

mean free path. The shear viscosity can be expected to behave in a way characteristic of infinitely extended superfluids only in a regime where the ratio Kn= λ_η/d of the mean free path and the typical cell dimension, the so-called Knudsen number, is negligibly small. For finite Knudsen numbers corrections have to be introduced into the theory which account for the presence of confining walls.

In this lecture I shall deal with mean free path effects which enter as *first order* corrections to the theory presented in the preceding lectures. At the beginning let me consider the case of stationary fluid flow past a plane wall (Couette flow). I assume the fluid to be restricted to the half space $z > 0$, bounded by a wall at $z = 0$. The wall is assumed to be plane on macroscopic, even on mesoscopic length scales. It may, however, have a corrugation on atomic scales which is thought to be the cause of diffuse scattering of the gas particles, which will be discussed below.

The flow of the gas is assumed to be purely transverse (along the x–direction). The quasiparticle transport equation, specialized to this case, assumes the simple form:

$$V_{kz}\frac{\partial h_k}{\partial z} = -\frac{1}{\tau_\eta(T)}\left(h_k - \varphi_k p_x v_x^{\mathrm{n}}\right) \tag{6.3}$$

Defining the deviation of the distribution function from local equilibrium $g_k \equiv h_k - \varphi_k p_x v_x^{\mathrm{n}}$ one may rewrite (6.3) in terms of g_k as

$$\left[1 + \tau_\eta(T)V_{kz}\frac{\partial}{\partial z}\right] g_k = -\tau_\eta(T)\varphi_k p_x V_{kz}\frac{\partial v_x^{\mathrm{n}}(z)}{\partial z} \tag{6.4}$$

The stress tensor component, relevant for the symmetry of our problem, reads:

$$\Pi_{xz}(z) = 2\sum_{\vec{k}} p_x V_{kz} g_k(z) = -2\sum_{\vec{k}} \varphi_k p_x^2 V_{kz}^2 \tau_\eta(T)\frac{\partial v_x^{\mathrm{n}}(z)}{\partial z} \tag{6.5}$$

where the prefactor of the gradient in the normal velocity is the shear viscosity $\eta(T)$ (cf. (5.11)). What precisely are the alterations which come into play through the presence of the wall? The wall has the effect of *distinguishing* between *incoming* (negative group velocity $V_{kz} < 0$) and *outgoing* (positive group velocity $V_{kz} > 0$) quasiparticles. In contrast to the impinging ones the outgoing gas particles have suffered interactions with the wall, which they can communicate via bulk collisions only over a typical distance λ_η near the wall. This defines a (so–called Knudsen) layer in which the quasiparticles move essentially ballistically. Let me decompose the distribution function g_k into contributions from incoming and outgoing quasiparticles:

$$g_k(z) = g_k^{>}(z)\Theta(V_{kz}) + g_k^{<}(z)\Theta(-V_{kz})$$

If the microscopic surface corrugation serves to scatter the quasiparticles diffusely, whereby they lose completely their memory about their earlier local equilibrium state with respect to the rest of the bulk fluid, we can state the following boundary condition for the distribution function $g_k^{>}$ of reemitted quasiparticles:

$$\begin{aligned} g_k^{>}(0) &= \nu_k\left[E_k - p_x v_x^{\mathrm{ext}} + p_x v_{x0}^{\mathrm{n}}\right] - \nu_k^0 \\ &= \varphi_k p_x\left[v_x^{\mathrm{ext}} - v_{x0}^{\mathrm{n}}\right] \end{aligned} \tag{6.6}$$

Our simple Couette–flow case is characterized by a linear variation $v_x^{\mathrm{n}}(z) = v'_\infty \cdot z + v_{x0}^{\mathrm{n}}$ of the macroscopic velocity field of the gas with the distance z from the wall. In such a case the velocity gradient $\partial v_x^{\mathrm{n}}(z)/\partial z = v'_\infty$ is a constant and the stress tensor is independent of z and can be written as

$$\begin{aligned}\Pi_{xz}(z) &= -\eta(T)v'_\infty = \Pi_{xz}(0) \\ &= 2\sum_{\substack{k \\ V_{kz}>0}} p_x V_{kz} g_k^>(z) + \sum_{\substack{k \\ V_{kz}<0}} p_x V_{kz} g_k^<(z)\end{aligned} \tag{6.7}$$

Here the first and second terms in the second equality of Eq. (6.7) refer to contributions from outgoing and incoming quasiparticles, respectively. The value of the stress tensor at the wall ($z = 0$) is the viscous force that the fluid exerts on the boundary:

$$\Pi_{xz}(0) = -2\sum_{\substack{k \\ V_{kz}>0}} \varphi_k p_x V_{kz}\left[v_{x0}^{\mathrm{n}} - v_x^{\mathrm{ext}}\right] + \frac{1}{2}\Pi_{xz}(0) \tag{6.8}$$

The first term on the r. h. s. of Eq. (6.8) originates from inserting the boundary condition (6.6) into Eq. (6.7) whereas the second term is the contribution of the incoming quasiparticles. The velocity difference $[v_{x0}^{\mathrm{n}} - v_x^{\mathrm{ext}}]$ can, with the aid of (6.8) now be expressed through moments of the k–dependent quasiparticle mean free path $\lambda_{kz} = V_{kz}\tau_\eta(T)$:

$$\left[v_{x0}^{\mathrm{n}} - v_x^{\mathrm{ext}}\right] = v'_\infty \frac{\langle p_x^2 V_{kz}\lambda_{kz}\rangle_{\varphi+}}{\langle p_x^2 V_{kz}\rangle_{\varphi+}}$$

where $< \ldots >_{\varphi+} = 2\sum_{\substack{k \\ V_{kz}>0}} \varphi_k \ldots$ denotes the summation over the half space $V_{kz} > 0$. A finite value of this velocity difference (the *velocity slip*) is therefore directly related to the mean free path λ_{kz} or its thermal average. As a consequence the quantity

$$\zeta_0(T) = \frac{v_{x0}^{\mathrm{n}} - v_x^{\mathrm{ext}}}{v'_\infty} = \frac{\langle p_x^2 V_{kz}\lambda_{kz}\rangle_{\varphi+}}{\langle p_x^2 V_{kz}\rangle_{\varphi+}} \tag{6.9}$$

is finite. It is referred to as *slip length*, because it is the length at which the velocity difference extrapolates to zero behind the wall. The assumption of a linear velocity profile everywhere is only approximately true, the slip length evaluated that way is sometimes referred to as Maxwell slip length [42]. It turns out that in a realistic situation the velocity profile aquires a positive curvature within the Knudsen layer, which is sketched in Fig. 8. Such profiles result, for example, from an exact treatment of the transport equation in the case of diffuse boundary scattering, for details of which I advise ref.[5] to the interested reader.

As a next step I generalize the above calculation of the normal velocity slip length ζ_0 to (i) the case that the plane wall at $z = 0$ moves with a velocity $v_x^{\mathrm{ext}}\exp(-i\omega t)$ and (ii) to

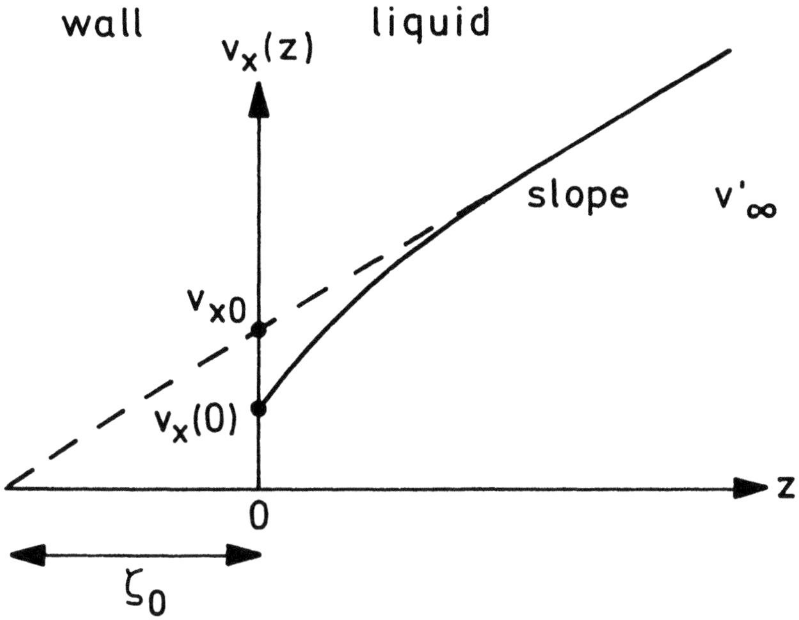

Fig. 8: Velocity profile for the transverse flow of a (quasiparticle) gas past a plane wall.

a more general class of wall scattering events. There are only two relevant observables in such a case:

(i) the transverse (normal) velocity field

$$v_x^{\rm n}(z,t) = \frac{1}{\rho_0^{\rm n}} 2 \sum_{\vec{k}} p_x h_k(z,t) \tag{6.10}$$

(cf. Eq. (4.15)), and

(ii) the shear component of the stress tensor field (cf. table 4)

$$\Pi_{xz}(z,t) = 2 \sum_{\vec{k}} p_x V_{kz} h_k(z,t) \tag{6.11}$$

It is straightforward to express the Landau expansion (5.2) of the change in the quasiparticle energy δE_k

$$\delta E_k(z,t) = \frac{F_1^{\rm s}}{3} \frac{m}{m^*} \frac{\rho^{\rm n}}{\rho} p_x v_x^{\rm n}(z,t) + \frac{F_2^{\rm s} Y_2/5}{1+F_2^{\rm s}/5} \frac{5 p_x V_{kz}}{n p_{\rm F} v_{\rm F} Y_2} \Pi_{xz}(z,t) \tag{6.12}$$

and of the conserving relaxation time approximation (4.29) of the collision integral (4.22)

$$\delta I_k(z,t) = -\frac{1}{\bar{\tau}_{\rm S}} \left\{ h_k(z,t) - \varphi_k \left[\lambda_1^+ p_x v_x^{\rm n}(z,t) + \lambda_2^+ \frac{5 p_x V_{kz}}{n p_{\rm F} v_{\rm F} Y_0} \Pi_{xz}(z,t) \right] \right\} \tag{6.13}$$

through the observables v_x^n and Π_{xz}. The transport equation for the distribution function h_k (cf. (5.5)), applied to the case of purely transverse flow with a harmonic time dependence $\partial/\partial t \to -i\omega$, reads

$$\left[\frac{\partial}{\partial z}+\frac{1}{\lambda_z}\right] h_k(z) = \frac{\varphi_k p_x}{\lambda_z}\left\{\frac{\Lambda_1(\omega)}{1-i\omega\bar{\tau}_S} v_x^n(z) + \frac{5\Lambda_2(\omega)\lambda_z}{np_F v_F Y_2 \bar{\tau}_S}\Pi_{xz}(z)\right\} \tag{6.14}$$

Here I have defined a renormalized complex quasiparticle mean free path $\lambda_z = V_{kz}\bar{\tau}_S/(1-i\omega\bar{\tau}_S)$ and frequency dependent coefficients

$$\Lambda_1(\omega) = \lambda_1^+ - i\omega\bar{\tau}_S \frac{F_1^s Y_0/3}{1+F_1^s Y_0/3}$$

and

$$\Lambda_2(\omega) = \lambda_2^+ \frac{Y_2}{Y_0} - i\omega\bar{\tau}_S \frac{F_2^s Y_2/5}{1+F_2^s/5}$$

The first order differential equation (6.14) can now be integrated with respect to the variable z to yield a formal solution

$$h_k(z) = C_k e^{-\frac{z}{\lambda_z}} + \frac{\varphi_k p_x}{\lambda_z}\int_0^z dz'\left[\frac{\Lambda_1(\omega)}{1-i\omega\bar{\tau}_S} v_x^n(z') + \frac{5\Lambda_2(\omega)\lambda_z}{np_F v_F Y_2\bar{\tau}_S}\Pi_{xz}(z')\right]\exp\left(-\frac{z-z'}{\lambda_z}\right) \tag{6.15}$$

Here C_k is an integration constant which is fixed by the boundary conditions for the distribution function, to be specified below. As before, let us split up the distribution function h_k into contribution from outgoing (>) and incoming (<) quasiparticles:

$$h_k(z) = h_k^>(z)\Theta(V_{kz}) + h_k^<(z)\Theta(-V_{kz})$$

and write the corresponding decompositions of the quasiparticle observables in terms of integrals over the half space $V_{kz} > 0$ as

$$v_x^n(z) = \frac{1}{\rho_0^n}\, 2 \sum_{\substack{k \\ V_{kz}>0}} p_x \left[h_k^>(z) + h_k^<(z)\right] \tag{6.16}$$

and

$$\Pi_{xz}(z) = 2 \sum_{\substack{k \\ V_{kz}>0}} p_x V_{kz} \left[h_k^>(z) - h_k^<(z)\right] \tag{6.17}$$

Next we have to specify the boundary conditions that the distribution function has to fulfill (i) at infinite distance from the wall ($z \to \infty$):

$$\lim_{z\to\infty} h_k^<(z) = 0 \tag{6.18}$$

and (ii) at the wall ($z = 0$, diffuse scattering):

$$h_k^{>}(0) = \nu_k \left[E_k - p_x v_x^{\mathrm{ext}}\right] - \nu_k^0 = \varphi_k p_x v_x^{\mathrm{ext}} \tag{6.19}$$

A more general form for the wall boundary condition (6.19) which is not restricted to the case of diffuse quasiparticle scattering can be written down by relating incoming and outgoing quasiparticle partial currents through a general elastic wall scattering probability $W_{kk'}$:

$$|V_{kz}|\delta\nu_k^{>}(0) = 2 \sum_{\substack{\vec{k} \\ V_{kz}<0}} W_{kk'}|V_{k'z}|\delta\nu_k^{<}(0) \tag{6.20}$$

The quantity $W_{kk'}$ is the probability that a particle hitting the wall with momentum $\hbar\vec{k}\,'$ leaves it in a quantum state $\hbar\vec{k}$. It represents a large class of *elastic* wall scattering laws which are thought to be superimposed the diffuse scattering processes. $W_{kk'}$ is therefore assumed to have the property

$$W_{kk'} = \tilde{W}_{kk'}\ \delta(E_k - E_{k'}) \tag{6.21}$$

Furthermore, I treat only the case when the scattering preserves particle number. This can be expressed as a normalization condition

$$2 \sum_{\substack{\vec{k} \\ V_{kz}>0}} W_{kk'} = 1 \tag{6.22}$$

Finally, the global equilibrium distribution ν_k^0 must obey (6.20). A sufficient condition for that is

$$|V_{k'z}|\ W_{kk'}\ \nu_{k'}^0 = |V_{kz}|\ W_{-k'-k}\ \nu_k^0 \quad \text{for} \quad V_{kz} > 0\ ;\ \ V_{k'z} < 0 \tag{6.23}$$

This is the reciprocity relation for boundary scattering (cf. ref. [43]). In the case of purely diffuse boundary scattering the wall scattering probability $W_{kk'}$ is an azimuthally isotropic function of $\hat{k}$ and $\hat{k}'$ which averages to zero upon integration with the weight function $\hat{p}_x$, characteristic for transverse flow. Special cases of elastic scattering events may include

(i) specular scattering

$$W_{kk'} = s\ \delta(V_{kz} - V_{k'z})\ \delta_{\vec{k}^*,\vec{k}'} \tag{6.24}$$

where $\vec{k}^*$ ia a vector with components $\{k_x, k_y, -k_z\}$, and

(ii) elastic backward scattering

$$W_{kk'} = -r\ \delta(V_{kz} - V_{k'z})\ \delta_{\vec{k},-\vec{k}'} \tag{6.25}$$

In pair correlated Fermi liquids a further elastic scattering process is known to exist, which is of quantum-mechanical origin. It is called Andreev scattering [44] or particle–hole conversion and is caused by a reduction of the (triplet) order parameter components near the boundary. This leads to a conversion of an incoming Bogolon into a "retro-flected" hole-like excitation which has all components of the group velocity reversed. The transverse momentum, transferred to the wall in this process, however, is negligibly small, eventually leading to a temperature dependent reduction of the frictional force of the normal fluid component on the surface. For a comprehensive review of Andreev scattering I refer to the work of Kurkijärvi and Rainer [45]. Near surfaces, the order parameter representing the B–phase of superfluid ^{3}He splits into components parallel and perpendicular with respect to the surface normal, which are in general suppressed in a different way. Here and in the following I restrict myself to discussing an order parameter profile which simply drops from its bulk value to zero near the surface. Such a step–like order parameter profile serves to provide an upper bound for the estimate of Andreev scattering effects, because it leads to the maximum possible reflection probability $R(E_k)$ [44]:

$$R(E_k) = \frac{E_k - \sqrt{E_k^2 - \Delta^2}}{E_k + \sqrt{E_k^2 - \Delta^2}} \tag{6.26}$$

The corresponding scattering probability $W_{kk'}$ reads in the case of

(iii) Andreev scattering

$$W_{kk'} = R(E_k)\delta(V_{kz} + V_{k'z})\ \delta_{\underline{k},\underline{k}'} \tag{6.27}$$

Let me now aim at a derivation of integral equations for the observables $v_x^{\rm n}$ and Π_{xz} for the general set of boundary conditions (6.20) which can also be written in terms of the distribution functions $h_k^>$ and $h_k^<$:

$$h_k^>(0) = \varphi_k p_x v_x^{\rm ext} + 2 \sum_{\substack{\underline{k} \\ V_{kz}<0}} W_{kk'} \frac{|V_{k'z}|}{|V_{kz}|} \left[h_{k'}^<(0) - \varphi_{k'} p'_x v_x^{\rm ext}\right] \tag{6.28}$$

In constructing expressions for the observables one generates moments of the mean free path λ_z which correspond to the diffuse part of the scattering process

$$L_n(z) = \frac{1}{\tilde{\tau}_S} 2 \sum_{\substack{\underline{k} \\ V_{kz}>0}} \varphi_k p_x^2 \lambda_z^n\ [\mathrm{sgn}(z)]^{n-1} \exp\left(-\frac{z}{\lambda_z}\right) \tag{6.29}$$

and to the elastic contribution:

$$K_{mn}(z,z') = \frac{1}{\tilde{\tau}_S} \sum_{\substack{\vec{k} \\ V_{kz}>0}} \sum_{\substack{\vec{k}' \\ V_{k'z}>0}} \varphi_k p_x p'_x \lambda_z^{m-1} \lambda'^n_z W_{k-k'} \exp\left(-\frac{z}{\lambda_z} - \frac{z'}{\lambda'_z}\right) \tag{6.30}$$

In terms of these mean free path moments the coupled integral equations for the observables $v_x^{\rm n}$ and Π_{xz} can be written in the compact form:

$$\begin{aligned}
2L_0(0)v_x^{\rm n}(z) \quad &- \quad \tilde{\Lambda}_1(\omega)\int_0^\infty dz'\,[L_{-1}(|z-z'|) - K_{00}(z,z')]\,v_x^{\rm n}(z') \\
&= \quad \frac{5\tilde{\Lambda}_2(\omega)}{np_f v_f Y_2 \tilde{\tau}_S}\int_0^\infty dz'\,[L_0(z-z') + K_{01}(z,z')]\,\Pi_{xz}(z') \\
&+ \quad [L_0(z) + K_{01}(z,0)]\,v_x^{\rm ext} \qquad\qquad (6.31)\\
\Pi_{xz}(z) \quad &- \quad \frac{5\tilde{\Lambda}_2(\omega)}{np_{\rm F} v_{\rm F} Y_2 \tilde{\tau}_S}\int_0^\infty dz'\,[L_1(|z-z'|) + K_{11}(z,z')]\,\Pi_{xz}(z') \\
&= \quad \tilde{\Lambda}_1(\omega)\int_0^\infty dz'\,[L_0(z-z') - K_{01}(z,z')]\,v_x^{\rm n}(z') \\
&+ \quad [L_1(z) + K_{11}(z,0)]\,v_x^{\rm ext}
\end{aligned}$$

Here I have abbreviated $\tilde{\Lambda}_\ell \equiv \Lambda_\ell/(1-i\omega\bar{\tau}_S)$; $\ell = 1,2$. We come now back to the case of stationary Couette flow which is obtained as a special case ($\omega\bar{\tau}_S \to 0$) from the second of the coupled integral equations (6.31):

$$\begin{aligned}
\Pi_{xz}(z) \quad &= \quad -\eta(T)\frac{\partial v_x^{\rm n}(z)}{\partial z} = -\eta(T)v'_\infty\,[1+\psi(z)] \\
&= \quad [L_1(z) + K_{11}(z,0)]\left[v_x^{\rm ext} - v_x^{\rm n}(0)\right] \qquad\qquad (6.32)\\
&- \quad \int_0^\infty dz'\,[L_1(|z-z'|) + K_{11}(z,z')]\,\frac{\partial v_x^{\rm n}(z')}{\partial z'}\,\frac{1}{1-\lambda_2^+ Y_2/Y_0}
\end{aligned}$$

Here $\psi(z) = [\partial v_x^{\rm n}(z)/\partial z - v'_\infty]/v'_\infty] > 0$ is a measure of the curvature of the velocity profile. From (6.32) one one may derive an integral equation for the slip length $\zeta_0 = [v_x^{\rm n}(0) - v_x^{\rm ext}]/v'_\infty + \int_0^\infty dz\psi(z)$:

$$\begin{aligned}
&\zeta_0\ \{L_1(z) + K_{11}(z,0)\} - \{L_2(z) - K_{12}(z,0)\} \\
&\quad = \int_0^\infty dz'\psi(z')\,\{[L_1(z) + K_{11}(z,0)] - [L_1(|z-z'|) + K_{11}(z,z')]\}
\end{aligned} \tag{6.33}$$

Here we put $z = 0$. Asuming the three expressions in curly brackets as well as the quantity $\psi(z)$ to be positive one may derive positive bounds for the slip length ζ_0 from (6.33):

(i) lower bound (Maxwellian slip, $\psi(z) \equiv 0$):

$$\zeta_0^{\mathrm{LB}} > \frac{L_2(0) - K_{12}(0,0)}{L_1(0) + K_{11}(0,0)} \; \frac{1}{1 - \lambda_2^+ Y_2/Y_0} \tag{6.34}$$

(ii) From an integration of Eq. (6.33) over z from 0 to ∞ one may also derive an upper bound:

$$\zeta_0^{\mathrm{UB}} < \frac{L_3(0) - K_{22}(0,0)}{L_2(0) + K_{21}(0,0)} \; \frac{1}{1 - \lambda_2^+ Y_2/Y_0} \tag{6.35}$$

(iii) Finally it is possible to derive an improved lower bound for the slip length:

$$\zeta_0^{\mathrm{ILB}} > \frac{[L_2(0) - K_{21}(0,0)]^2 + [L_1(0) + K_{11}(0,0)][L_3(0) - K_{22}(0,0)]}{2L_2(0)[L_1(0) + K_{11}(0,0)]\,[1 - \lambda_2^+ Y_2/Y_0]} \tag{6.36}$$

These bounds have, for the case of diffuse scattering, first been derived by Jensen et al. [46]. Let me discuss this special case first. Applying the bounds (6.34), (6.35) and (6.36) to the case of purely diffuse scattering, the slip length of a normal Fermi liquid is bounded by

$$\frac{8}{15}\lambda_\eta < \zeta_0 < \frac{5}{8}\lambda_\eta \quad ; \quad \zeta_0 < \frac{1}{2}\left[\frac{8}{15} + \frac{5}{8}\right] = 0.579...$$

In a superfluid Fermi liquid the ratio of the slip length to the mean free path varies slightly with temperature in the case of diffuse scattering as is shown in part a) of Fig. 9 for ^{3}He–B at 20 bar pressure. The three bounds (LB, UB, ILB, dashed lines) are shown together with the result of an exact treatment (EX, full line) [5].

The improved lower bound for the slip length is hence seen to be an excellent approximation in the case of diffuse scattering. It can be anticipated to work as well in the more general case of additional elastic wall scattering processes, which I would now like to discuss. Inserting the scattering probabilities (6.24), (6.25) into the mean free path moments (6.30) one finds that

$$K_{mn}(z, z') = -tL_{m+n-1}(z + z')$$

where $t = s$ in the case of specular scattering and $t = -r$ in the case of elastic backward scattering. The lower bound for the slip length has in these cases the form

$$\zeta_0^{\mathrm{LB}}(t) = \zeta_0^{\mathrm{LB}}(0)\,\frac{1+t}{1-t} \tag{6.37}$$

and is hence seen to be enhanced for specular scattering and reduced for elastic backscattering. The case of Andreev scattering in a Fermi superfluid has to be discussed separately because the functions $K_{mn}(z, z')$ have to be evaluated with the amplitude $R(E_k)$ for

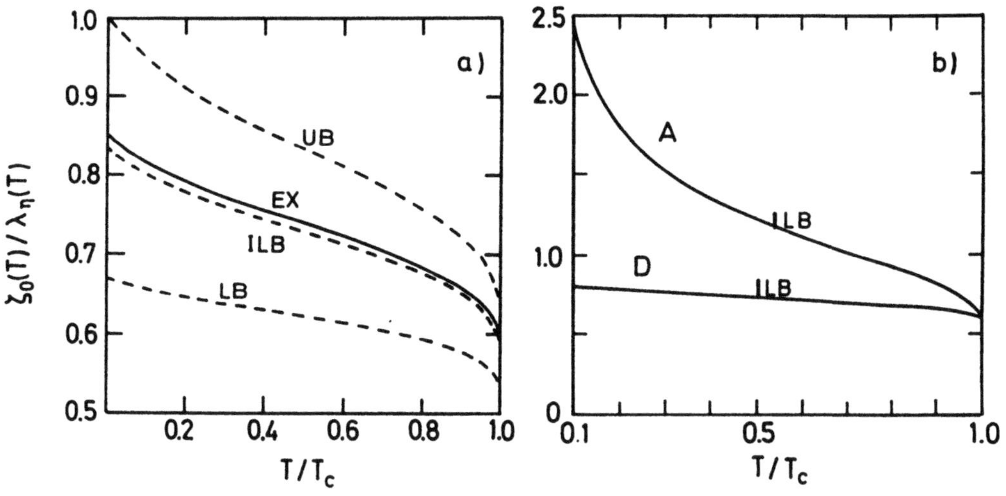

Fig. 9: a) Slip length ζ_0 of an isotropic Fermi superfluid, normalized to the viscous mean free path λ_η. Dashed lines: bounds on the slip length; full line: exact result (see text). b) Comparison of the slip lengths normalized to the viscous mean free path for the cases of diffuse (D) and Andreev (A) scattering (step profile).

particle–hole conversion:

$$K_{mn}(z,z') = -\frac{1}{\tilde{\tau}_S} 2 \sum_{\substack{k \\ V_{kz}>0}} \varphi_k p_x \lambda_z^{m+n-1} R(E_k) e^{-\frac{z+z'}{\lambda_z}} \equiv -R_{m+n-1}(z+z')$$

Inserting this into the result (6.34) for the Maxwellian slip length one finds that Andreev scattering leads to an enhancement of the slip length over its value for purely diffuse scattering:

$$\zeta_0^{\mathrm{LB}}(R) = \zeta_0^{\mathrm{LB}}(0)\, \frac{1 + R_2(0)/L_2(0)}{1 - R_1(0)/L_1(0)} \tag{6.38}$$

For the special case of a step–like order parameter profile near the wall $\Delta(z) = \Delta(T)\Theta(z - z_0)$ on may show from (6.38) that the ratio $\zeta_0^{\mathrm{LB}}(R)/\zeta_0^{\mathrm{LB}}(0)$ diverges as $\propto T^{-1/2}$ in the low temperature limit. The temperature dependence of the slip enhancement in the presence of Andreev scattering over that for diffuse scattering is illustrated in part b) of Fig. 9. The dramatic increase which is visible there has been termed "quantum slip effect" because it is the quantum–mechanical scattering of Bogoliubov quasiparticles which is responsible for the loss of transverse momentum transfer to the container walls.

Being aware of the fact that the length ζ_0 enters the description of fluid flow as an additional temperature dependent parameter and being able to compute it for various wall scattering probabilities, we now have to investigate in detail where the slip length enters the hydrodynamic description of fluid flow as a finite size correction. This will be the topic of the next lecture.

7 Phenomenological theory of flow in restricted geometry

In this lecture I shall deal with the incorporation of experimental finite size constraints on the determination of the shear viscosity. These considerations will be restricted to certain classes of torsional oscillator and sound propagation experiments which use a parallel plate geometry. The treatment will be entirely on a phenomenological level. A phenomenological treatment of mean free path effects, i.e., the incorporation of the slip length ζ_0 into the theory, can be achieved by postulating a boundary condition for the macroscopic velocity field v_x^{n} at the container wall, which allows for velocity slip. Such a boundary condition is particularly simple in the case of one ideally plane wall, located at $z = 0$, which confines the fluid to the half space $z > 0$, as discussed in the last lecture. For such a geometry we obtain a simple slip boundary condition for transverse flow:

$$v_x^{\mathrm{n}}(0) = \zeta_0 \frac{\partial v_x^{\mathrm{n}}}{\partial z}(0) \tag{7.1}$$

Let me now consider an experimental arrangement which is typical for the determination of the shear viscosity, namely the torsional oscillator.

Torsion oscillator experiments

It consists of two circular parallel plates of radius R, a distance d apart, located at $z = \pm d/2$ at the top and bottom, and a cylindrical sidewall. The influence of the sidewall is negligible for $R \gg d$ and turns out to be trivial for $R \approx d$ and will therefore be omitted in the calculations to follow. This cell entrains the Fermi liquid under investigation and oscillates with a frequency ω about the cylinder axis. In such a geometry the macroscopic velocity field is purely azimuthal

$$v_\phi^{\mathrm{n}}(r, z, t)$$

where r measures the distance from the cylinder axis and $-d/2 < z < d/2$. The hydrodynamic form of the momentum current reads then (cf. (5.10)):

$$\Pi_{\phi z}(r,z,t) = -\,\eta(T)\,\frac{\partial v_\phi^{\rm n}(r,z,t)}{\partial z} \tag{7.2}$$

Splitting off the harmonic time dependence of the velocity field

$$v_\phi^{\rm n}(r,z,t) = v_\phi^{\rm n}(r,z)\; e^{-i\omega t} \tag{7.3}$$

one may write the Navier–Stokes equation in cylindrical coordinates as

$$-\,i\omega\rho^{\rm n} v_\phi^{\rm n}(r,z) = \eta\,\left[\frac{\partial^2}{\partial r^2} + \frac{1}{r}\frac{\partial}{\partial r} - \frac{1}{r^2} + \frac{\partial^2}{\partial z^2}\right] v_\phi^{\rm n}(r,z) \tag{7.4}$$

Dividing (7.4) by the viscosity η one observes that the length scale for the spatial variation in r and z is set by the quantity

$$\delta = \sqrt{\frac{2\,\eta}{\rho^{\rm n}\,\omega}} \tag{7.5}$$

which has the physical meaning of a viscous penetration (or skin) depth, i.e. the characteristic thickness of the fluid layer that is dragged along by the motion of the wall. In Fig. 10 I have plotted the viscous penetration depth for normal and superfluid ^{3}He (–B) at 0 bar pressure, normalized to its value at the transition, as a function of the inverse reduced temperature T_c/T. Depending on the frequency of the oscillator, the viscous penetration depth specifies different regimes of viscous confinement. If $\delta \gg d$ the fluid is "clamped" and moves as a whole with the oscillator whereas in the opposite limit $\delta \ll d$ only a sheet of thickness δ is dragged along by the surfaces. The purpose of the following considerations is to study the fluid response to oscillations of the confining walls in the whole regime of parameters d/δ. Defining a complex wavenumber $k = (1+i)/\delta$, the differential equation for the velocity field assumes the form:

$$\left[\frac{\partial^2}{\partial r^2} + \frac{1}{r}\frac{\partial}{\partial r} - \frac{1}{r^2} + \frac{\partial^2}{\partial z^2} + k^2\right]\; v_\phi^{\rm n}(r,z) = 0 \tag{7.6}$$

The r– and z dependence of the solution of (7.6) factorizes

$$v_\phi^{\rm n}(r,z) = \left[ar + \frac{b}{r}\right]\cos kz \tag{7.7}$$

and only the choice $b = 0$ for the second integration constant guarantees the regular behavior of the solution at the origin. The remaining integration constant is fixed by the slip boundary condition (7.1), which, applied to the double plate geometry, reads

$$v_\phi^{\rm n}\left(r, \pm\frac{d}{2}\right) - \dot{\Theta} r = \mp\zeta_0\,\frac{\partial v_\phi^{\rm n}}{\partial z}\left(r, \pm\frac{d}{2}\right) \tag{7.8}$$

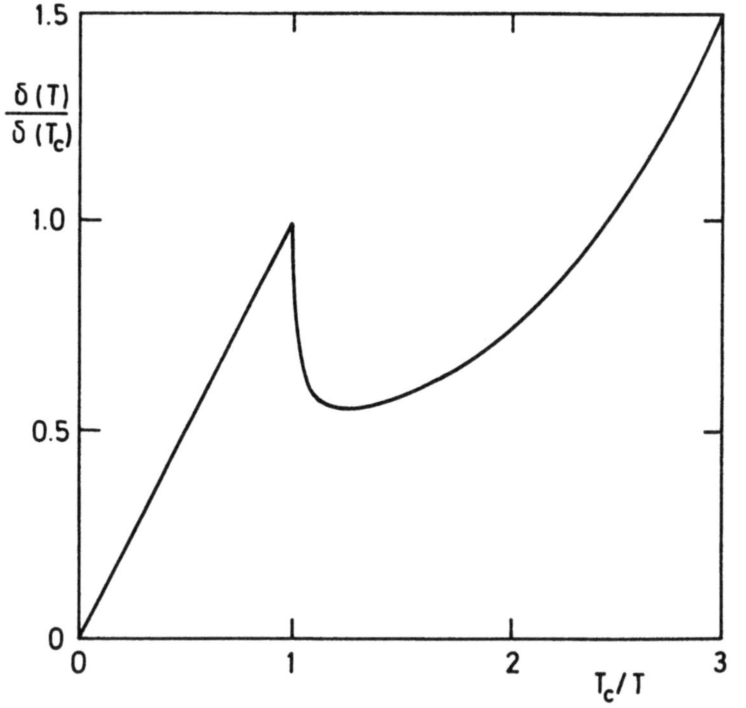

Fig. 10: The viscous penentration depth δ of normal and superfluid ^{3}He(–B) at 0 bar pressure, normalized to its value at T_c vs. inverse reduced temperature T_c/T.

where $\dot{\Theta}$ denotes the angular frequency of the torsional oscillator. Inserting the solution for the velocity field (7.7) into the boundary condition (7.8) fixes the integration constant a to be:

$$a = \frac{\dot{\Theta}}{\cos(kd/2) - k\zeta_0 \sin(kd/2)} \tag{7.9}$$

For the torsional oscillator experiment it turns out that two quantities are important:

(i) the cross sectional average of the velocity field v_ϕ^{n}

$$\left\langle v_\phi^{\mathrm{n}} \right\rangle_z \equiv \frac{1}{d} \int_{-d/2}^{d/2} dz \; v_\phi^{\mathrm{n}}(r,z) = -\dot{\Theta} r \left\{ \frac{2}{kd} \; \frac{\tan(kd/2)}{1 - k\zeta_0 \tan(kd/2)} \right\} \tag{7.10}$$

and

(ii) the viscous shear force on the walls which reads for the upper wall

$$\Pi_{\phi z}(r, \frac{d}{2}) = \dot{\Theta} \; r \; \eta \; \frac{k^2 d}{2} \; \left\{ \frac{2}{kd} \; \frac{\tan(kd/2)}{1 - k\zeta_0 \tan(kd/2)} \right\} \tag{7.11}$$

Both these quantities are described by the same function in curly brackets which determines the amount of viscous confinement of the fluid in the presence of velocity slip:

$$F = F(d, \delta, \zeta_0) = \frac{2}{kd} \frac{\tan(kd/2)}{1 - k\zeta_0 \tan(kd/2)} \tag{7.12}$$

It is illustrative to evaluate the viscous confinement function F in the limits where the viscous penetration depth is either small or large with respect to the plate spacing d:

$$\begin{aligned} \lim_{\delta \gg d} F(d, \delta, \zeta_0) &= 1 + \frac{i}{3}\left(\frac{d}{\delta}\right)^2 \left[1 + 6\frac{\zeta_0}{d}\right] + \ldots \\ \lim_{\delta \ll d} F(d, \delta, \zeta_0) &= (1+i)\frac{\delta}{d}\left[1 - (1-i)\frac{\zeta_0}{\delta}\right] + \ldots \end{aligned} \tag{7.13}$$

The behavior of the real and imaginary parts of the viscous confinement function for arbitrary values of the parameter d/δ is shown in the insert of Fig. 11 for two different ratios ζ_0/d of the slip length with respect to the plate spacing.

From the viscous shear force (7.11) on the plates one may now compute the torque that the fluid exerts on the top and bottom plates:

$$M = 2\int_0^R drr(2\pi r)\Pi_{\phi z}(r, z) = \frac{\pi}{2} R^4 \eta \dot{\Theta} k^2 d\, F(d, \delta, \zeta_0) \tag{7.14}$$

The complex viscous torque (7.14) specifies influence of the quasiparticle gas on the equation of motion of the torsional oscillator

$$I_s \ddot{\Theta} + \left[\frac{M}{\dot{\Theta}} + N\right] \dot{\Theta} + R\Theta = D_0 e^{-i\omega t} \tag{7.15}$$

in which I_s denotes the moment of inertia of the empty cell, N is the nuisance damping, R the torsion constant and D_0 the driving force. Calling

$$I_f = \frac{1}{2}\pi R^2 d \cdot \rho R^2$$

the fluid moment of inertia, the characteristic frequency of the oscillator can be found to be

$$\omega = \omega_0 \left\{1 + \frac{1}{2}\frac{\rho^n}{\rho}\frac{I_f}{I_s} F(d, \delta, \zeta_0) + \frac{iN}{2I_s\omega_0}\right\} \tag{7.16}$$

where $\omega_0^2 = R/I_s$ is the resonance frequency of the oscillator. From the result (7.16) for the complex characteristic frequency one may obtain the dissipation as the imaginary part

$$Q^{-1} = 2\ \mathrm{Im}\left(\frac{\omega}{\omega_0}\right) = \frac{N}{I_s\omega_0} + \frac{\rho^n}{\rho}\frac{I_f}{I_s}\ \mathrm{Im}\ F(d, \delta, \zeta_0) \tag{7.17}$$

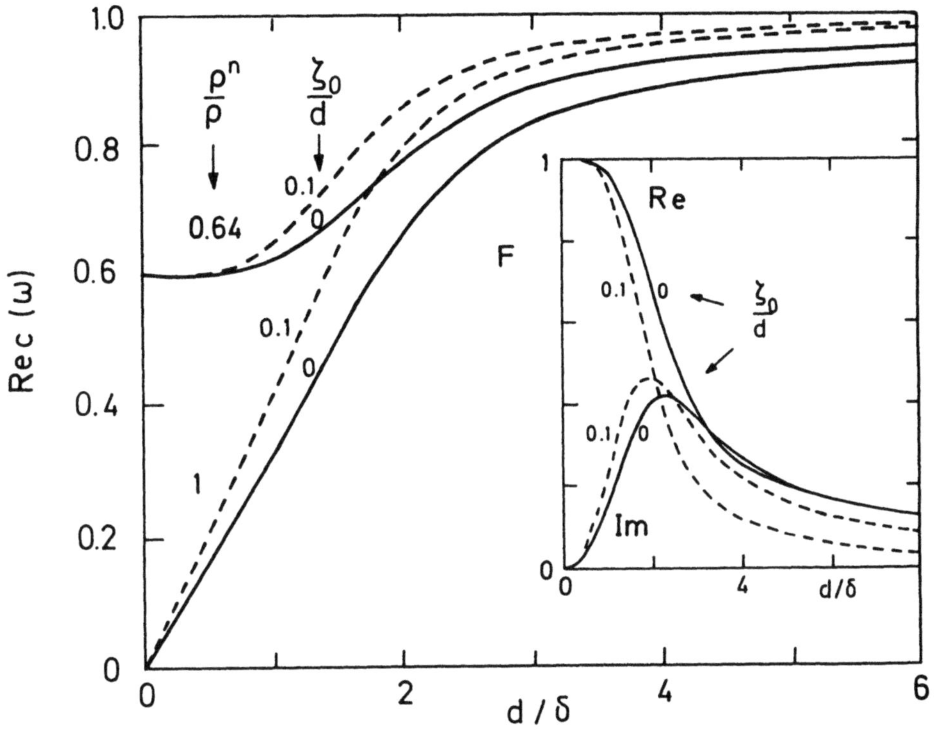

Fig. 11: The transition from fourth to first sound in a parallel plate resonator for two values of ρ^n/ρ and ζ_0/d as a function of d/δ (after ref. [47]). The insert shows real and imaginary part of the viscous confinement function F as a function of d/δ for two values of ζ_0/d.

Using the asymptotic behavior of the F–function (cf. (7.13)) one may evaluate the dissipation in the limits of small and large values of the viscous confinement parameter d/δ

$$\begin{aligned} \lim_{\delta \gg d} Q^{-1} &= \frac{N}{I_s\omega_0} + \frac{\rho^n}{\rho}\frac{I_f}{I_s}\frac{\rho^n \omega d^2}{12\eta}\left[1 + 6\frac{\zeta_0}{d}\right] \\ \lim_{\delta \ll d} Q^{-1} &= \frac{N}{I_s\omega_0} + \frac{\pi R^4}{I_s\omega_0}\,\mathrm{Re}\, Z_\perp(\omega;\zeta_0) \end{aligned} \qquad (7.18)$$

The frequency (or period) shift of the oscillator is obtained as the real part of the complex characteristic frequency:

$$2\frac{\Delta\omega}{\omega_0} = \frac{\rho^n}{\rho}\frac{I_f}{I_s}\,\mathrm{Re}\, F\,(d,\delta,\zeta_0) \qquad (7.19)$$

which again can be treated in the two asymptotic limits

$$2\lim_{\delta \gg d} \frac{\Delta\omega}{\omega_0} = \frac{\rho^n}{\rho}\frac{I_f}{I_s}\left[1 + O\left(\frac{\delta^4}{d^4}\right)\right]$$
$$2\lim_{\delta \ll d} \frac{\Delta\omega}{\omega_0} = \frac{\pi R^4}{I_s\omega_0}\,\mathrm{Im}\; Z_\perp(\omega;\zeta_0) \tag{7.20}$$

The above calculations may be summarized as follows: There are two quantities which emerge from a torsional oscillator measurement in the two limits of small and large d/δ: In the "clamped" limit $\delta \gg d$ the experiment sees an effective viscosity

$$\eta_{\mathrm{eff}}(T) = \eta(T)\frac{d}{d + 6\zeta_0(T)} \tag{7.21}$$

in the dissipation, with the normal fluid density ratio that can be taken from the simultaneous measurement of the frequency shift. In the opposite limit $\delta \ll d$ the same experiment measures the complex transverse surface (or shear) impedance $Z_\perp$:

$$Z_\perp(\omega;\zeta_0) = \frac{(1-i)\;\eta(T)}{\delta(T) + (1-i)\;\zeta_0(T)} \tag{7.22}$$

The equations (7.21) and (7.22) for the effective viscosity and the transverse surface impedance are the central results of this lecture. They show that both quantities are affected by mean free path effects which, in our phenomenological treatment, enters as a first order slip correction to the hydrodynamic result for infinitely extended systems. In the case of the effective viscosity the importance of the slip correction is given by the ratio ζ_0/d, i.e., essentially the Knudsen number Kn, whereas in the case of the surface impedance the importance of the slip effect scales with ζ/δ, i.e., with the inverse of the frequency dependent viscous penetration depth $\delta^{-1} \propto \omega^{1/2}$. Therefore the slip correction to the transverse surface impedance vanishes in the hydrodynamic limit $\omega \to 0$.

Sound propagation experiments

Let me now consider another kind of a clean experimental arrangement for the determination of viscous transport as influenced by mean free path effects. It is a parallel plate sound resonator the discusssion of which will be to some extent along the lines of a treatment by Jensen et al. [47]. Let us consider the idealized case of a parallel plate flow channel with height d in which sound is generated on the one side, propagates in the x–direction and is detected at the other end. The relevant two–fluid hydrodynamic equations for this case are the continuity equation

$$\frac{\partial \rho}{\partial t} + \rho^{\mathrm{n}} \frac{\partial v_x^{\mathrm{n}}}{\partial x} + \rho^{\mathrm{s}} \frac{\partial v_{x0}^{\mathrm{s}}}{\partial x} = 0 \tag{7.23}$$

the Navier–Stokes equation

$$\rho^{\mathrm{n}} \frac{\partial v_x^{\mathrm{n}}}{\partial t} + \rho^{\mathrm{s}} \frac{\partial v_{x0}^{\mathrm{s}}}{\partial t} = -\frac{\partial \delta P}{\partial x} + \eta \left[\frac{4}{3} \frac{\partial^2}{\partial x^2} + \frac{\partial^2}{\partial z^2} \right] v_x^{\mathrm{n}} \tag{7.24}$$

and the acceleration equation for the superfluid

$$\frac{\partial v_{x0}^{\mathrm{s}}}{\partial t} = -\frac{1}{\rho} \frac{\partial \delta P}{\partial x} + \zeta_3 \frac{\partial^2}{\partial x^2} [g_x - \rho v_x^{\mathrm{n}}] \tag{7.25}$$

The dispersion of sound can be studied by differentiating the continuity equation with respect to the time and inserting the Navier–Stokes equation. Assuming a spatial variation in the x–direction $\propto \exp(iqx)$, expressing the pressure by the mass density change $\delta P = c_1^2 \delta\rho$ via the first sound velocity $c_1 = (\rho\kappa)^{-1} = (1 + F_0^{\mathrm{s}})(1 + F_1^{\mathrm{s}}/3)v_{\mathrm{F}}^2/3$, and going over to quantities averaged over the cross section of the flow channel, one eventually arrives at the following sound dispersion equation:

$$\left[\frac{\partial^2}{\partial t^2} - c_1^2 \frac{\rho^{\mathrm{s}}}{\rho} \frac{\partial^2}{\partial x^2} \right] \langle \delta\rho \rangle_z = -\rho^{\mathrm{n}} \frac{\partial^2}{\partial x \partial t} \langle v_x^{\mathrm{n}} \rangle_z + \ldots \tag{7.26}$$

In the derivation of (7.26) terms of the order $O(q\delta)$ have been ignored for the moment. Let us first assume that the viscous confinement parameter d/δ is very small so that the right hand side of Eq. (7.26) is negligible, i.e., that the normal fluid velocity field is clamped. Then we obtain from (7.26) a dispersion relation of fourth sound with a complex frequency

$$c_4^2(\omega) = (c_4^0)^2 \left[1 - \frac{i}{Q_4} \right] \tag{7.27}$$

from which one identifies the fourth sound velocity

$$c_4^0 = c_1 \sqrt{\frac{\rho^{\mathrm{s}}}{\rho}} \tag{7.28}$$

and the damping or Q–factor

$$Q_4^{-1} = \frac{\rho\omega}{c_1^2} \zeta_3 \tag{7.29}$$

Here ζ_3 is the coefficient of bulk viscosity, which has been discussed in lecture 5. Let us next relax the assumption of small d/δ and investigate the dispersion relation for arbitrary

d/δ. For this purpose we need the normal fluid velocity profile v_x^n which is calculated from the differential equation

$$\left[\frac{\partial^2}{\partial z^2}+k^2\right]v_x^n(z)=\frac{\rho^n}{\rho}\frac{iq}{\eta}c_1^2\delta\rho \tag{7.30}$$

to be of the form

$$v_x^n(z)=\frac{q}{\rho\omega}c_1^2\delta\rho\,[1+A\cos kz] \tag{7.31}$$

As in the torsional oscillator problem, the slip boundary condition fixes the integration constant A and after a straightforward calculation we arrive a the desired result for the cross sectional average of the velocity field

$$\langle v_x^n\rangle_z=\frac{q}{\omega}\frac{c_1^2}{\rho}\delta\rho\,[1-F(d,\delta,\zeta_0)] \tag{7.32}$$

This result, when inserted into the r. h. s. of Eq. (7.26) leads to the following general expression for the dispersion of sound in a parallel plate geometry:

$$c^2(\omega)=c_1^2\left\{\frac{\rho^s}{\rho}\left[1-\frac{i\omega\rho\zeta_3}{c_1^2}\right]+\frac{\rho^n}{\rho}\left[1-\frac{4i\omega\eta(\omega)}{3\rho^n c_1^2}-F(d,\delta,\zeta_0)\right]\right\} \tag{7.33}$$

Here $\eta(\omega)=\eta/(1-i\omega\tau_\eta/\phi_2(T))$ is the dynamic shear viscosity which characterizes the constitutive relation (5.10) for the dissipative part of the stress tensor at finite frequencies. Not surprisingly, it is the viscous confinement function F that governs also the finite size effects on the dispersion of sound in a parallel plate geometry. Here the limiting cases of small and large viscous confinement parameter d/δ correspond to the propagation of fourth ("clamped" regime) and first ("open" regime) sound, respectivly. Let me discuss the two limits separately. The dispersion of fourth sound is obtained from (7.33) in the limit $d/\delta\rightarrow 0$ as

$$c_4^2(\omega)=c_1^2\frac{\rho^s}{\rho}\left\{1+\frac{1}{30}\frac{\rho^n}{\rho^s}\left(\frac{d}{\delta}\right)^4\left[1+10\frac{\zeta_0}{d}\right]-\frac{i}{Q_4}\right\} \tag{7.34}$$

The imaginary part in the dispersion relation (7.34) has now acquired a surface contribution, which, as in the torsional oscillator case, is characterized by the inverse effective (i.e. slip–corrected) viscosity:

$$Q_4^{-1}=\omega\frac{\rho\zeta_3}{c_1^2}+\frac{1}{6}\frac{\rho^n}{\rho^s}\frac{\rho^n\omega d}{2\eta}\left[1+6\frac{\zeta_0}{d}\right] \tag{7.35}$$

In the opposite limit, $d/\delta\gg 1$ Eq. (7.33) describes the dispersion of first sound

$$c_1^2(\omega)=c_1^2\left\{1-i\frac{4\omega\eta(\omega)}{3\rho c_1^2}-i\frac{2Z_\perp(\omega)}{\rho\omega d}\right\} \tag{7.36}$$

from which one finds the velocity of first sound

$$\text{Re } c_1(\omega) = \left\{1 + \frac{2\omega \text{Im } \eta(\omega)}{3\rho c_1^2} + \frac{\text{Im } Z_\perp(\omega)}{\rho\omega d}\right\} \tag{7.37}$$

and the attenuation

$$\alpha(\omega) = \text{Im } \left(\frac{\omega}{c_1(\omega)}\right) = \frac{2\omega^2 \text{Re } \eta(\omega)}{3\rho c_1^3} + \frac{\text{Re } Z_\perp(\omega)}{\rho d c_1} \tag{7.38}$$

The transition from fourth to first sound in a resonator with parallel plate geometry as a function of d/δ is shown in Fig. 11 for two different temperatures (corresponding to two values for ρ^n/ρ) and two different slip lengths. It turns out that the surface contributions to the sound dispersion relations dominate the bulk ones in the hydrodynamic limit, as an inspection of the frequency dependence shows. It should be noted that similar dispersion relations for fourth and first sound hold in the case that the shape of the flow channel cross section is arbitrary. If the cross section is characterized by an area S and a circumference L one simply has to replace $d/2$ by S/L. For cylindrical flow channels of radius R, for example, this implies the replacement $d \rightarrow R$.

8 Poiseuille flow and surface impedance

In the last lecture I have shown that the flow channel width d or the viscous penetration depth δ, typical sizes characterizing fluid flow experiments, are length scales on which effects of finite quasiparticle mean free path may become important. The slip length ζ_0 appears as an additional temperature dependent parameter of the theory which renormalizes the shear viscosity and the transverse surface impedance and turns them into an effective viscosity and an effective surface impedance, respectively, which explicitly depend on the geometry of the experimental arrangement for its measurement.

Shear viscosity and Knudsen flow

It is the aim of this section to calculate the effective viscosity, as determined in a parallel plate geometry in the limit $d/\delta \rightarrow 0$, from the scalar Landau–Boltzmann equation for quasiparticles at arbitrary Knudsen numbers Kn= λ_η/d. We start by applying the Landau–Boltzmann equation (6.14) for the distribution function h_k to the case of stationary flow of a fluid, which is confined to the space $-d/2 \leq z \leq d/2$ between two plane walls at $z = \pm d/2$ (double plane boundary problem):

$$\left[\frac{\partial}{\partial z} + \frac{1}{\lambda_z}\right] h_k = \frac{\varphi_k p_x}{\lambda_z}\left\{\frac{\bar{\tau}_S F_x^{\mathrm{ext}}}{m(1+F_1^s Y_0/3)} + \frac{1}{\lambda_z} v_x^{\mathrm{n}} + \lambda_2^{+}\frac{5V_{kz}}{n p_{\mathrm{F}} v_{\mathrm{F}} Y_0}\Pi_{xz}\right\} \tag{8.1}$$

Here λ_z is the $\omega \rightarrow 0$–limit of the k–dependent quasiparticle mean free path introduced in context with Eq. (6.14). The external force appearing in this first order differential equation can be thought of as originating from a time–dependent wall velocity (cf. torsional oscillator) or, equivalently, from a pressure gradient (stationary Poiseuille flow):

$$\frac{F_x^{\mathrm{ext}}}{m} = i\omega v_x^{\mathrm{ext}} \equiv \frac{1}{\rho^{\mathrm{n}}}\frac{\partial \delta P}{\partial x}$$

This external force term appears also in the conservation law for the quasiparticle momentum (quasiparticle Navier–Stokes equation)

$$-i\omega v_x^{\rm n} + \nabla_z \Pi_{xz} = \frac{\rho^{\rm n}}{m} F_x^{\rm ext}$$

For stationary flow ($\omega \to 0$) the stress tensor varies linearly with z between the plates

$$\Pi_{xz}(z) = \frac{\rho^{\rm n}}{m} F_x^{\rm ext} \cdot z$$

The effect of the stress tensor term in (8.1) is essentially to turn the averaged quasiparticle lifetime $\bar{\tau}_{\rm S}$ into the proper transport time τ_η. In what follows, I shall therefore omit the stress tensor term from the kinetic equations and do the replacement, mentioned above, at the end of the calculation. The kinetic equation can be formally integrated to yield

$$h_k(z) = h_k(\pm\frac{d}{2}) e^{-\frac{z}{\lambda_z}} + \frac{\varphi_k p_x}{\lambda_z} \int\limits_{-d/2}^{z} dz' \left[v_x^{\rm n}(z') + \frac{\bar{\tau}_{\rm S} F_x^{\rm ext}}{m} \right] \exp\left(-\frac{z - z'}{\lambda_z} \right) \tag{8.2}$$

The values of the distribution function h_k at the boundaries $z = \pm d/2$, required in Eq. (8.2) are taken from the boundary condition (cf. (6.28)):

$$|V_{kz}| h_k^{\gtrless} \left(\mp\frac{d}{2} \right) = 2 \sum_{\substack{k' \\ V_{k'z} < 0}} W_{kk'} \, |V_{k'z}| \, h_k^{\lessgtr} \left(\mp\frac{d}{2} \right) \tag{8.3}$$

We proceed by deriving an integral equation for the quasiparticle velocity field $v_x^{\rm n}$ by inserting the solutions $h_k^{>}$ and $h_k^{<}$ of (8.2) into its decomposition (6.16). Besides the mean free path moments $L_n(z)$, introduced through Eq. (6.29), characteristic of purely diffuse scattering, in the cases of specular, backward and Andreev scattering one encounters integrals of the form

$$R_n(z, z') = \frac{1}{\bar{\tau}_{\rm S}} 2 \sum_{\substack{k \\ V_{kz} > 0}} \varphi_k p_x^2 \lambda_z^n \frac{2R \exp(-d/\lambda_z)}{1 - R \exp(-d/\lambda_z)} \cosh(\frac{z}{\lambda_z}) \cosh(\frac{z'}{\lambda_z}) \tag{8.4}$$

where $R = s$, $-r$ and $R(E)$ in the cases of specular, backward and Andreev scattering, respectivly. With the definition (8.4) the desired integral equation for the velocity field reads

$$v_x^{\rm n}(z) = \frac{1}{2L_0(0)} \int\limits_{-d/2}^{d/2} dz' \left[L_{-1}(|z - z'|) + R_{-1}(z, z') \right] \left[v_x^{\rm n}(z') + \frac{\bar{\tau}_{\rm S} F_x^{\rm ext}}{m} \right] \tag{8.5}$$

If one introduces a dimensionless velocity field $w(z)$

$$v_x^{\rm n}(z) = \frac{\bar{\tau}_{\rm S} F_x^{\rm ext}}{m} \left[w(z) - 1 \right] \tag{8.6}$$

and the integral kernel

$$H(z,z') = \delta(z-z') - \frac{1}{2L_0(0)}\left[L_{-1}(|z-z'|) + R_{-1}(z,z')\right] \tag{8.7}$$

the integral equation can be rewritten in the compact form

$$Hw \equiv \int_{-d/2}^{d/2} dz' H(z,z')w(z') = 1 \tag{8.8}$$

An inspection of Eq. (8.4) shows that the integral operator $H(z,z')$ is symmetric

$$H(z,z') = H(z',z)$$

and positive semidefinite

$$(\phi, H\phi) \equiv \int_{-d/2}^{d/2} dz\phi(z) \int_{-d/2}^{d/2} dz' H(z,z')\phi(z') \geq 0$$

This allows us to use a variational procedure for the solution of (8.8): For a set of trial functions $\lambda\phi$ with amplitudes λ one may use the positivity of

$$(w - \lambda\phi, H[w - \lambda\phi]) \geq 0$$

for an estimate of the cross sectional average of the quasiparticle flow velocity field $< v_x^{\mathrm{n}} >$. This can be expressed through the variational functional of ϕ in the following way:

$$\langle v_x^{\mathrm{n}} \rangle_z = \frac{\bar{\tau}_{\mathrm{S}} F_x^{\mathrm{ext}}}{m}\left[(w, Hw) - 1\right] \geq \frac{\bar{\tau}_{\mathrm{S}} F_x^{\mathrm{ext}}}{m}\left[\frac{(\phi,1)^2)}{(\phi, H\phi)} - 1\right] \tag{8.9}$$

As a final step one may write the averaged velocity field $< v_x^{\mathrm{n}} >$ in the form of a generalization of Hagen–Poiseuille's law

$$\langle v_x^{\mathrm{n}} \rangle = \frac{\rho^{\mathrm{n}} d^2 F_x^{\mathrm{ext}}}{12 m \eta_{\mathrm{eff}}\{\mathrm{Kn}; \phi, W\}} \tag{8.10}$$

and hence express it through an effective shear viscosity functional, which depends on the Knudsen number Kn$= \lambda_\eta / d$, on the choice for the trial function ϕ and on the details of the wall scattering processes, described by the probability $W_{kk'}$:

$$\eta_{\mathrm{eff}}\{\mathrm{Kn}; \phi, W\} = \eta(T)\frac{d}{d + a\{\mathrm{Kn}; \phi, W\}\lambda_\eta(T)} \tag{8.11}$$

One immediately recognizes that our result for the effective viscosity is of a form reminiscent of the slip corrected viscosity (cf. Eq. (7.21)). It is, in fact, a generalization of it in that it is valid for *all* Knudsen numbers Kn. The Knudsen coefficient a is a functional of ϕ and the wall scattering probability $W_{kk'}$:

$$a\{\mathrm{Kn};\phi,W\} = \frac{12}{5}\mathrm{Kn}^0\left[\frac{(\phi,1)^2}{(\phi,H\phi)}-1\right]-\frac{1}{\mathrm{Kn}^0} \tag{8.12}$$

Here Kn^0 is the Knudsen number taken for $\lambda_2^+ = 0$. It is well known from exact solutions of the kinetic equation for classical gases that the velocity profile is quadratic in z for small Kn and essentially flat in the Knudsen limit [43]. This suggests the use of the simplest set of trial functions of the form

$$\phi(z) = A - z^2$$

that allows at least a numerical treatment of the integrals in (8.12) for arbitrary Knudsen numbers. In the limit of small Kn an analytical result for the Knudsen coefficient can be obtained:

$$\lim_{\mathrm{Kn}\to 0} a\{\mathrm{Kn}; A-z^2, W\} = 6\,\frac{\zeta_0^{\mathrm{ILB}}}{\lambda_\eta} \tag{8.13}$$

with ζ_0^{ILB} the improved lower bound for the slip length (cf. (6.36)), which I have derived already in lecture 6. The variational treatment does hence reproduce the best estimate for the slip length in the presence of elastic wall scattering processes. The full result for the effective viscosity of superfluid ^{3}He–B at 20 bar pressure, normalized to its value at the transition, is plotted on a logarithmic scale vs. the inverse reduced temperature T_c/T in Fig. 12. The dashed line shows the result for the infinitely extended system. The upper of the pair of curves (1), which show the characteristic "droop" has been evaluated with the assumption of purely diffuse scattering of quasiparticles off the wall. The lower of these curves (2) has been evaluated by assuming in addition Andreev scattering originating from a step–like order parameter drop near the wall. As discussed already in lecture 6, this assumption leads to an overestimation of the effect and hence to an upper bound for the estimate of its influence. The points stem from a more recent set of torsional oscillator data by Jeevak Parpia [8]. As one should expect, the data points lie nicely between the two bounds of diffuse and Andreev scattering at an order parameter step and hence form an experimental manifestation of the "quantum slip effect". It should be noted, however, that the agreement is less good at the lowest pressures, though still qualitative. It has been proposed that this might be due to surface roughness on mesoscopic rather than microscopic length scales. For a detailed discussion the reader is advised to see refs. [8]

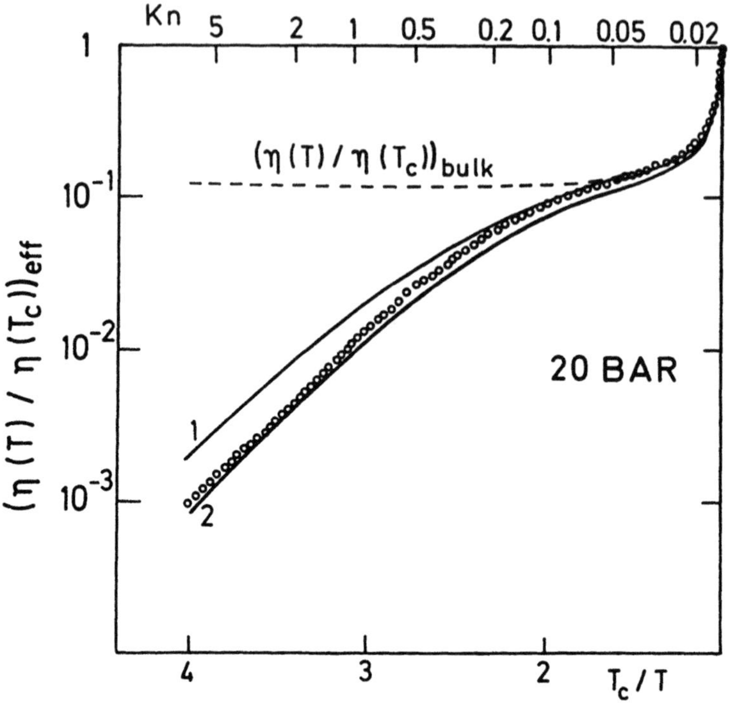

Fig. 12: Normalized effective shear viscosity of superfluid ^{3}He–B at 20 bar pressure vs. inverse reduced temperature. The dashed line represents the bulk theoretical result. The solid lines are evaluated for diffuse (1) and diffuse + Andreev (2) scattering (order parameter step). The experimental data from a 135 μm torsional oscillator [8] are shown as circles.

and [48]. The upper horizontal scale in Fig. 12 shows the Knudsen numbers appropriate for the plate spacing of $d = 135\mu m$. It indicates that the data points extend in fact into a regime of quasiparticle Knudsen flow. Finally I should like to emphasize that a similar treatment brings also the torsional oscillator data by Archie et al. [49] into agreement with the full theoretical treatment at arbitrary Knudsen numbers. The vibrating wire data by Carless et al. [7], on the other hand, have only been analyzed using a first order slip approximation. A treatment for arbitrary Knudsen numbers has not yet been attempted.

Transverse surface impedance

As already explained in lecture 7, a torsional oscillator experiment determines the shear viscosity directly only if the viscous penetration depth δ is large compared to the size d of the flow channel. In the opposite limit, one can rather observe a viscous force, acting on the surfaces of the container, which is exerted by a fluid layer of thicknes δ only. The ratio of this viscous force and the wall velocity $v_x^{\rm ext}$ is called the *transverse surface impedance* $Z_\perp$ of the quasiparticle gas:

$$Z_\perp(\omega) = \frac{\Pi_{xz}}{v_x^{\rm ext}}(\text{wall}) \tag{8.14}$$

My aim is to generalize the result (7.22) for the transverse surface impedance in the slip approximation to arbitrary values of $\omega\bar{\tau}_S$ and ratios ζ_0/δ and to include the elastic wall scattering events represented by the probability $W_{kk'}$. For that purpose it is sufficient to restrict onself to just one oscillating plane wall, which we assume to be located at $z = 0$, restricting the fluid to the half space $z > 0$. This is, however, precisely the geometry that I have treated already in lecture 6, namely the frequency dependent single plane boundary problem. The result was the set (6.31) of coupled integral equations for the velocity field $v_x^{\rm n}$ and the stress tensor component Π_{xz}. The method of solution of these integral equations, again a variational one, is largely along the lines of the preceding treatment of the effective shear viscosity and will therefore not be discussed in detail. The interested reader may see ref. [50] for a comprehensive treatment of the problem. At this stage I would like to restrict myself to mentioning only a few important points. In contrast to the Poiseuille flow case, the trial functions for velocity and stress tensor are chosen to decay exponentially away from the surface,

$$v_x^{\rm n}(z) = A_v\, e^{-qz} \quad ; \quad \Pi_{xz}(z) = A_\Pi e^{-qz}$$

characterized by a complex wave number q, which, besides the amplitudes A_v and A_Π serves as a variational parameter. The general form of the resulting surface impedance functional is rather complicated and will be omitted here. Nevertheless there are a few limits in which the resulting expressions for the variational surface impedance are simple. The first of these is the hydrodynamic limit for which $q \propto \omega^{1/2} \to 0$. The optimal wavenumber $q^{\rm var}$ is in this case obtained as

$$\lim_{\omega\tau_S\to 0} q^{\rm var} = \frac{1-i}{\delta}\left[1 + \frac{1-i}{\delta}\,\frac{L_3 - K_{22} + (K_{12}^2 - L_2^2)/(L_1 + K_{11})}{L_2(1 - \lambda_2^+ Y_2/Y_0)}\right]$$

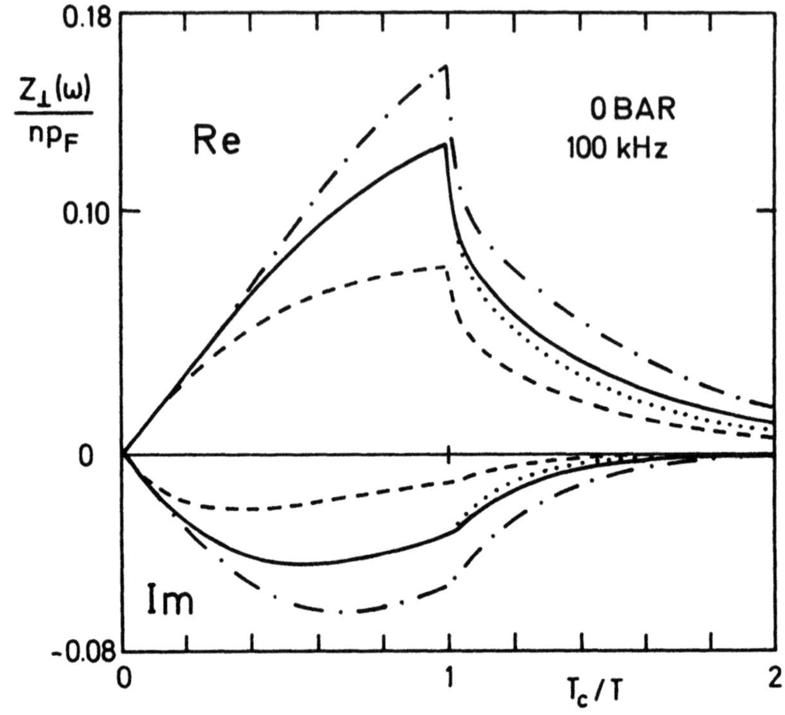

Fig. 13: Real and imaginary part of the normalized surface impedance $Z_\perp(\omega)/np_F$ of normal and superfluid ^{3}He(–B) at 0 bar pressure and 100 kHz vs. inverse reduced temperature for four different elastic scattering laws. Full lines: diffuse scattering; dashed lines: specular scattering (s=0.5); dashed–dotted lines: backward scattering (r=0.5) and dotted lines: Andreev scattering from a steplike order parameter profile.

Here I have used the shorthand notation $L_\mu = L_\mu(0)$ and $K_{\mu\nu} = K_{\mu\nu}(0,0)$. In the limit $\delta \to \infty$ q^{var} is just the wavenumber of the diffusive shear mode, present in all (thermal excitation) gases. Not unexpectedly, the impedance turns out to be of the slip corrected form (7.22)

$$\lim_{\omega\bar{\tau}_S \to 0} Z_\perp(\omega) = (1-i)\frac{\eta}{\delta}\left[1-(1-i)\frac{\zeta_0^{\text{ILB}}}{\delta}\right] \tag{8.15}$$

with the improved lower bound result for the slip length emerging again from the variational treatment.

In order to get some information about the situation at arbitrary $\omega\bar{\tau}_S$ one has to stick to the special cases (6.24), (6.25) and (6.27) for the elastic scattering laws. An approximation

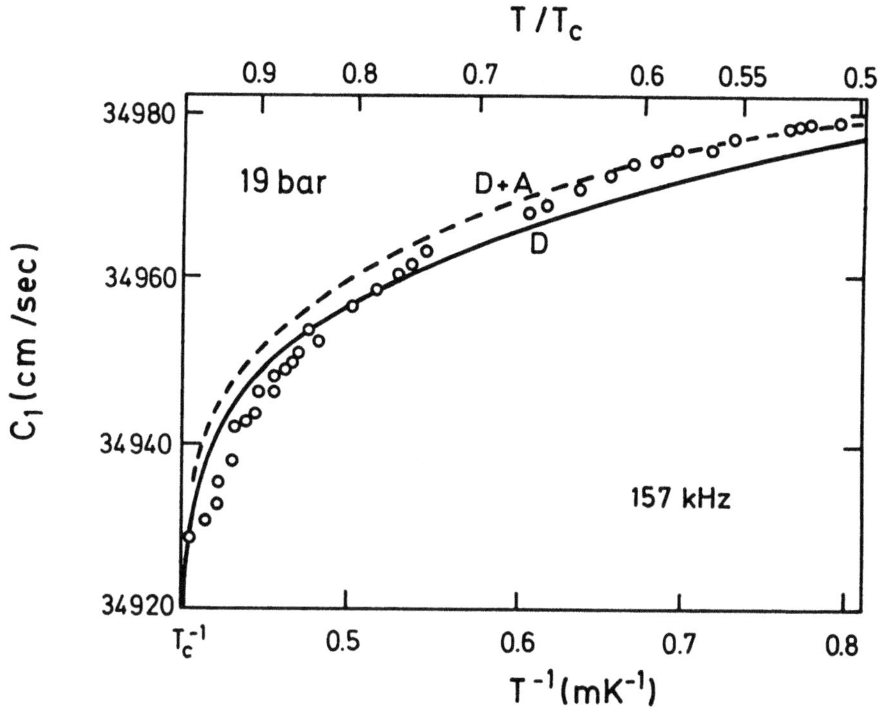

Fig. 14: First sound velocity c_1 of superfluid ^{3}He–B at 19 bar pressure and 157 kHz vs. inverse temperature. The full lines represent the theoretical result for diffuse (D) and diffuse + Andreev (D+A) scattering (order parameter step). The circles are sound resonator data taken from Eska et al. [51].

for the optimal wavenumber q^{var} is then obtained as

$$q^{\text{var}}(\omega) = \frac{1-i}{\delta}\sqrt{\frac{1-i\omega\bar{\tau}_S}{1-i\omega\bar{\tau}_S a(T)}\left(1-i\omega\frac{\tau_\eta(T)}{\phi_2(T)}\right)}$$

where $a(T) = [F_1^s Y_0/3 + L_0(0)L_2(0)/L_1^2(0)]/[1 + F_1^s Y_0/3]$ and $\phi_2(T) = [1 + F_2^s/5]/[1 + F_2^s(1-Y_2)/5]$. The corresponding approximate form of the variational impedance reads then

$$Z_\perp(\omega) = q^{\text{var}}(\omega)\left\{1 + q^{\text{var}}(\omega)\zeta_0^{\text{LB}}\frac{1+t}{1-t} - i\omega\frac{\tau_\eta(T)}{\phi_2(T)}\right\}^{-1} \tag{8.16}$$

The dependence on inverse reduced temperature T_c/T of the transverse surface impedance $Z_\perp(\omega)$ of normal and superfluid ^{3}He(-B) at 0 bar pressure, is shown for a frequency of 100 kHz in Fig. 13. Only the initial linear increase of Re $Z_\perp$ and Im $Z_\perp$ at the high temperature side of the normal phase indicates hydrodynamic behavior. At lower temperatures a deviation from hydrodynamic behavior sets in. As the elastic scattering processes become visible exclusively through the slip correction to the impedance, they can be detected only in the deviations from hydrodynamic behavior which is seen to be amplified by specular (dashed line, s=0.5) and Andreev scattering (dotted line, step profile) and reduced by backward scattering (dashed–dotted line, r=0.5). Surface contributions to the propagation of first sound have been detected in an experiment by Eska et al. [51] where they enter most prominently in the sound velocity change (cf. (7.37)). The change in the velocity of first sound as a function of inverse temperature T^{-1}, as expected from (7.37) together with (8.16) in the cases of diffuse (D) and Andreev (D+A) scattering is compared with these data in Fig. 14. The comparison of experiment and theory seems to indicate that this experiment has most probably seen the effects of Andreev scattering on the transverse surface impedance. The data of ref. [51] allow, however, only for a qualitative conclusion in this direction. A quantitative analysis would become possible only on the basis of data with much higher resolution in the superfluid phase [52].

9 Nonhydrodynamic spin transport

It is the purpose of this lecture to reconsider in some detail the spin dependent transport parameters of normal and superfluid Fermi liquids (spin diffusion and intrinsic spin relaxation). Particular emphasis will here be on the nonhydrodynamic effects originating from the precession terms involving the exchange frequency $\vec{\omega}^{\rm exch}$ in the corresponding kinetic equations.

The Leggett–Rice effect

Let me first come back to the calculation of the spin diffusion coefficient of a normal Fermi liquid. The evaluation of the magnetization current differs from their spin independent counterparts through the existence of the precession term in the spin part of the Landau–Silin equation [53]:

$$\frac{\partial}{\partial t}\vec{J}_i^M(\vec{r},t)+\ldots=\vec{J}_i^M(\vec{r},t)\times[\vec{\omega}_{\rm L}+\vec{\omega}^{\rm exch}]-\frac{1}{\tau_{\rm D}(T)}\vec{J}_i^M(\vec{r},t)\;\;;\;\;\tau_{\rm D}(T)=\bar{\tau}_{\rm N}(T)/(1-\lambda_1^{\rm a}) \quad (9.1)$$

with a spin diffusion transport (or magnetization current decay) time $\tau_{\rm D}(T)$ which has been defined by (2.25).

As a consequence the magnetization current precesses not only about the external magnetic field (Larmor precession), but also about the exchange part of the Landau molecular field (exchange precession),

$$\vec{\omega}^{\rm exch}=\gamma\lambda\vec{M}/\chi_M \quad (9.2)$$

with exchange coupling parameter

$$\lambda=\frac{1}{1+F_0^{\rm a}}-\frac{1}{1+F_1^{\rm a}/3} \quad (9.3)$$

A steady state solution of (9.1) can be found in a straightforward manner (cf. ref. [9]). It turns out that in spite of the exchange precession the magnetization current can be expressed via a constitutive relation through a magnetization gradient, however, with a modified spin diffusion coefficient:

$$\vec{J}_i^M(\vec{r},t) = -\overleftrightarrow{D}\,\nabla_i\delta\vec{M}(\vec{r},t) \;\; ; \;\; \delta\vec{M}(\vec{r},t) = \vec{M}(\vec{r},t) - \chi_M\vec{H}^{\rm ext}(\vec{r},t) \qquad (9.4)$$

As a consequence of the exchange precession, even an ordinary isotropic Fermi liquid subject to a space and time–dependent magnetic field is characterized by an anisotropic (tensorial) coefficient of spin diffusion:

$$D_{\alpha\beta}(T) = D_0(T)\;\frac{\delta_{\alpha\beta} + \tau_{\rm D}(T)\,\epsilon_{\alpha\beta\gamma}\,\omega_\gamma^{\rm exch} + \tau_{\rm D}^2(T)\,\omega_\alpha^{\rm exch}\,\omega_\beta^{\rm exch}}{1+[\vec{\omega}^{\,\rm exch}\tau_{\rm D}(T)]^2} \qquad (9.5)$$

Here $D_0(T)$ denotes the ordinary hydrodynamic spin diffusion coefficient of a Fermi liquid (cf. (2.25))

$$D_0(T) = \frac{v_{\rm F}^2}{3}\,(1+F_0^a)\,\tau_{\rm D}(T) \qquad (9.6)$$

Eq. (9.5) may be rewritten in a more compact and physically transparent form as:

$$\overleftrightarrow{D}(T) = D_0(T)\,\overleftrightarrow{L}\,[\hat{M},s(T)] \;\; ; \;\; \overleftrightarrow{L}\,[\hat{M},s] = \frac{1}{2}\left\{\overleftrightarrow{1} + \overleftrightarrow{R}\left[\hat{M},2\ \arctan(s)\right]\right\} \qquad (9.7)$$

where the parameter $s(T) = \omega^{\rm exch}\tau_{\rm D}(T)$, the ratio of the exchange frequency and the collision rate, measures the strength of the nonhydrodynamic effect.

Eq. (9.7) is the most general representation of a nonhydrodynamic effect on the spin transport in a Fermi liquid, which was originally discovered by Leggett and Rice [9] and named after them. The result (9.7) implies that

(i) the component of the spin diffusion tensor $\overleftrightarrow{D}$ *perpendicular* to the local magnetization direction $\hat{M}$ is strongly reduced in the low temperature limit (Leggett–Rice effect, "diffusion blocking"), and that

(ii) the spin current acquires, besides the purely dissipative (blocked) contribution, a reactive (rotated) component, which scales with the amplitude λ of the exchange interaction (spin current rotation effect).

The Leggett–Rice effect leads to a variety of exotic consequences for the NMR–properties of a normal Fermi liquid such as the experimentally verified occurrence of spin waves [54] and multiple spin echoes [11] in normal liquid ^{3}He.

At the beginning of the eighties E. P. Bashkin as well as C. Lhuillier and F. Laloe [55] demonstrated that the spin current rotation aspect of the Leggett–Rice effect, described by (9.7), is neither restricted to the degenerate temperature regime of Fermi liquids, nor to Fermi systems at all and that it occurs very generally as a consequence of quantum mechanical indistinguishability of identical particles ("identical spin rotation effect"). The mechanism should actually work with *any* paramagnetic system for both electronic and nuclear spins, provided the spin diffusion transport time τ_D is long enough to provide $\omega_L \tau_D \sim 1$. Candidate systems, in which related effects have already been observed, are spin polarized gaseous ^3He [56], mixtures of ^3He and ^4He [57] and spin polarized hydrogen [58]. Despite the fact that these sytems may behave classically with respect to their momentum distribution, due to the quantum nature of the spins, they possess a kind of Landau molecular field as well. Our understanding of the relations between spin rotation phenomena in Boltzmann gases and degenerate Fermi liquids, has been poor for quite some time; some clarification was provided by ref. [59].

Nonhydrodynamic spin diffusion in Fermi Superfluids

Let me now turn to a generalization of the Leggett–Rice type of nonhydrodynamic effect to s– and p–wave superfluids. To keep the calculations on a tractable level, I shall put the spin–antisymmetric Landau parameter $F_1^a \equiv 0$ in the following. For the calculation of the spin diffusion of a Fermi superfluid we need to reconsider the transport equation (cf. Eq. (5.7)) for the distribution function $\vec{h}_k^{(-)}$ which can be written in the form

$$\left[\frac{\partial}{\partial t} + \overleftrightarrow{\tau}_D^{-1}\right] \vec{h}_k^{(-)} = -\varphi_k \overleftrightarrow{V}_{kz} \nabla_z \delta\vec{\mu} - \frac{2}{\hbar}\left[\vec{h}_k^{(-)} \times \delta\vec{E}_k^{(+)}\right] \tag{9.8}$$

where $\overleftrightarrow{V}_{kz} = v_{kz} \overleftrightarrow{S}_k^{(-)}$ is the tensorial group velocity which has been introduced in context with the derivation of the two–fluid picture for the spin transport (cf. table 4). Assuming the magnetization to lie along the field direction, the quasiparticle energy change occuring in the precession term of (9.8) can be written as

$$\delta\vec{E}_k^{(+)} = \frac{\hbar}{2}\vec{\omega}_L' \quad ; \quad \vec{\omega}_L' = \vec{\omega}_L + \vec{\omega}^{exch}$$

where a dissipative term $\propto h_0(\vec{k})$ has been neglected. The quantity $\vec{\omega}^{exch}$ is the superfluid analogue of the normal Fermi liquid exchange frequency (9.2), and is defined by

$$\vec{\omega}^{exch} = \gamma\lambda_s(T)\frac{\vec{M}}{\chi_M(T)} \quad ; \quad \lambda_s(T) = -F_0^a \chi(T)$$

with $\lambda_s(T)$ a temperature dependent exchange coupling parameter (cf. (9.3)) taken in the limit $F_\ell^a = 0 \; \forall \, \ell \geq 1$. Here $\chi(T) = \chi_0(T)/(1+F_0^a\chi_0(T))$ is the dimensionless renormalized spin susceptibility of the superfluid Fermi system. Let us now try and solve the transport equation (9.8) with the ansatz

$$\vec{h}_k^{(-)} = \varphi_k \; \overleftrightarrow{V}_{kz} \; \overleftrightarrow{\tau}_{\rm D} \; \vec{\Phi}(E_k) \tag{9.9}$$

with a vector function $\vec{\Phi}$ which is assumed to depend only on the energy variable E_k. This ansatz is exact in the s–wave case. In the case of p–wave pairing it is an approximation in that the dependence of the distribution function $\overleftrightarrow{S}_k^{(-)-1} \; \vec{h}_k^{(-)}$ on the azimuthal angle is ignored. It is nevertheless sufficiently general to display all details of the spin current rotation effect in p–wave superfluids and leads to results, as we shall see, which are physically particularly transparent. The kinetic equation for $\vec{\Phi}$ aquires a simple form after the integration with respect to the azimuthal angle ϕ:

$$\left[\frac{\partial}{\partial t} + \overleftrightarrow{\tau}_{\rm D}^{-1}\right] \vec{\Phi} = -\nabla_z \delta\vec{\mu} \; \overleftrightarrow{\tau}_{\rm D}^{-1} + a_{k\perp} \vec{\Phi} \times \vec{\omega}_{\rm L}' \tag{9.10}$$

Here I have defined $a_{k\perp} = 1-(\Delta/2E_k)^2(1-\hat{k}_z^2)$. In terms of $\vec{\Phi}$ the macroscopic quasiparticle magnetization current (cf. table 4) reads

$$\vec{J}_z^{M\prime} = \frac{\gamma\hbar}{2} 2\sum_{\vec{k}} \varphi_k \; \overleftrightarrow{V}_{kz}^2 \; \overleftrightarrow{\tau}_{\rm D} \; \vec{\Phi}(E_k) = - \overleftrightarrow{D}_{zz} (T;\omega_{\rm L}) \nabla_z \delta\vec{M} \tag{9.11}$$

Hence, again, even in the presence of the precession term, the spin current has the form of a constitutive relation, however, with a modified spin diffusion tensor, which explicitly depends on the Larmor frequency:

$$\overleftrightarrow{D}_{zz} (T;\omega_{\rm L}) = D_{zz}^{\parallel}(T;\omega_{\rm L}) \; \hat{z} : \hat{z} + \overleftrightarrow{D}_{zz}^{\perp} (T;\omega_{\rm L})[\overleftrightarrow{1} - \hat{z} : \hat{z}] \tag{9.12}$$

As in a normal Fermi liquid, the longitudinal component of the spin diffusion tensor is found to remain unaffected by the nonhydrodynamic effect, viz $D_{zz}^{\parallel}(T;\omega_{\rm L}) = D_{zz}^{\parallel\,{\rm hyd}}(T)$. Only the transverse component is modified according to

$$\begin{aligned} \overleftrightarrow{D}_{zz}^{\perp} (T;\omega_{\rm L}) &= \frac{v_{\rm F}^2}{3\chi(T)} \tau_{\rm D}^{\perp}(T) \int_{-\infty}^{\infty} d\xi_k \varphi_k \; 3 \int \frac{d\Omega_k}{4\pi} \hat{k}_z^2 S_{kzz}^{(-)2} \; \overleftrightarrow{L} \; [s_k^-(T)] \\ L_{\mu\nu}(s) &= \frac{\delta_{\mu\nu} + s\epsilon_{\mu\nu z}}{1+s^2} = \frac{1}{2} \{\delta_{\mu\nu} + R_{\mu\nu}[2 \arctan s]\} \end{aligned} \tag{9.13}$$

Here $L_{\mu\nu}$ is a two-dimensional matrix which describes the nonhydrodynamic modification of the transverse spin diffusion and $\overleftrightarrow{R}_{\mu\nu}(\theta) = \cos\theta\delta_{\mu\nu} + \sin\theta\epsilon_{\mu\nu z}$ is a two dimensional rotation matrix. In the case of s-wave pairing (9.13) has to be evaluated using $S_{kxx}^{(-)2} = (\xi_k/E_k)^2$ and $s_k^{(-)}(T) = \omega^{\rm exch}\tau_{\rm D}(T)$. This result can be physically interpreted as a Leggett–Rice effect for Bogoliubov quasiparticles: the exchange interaction causes a rotation of the macroscopic quasiparticle spin current (reactive effect) and the blocking of the transverse spin diffusion (dissipative effect). As in the normal Fermi liquid, the size of the effect scales with the exchange interaction parameter $\lambda_s(T)$; the effect is absent in the limit $\lambda_s(T) \to 0$.

In the case of p–wave pairing, however, the situation is quite different. Due to the anisotropic structure of the coherence tensors $\overleftrightarrow{S}_k^{(+)}$ and $\overleftrightarrow{S}_k^{(-)}$, which enter the precession term of (9.8), the quasiparticle system experiences only the projection $\overleftrightarrow{S}_k^{(+)}\vec{H}^{\rm ext}$ of the external field. The Larmor transformation does therefore not eliminate the external field completely. As a consequence, a spin current rotation effect is present in p–wave superfluids even in the absence of the exchange interaction ($\lambda_s(T) \to 0$). To be specific, the quantity $s_k^-(T)$ is in the p-wave case obtained as $s_k^-(T) = \delta\omega_k\tau_{\rm D}^{\perp}(T)$ with $\delta\omega_k = \omega_{\rm L}[S_{kxx}^{(-)2} - 1] + \omega^{\rm exch}$. Another convenient way of writing the transverse part of the constitutive relation (9.11) is in the complex notation

$$(J_z^{M'})_+(\vec{r},t) = (J_z^{M'})_x + i(J_z^{M'})_y = -D_\perp(T;\omega_{\rm L})\nabla_z\delta\vec{M}(\vec{r},t) \tag{9.14}$$

with a very simple result for the complex diffusion constant

$$\begin{aligned} D_\perp(T;\omega_{\rm L}) &= D_{zz}^{\perp xx} - iD_{zz}^{\perp xy} \\ &= \frac{v_{\rm F}^2}{3\chi(T)}\tau_{\rm D}^{\perp}(T)\int_{-\infty}^{\infty} d\xi_k\varphi_k\, 3\int\frac{d\Omega_k}{4\pi}\hat{k}_z^2\frac{S_{kxx}^{(-)2}}{1+is_k^-(T)} \end{aligned} \tag{9.15}$$

It should be noted that a full solution of Eq. (9.8) without making the approximate ansatz (9.9) shows that the complex transverse diffusion has a form almost as simple as (9.15) ([41], [60]):

$$D_\perp(T;\omega_{\rm L}) = \frac{v_{\rm F}^2}{3\chi(T)}\frac{\tau_{\rm D}^{\perp}(T)}{1-i\omega_{\rm L}\tau_{\rm D}^{\perp}(T)}\int_{-\infty}^{\infty} d\xi_k\varphi_k\, 3\int\frac{d\Omega_k}{4\pi}\hat{k}_z^2\frac{S_{kxx}^{(-)2} - is(T)S_{kzz}^{(+)2}}{1+s^2(T)S_{kzz}^{(+)2}} \tag{9.16}$$

which, however, cannot be written in the form (9.13), reminiscent of the Leggett–Rice effect in the normal state, because of the more complicated dependence on the coherence factors $S_{kxx}^{(-)2}$ and $S_{kzz}^{(+)2}$. In (9.16) I have defined $s(T) = [\omega_L + \omega^{\rm exch}(T)]\tau_{\rm D}^{\perp}(T)/[1 - i\omega_L\tau_{\rm D}^{\perp}(T)]$.

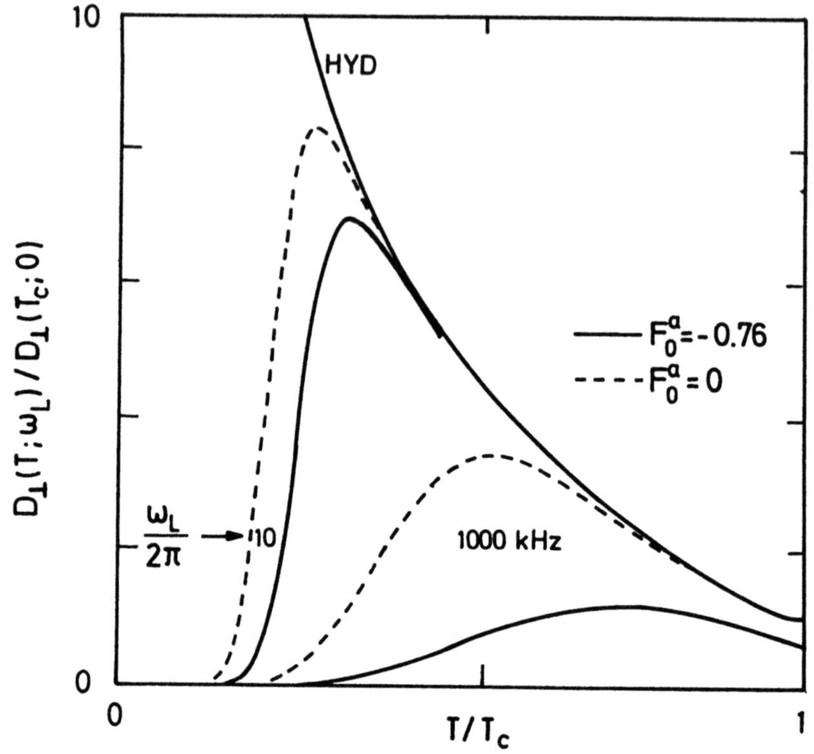

Fig. 15: The normalized transverse spin diffusion of superfluid ^{3}He–B at 20 bars vs. reduced temperature T/T_c for two different Larmor frequencies and two values for the (exchange–) Landau parameter F_0^a. Also shown is the hydrodynamic limiting form (HYD).

The more general result (9.16) for the transverse spin diffusion shares all the properties of (9.13), described above and will be used for the numerical evaluation. In order to visualize the deviation of the transverse diffusion from its hydrodynamic behavior I have plotted in Fig. 15 the numerical result for the dissipative part Re $D_\perp$ of the complex transverse component of the diffusion tensor of superfluid ^{3}He–B (cf. Eq. (9.16)) at 20 bar pressure for two different Larmor frequencies and the exchange Landau parameter F_0^a set equal to its value of -.76 at 20 bar, and to 0. The effect of spin diffusion blocking is, due to the exponential increase of the spin current decay time τ_D in the low temperature limit, more pronounced than in the normal Fermi liquid, and it is there even in the absence of the exchange interaction. In the limit as $T \to T_c$ the quantity $S_{kxx}^{(-)2} - 1 = -(\Delta^2(T)/2E_k^2)(1 - \hat{k}_z^2)$ vanishes,

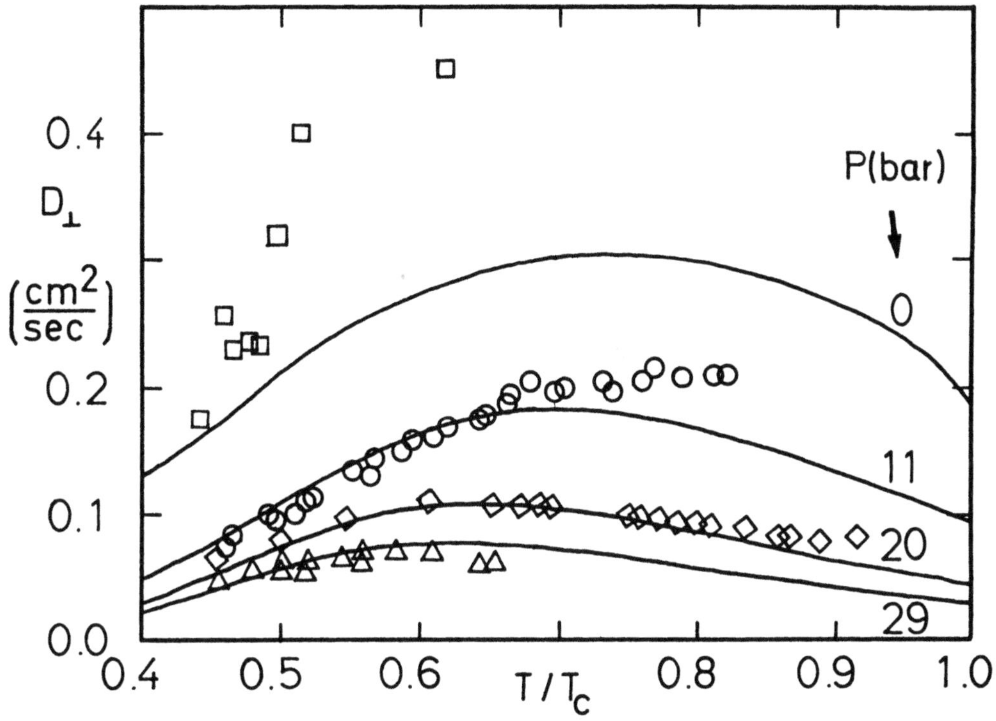

Fig. 16: The transverse spin diffusion of superfluid ^{3}He–B at a frequency of 460 kHz and four pressures as a function of reduced temperature T/T_c. Triangles: 29 bar, diamonds: 20 bar, circles: 11 bar and squares: 0 bar. The solid lines represent the theoretical results evaluated with Eq. (9.16) in the text.

the coherence factor $S_{k\pm\pm}^{(+)2} \to 1$ and the normal state Leggett–Rice effect (9.7) is recovered. In Fig. 16 I compare the results of a numerical evaluation of the real (dissipative) part of $D_\perp$ from (9.16) for superfluid ^{3}He–B at four different pressures with recent experimental data on the transverse spin diffusion [41]. The conventional spin echo method for the experimental determination of the spin diffusion has turned out not to be useful because the dissipative spin transport is masked by the reactive spin supercurrents (cf. table 4, which, for example, may give rise to the occurrence of (multiple) spin echoes).

Fomin [61] has shown, however, that the helicoidal magnetization distribution occurring in a typical NMR experiment becomes redistributed by spin supercurrents in the presence of a field gradient. The result is a two–domain structure of homogeneous magnetization pre-

cession in a closed chamber with $\vec{M} \parallel \vec{H}^{ext}$ on the side of higher fields and $\vec{M}$ tilted away from $\vec{H}^{ext}$ by the Leggett angle $\theta_0 = \arccos(-1/4)$ (cf. Eq. (3.2)) on the lower field side (the so-called "homogeneous precession domain", HPD). The two domains are separated by a domain wall. This domain wall moves with a velocity proportional to a combination of the transverse spin diffusion Re $D_\perp$ and the intrinsic spin relaxation parameter Ξ towards the lower field side such as to remove the domain with tilted magnetization. Due to their different dependence on the volume of the HPD the two spin transport parameters can be extracted separately from the data for the domain wall velocity.
The agreement between theory and the experimental data on the transverse spin diffusion is quantitative, except at the lowest pressure, where the applicability of Fomin's theory breaks down.

Nonhydrodynamic intrinsic spin relaxation

In principle, nonhydrodynamic effects on the intrinsic quasiparticle spin relaxation exist as well and can be treated in the same way as the spin diffusion. As a general result, at finite Larmor frequencies the intrinsic spin relaxation parameter Ξ becomes a tensor quantity:

$$\overleftrightarrow{\Xi}(T;\omega_L) = \Xi_\parallel(T;\omega_L)\, \hat{z}:\hat{z} + \overleftrightarrow{\Xi}_\perp(T;\omega_L)[\overleftrightarrow{1} - \hat{z}:\hat{z}] \tag{9.17}$$

The longitudinal component of the intrinsic spin relaxation tensor is found to remain unaffected by the nonhydrodynamic effect, viz $\Xi_\parallel(T;\omega_L) = \Xi^{\,hyd}(T)$. Work on the evaluation of the transverse component $\overleftrightarrow{\Xi}_\perp$ based on the kinetic equation for Bogoliubov quasiparticles is in progress. Here I restrict myself to discussing the result of a phenomenological treatment by Markelov [62], which was derived from the phenomenological Leggett–Takagi equations for the spin dynamics of p–wave superfluids. It reads:

$$\overleftrightarrow{\Xi}_\perp(\omega_L, T) = \Xi(0,T)\ \overleftrightarrow{L}\,[s_A(T)] \tag{9.18}$$

where the two–dimensional matrix $\overleftrightarrow{L}$ is defined through (9.13). Its argument $s_A(T) = A\omega_L\tau_{LT}(T)$ involves the Leggett–Takagi relaxation time τ_{LT} (cf. (4.35)) and a phenomenological parameter A which vanishes in the limit as $F_0^a \to 0$. It should be noted that the purely phenomenological model leading to (9.18) implies that we can expect only a

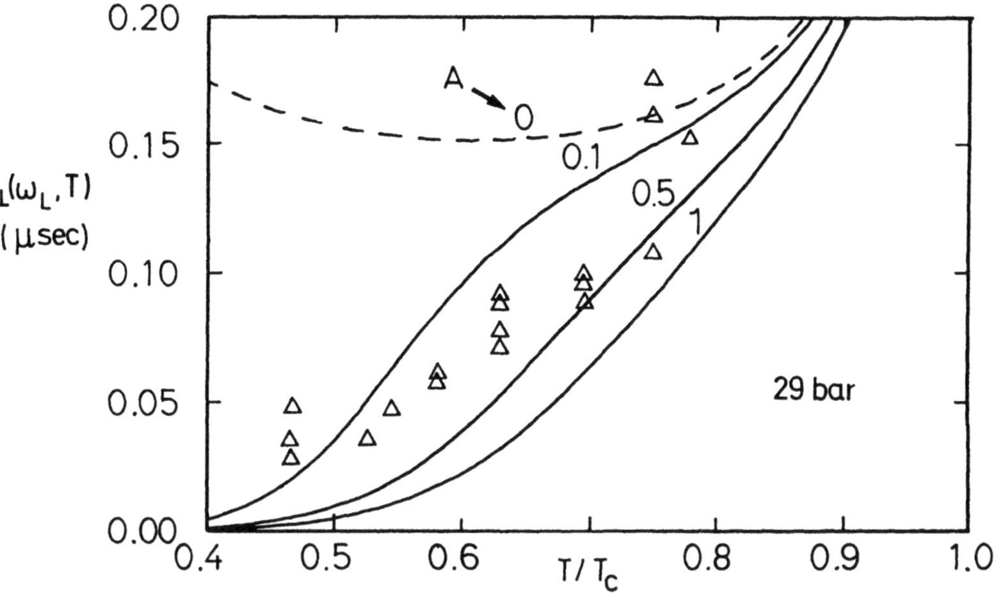

Fig. 17: The intrinsic spin relaxation parameter $\Xi_{\perp xx}(\omega_L, T)$ of ^{3}He–B at 29 bar and 920 kHz vs. reduced temperature. The dashed line is the result from hydrodynamic theory and the solid lines represent the result of the phenomenological treatment (cf. Eq. (9.18)) evaluated with parameter values as indicated.

quantitative understanding of the experimental data. In Fig. 17 I have plotted a numerical result for the intrinsic spin relaxation parameter $\Xi_{\perp xx}$ for superfluid ^{3}He–B at 29 bar pressure and a frequency of 920 kHz vs. reduced temperature. The data points seem to be best described for values of the phenomenological parameter A between 0.1 and 0.5. The agreement is qualitative and shows that the nonhydrodynamic modification is really present in superfluid ^{3}He–B at high Larmor frequencies.

10 Summary and conclusions

In this series of lectures I have discussed the dynamic behavior of normal Fermi liquids and pair–correlated Fermi superfluids in various experimental situations. As far as superfluids are concerned I have limited all considerations to the case, where the dynamics of the superfluid order parameter is governed by variations in time and space of a few phase variables, which represent the broken symmetries of the superfluid ground state and the resulting macroscopic quantum coherence. The associated regime of frequencies, at which external perturbations can vary, is therefore sometimes referred to as "macroscopic". In such a situation the response of the condensate to external perturbations is instantaneous and local in space. It turns out that it is only in such a case that a two–fluid description of pair–correlated superfluids is possible.

The dynamics of the Fermi systems was treated on the level of quasiclassical kinetic equations for quasiparticle distribution functions in phase space in the spirit of Landau's Fermi liquid theory and its generalizations to the presence of pairing–correlations.
The general emphasis was on the description of dissipative transport in the hydrodynamic limit and beyond. Hydrodynamic results for the transport parameters of normal and superfluid Fermi systems emerge in situations where the size of the measuring cell can be taken to be infinite and the amplitudes of the perturbing fields can be taken to be negligibly small. Deviations from hydrodynamic behavior of the transport parameters were shown to become essential when either one of the two experimental conditions mentioned above can no longer be met. The derivation of the hydrodynamic transport parameters was imbedded in a general discussion of the conservation and relaxation properties of the Fermi liquid under consideration. Only the dynamics of the normal component is nontrivial in the macroscopic regime. It is governed by scalar kinetic equations for spin–independent and the spin–dependent part of the distribution function representing the

dilute gas of Bogoliubov quasiparticles. The spin dependent dynamics is characterized, in addition, by nonlinear precession terms which turned out later to be the origin of nonhydrodynamic modification of the spin transport parameters.

The exact and in particular the approximate treatment of quasiparticle collisions at low temperature has proven to be of central importance for the discussion of the conservation and relaxation properties of Fermi liquids.
Relaxation time type of approximations for the collision term in the kinetic equations have been chosen such as to reproduce the conservation laws and to guarantee good estimates for all relevant transport times, i.e. the decay times of those macroscopic observables, which are not conserved.

In Fermi superfluids the interesting situation can arise that a macroscopic observable is conserved in the normal phase, but, regarded as a property of the gas of Bogoliubov quasiparticles in the superfluid, it is not. This happens precisely when the condensate can contribute a certain fraction to this quantity. Although the sum of quasiparticle and condensate contributions to this observable remains constant there can arise a *mutual* disequilibrium between the normal and superfluid part, which relaxes on the scale set by new transport times, which diverge in the limit $T \rightarrow T_c^-$. Examples of such intrinsic relaxation processes, which are well understood, are that of the normal mass and magnetization density in superfluid Fermi superfluids. The intrinsic spin relaxation mechanism is present, however, only in systems with p–wave pairing correlations, because in ordinary superconductors the total spin polarization originates entirely from thermal excitations and has no condensate contribution.
The relaxation time approximation chosen for the collision term in the kinetic equation for Bogoliubov quasiparticles reproduces also the abovementioned features of intrinsic relaxation, particularly the correct divergences of the intrinsic relaxation times near the transition.
In the hydrodynamic limit, relaxing currents (which may be connected with the normal component in the superfluid phase) can, as a result of a gradient expansion, be expressed via constitutive relations through gradients in the thermodynamically conjugate density

fields. The temperature dependent prefactors in these constitutive relations are identified as the hydrodynamic transport parameters of the Fermi liquid under consideration.
A comparison of these transport parameters with experimental data for normal and superfluid ^{3}He(–B) uncovered discrepancies in the cases of shear viscosity and the spin transport parameters which were particularly pronounced at low temperature. This led us to reconsider the experimental situation in these particular cases.

In the case of the shear viscosity the discrepancy could be traced back to the narrow size of the flow channels used in the experiment. The finiteness of the mean free path of thermal excitations, or equivalently, the Knudsen number, which is the ratio of the mean free path and the typical size of the flow channel, turned out to become important in such a case. The presence of oscillating surfaces, typical for a large class of flow experiments, introduces a new phenomenological length into the problem, the frequency dependent skin– or viscous penetration depth. Accordingly, the treatment of the viscosity problem had to be divided into a separate discussion of Poiseuille flow in narrow flow channels and the surface impedance of a transversely oscillating wall, which, respectively, represent the limiting cases of very large and very small skin depth.

On the level of the microscopic description of Fermi liquids, the presence of walls was generally acounted for by amending the kinetic equation by boundary conditions for the quasiparticle distribution function which, in addition to the two–particle collisions in the bulk, described quasiparticle collisions with the container surfaces. Wall collision processes were taken to include, besides the usual diffuse scattering, elastic scattering processes of quasiparticles, such as specular, backward and Andreev scattering.
Integral equations for the relevant quasiparticle observables (velocity field and stress tensor) were derived and solved by well controlled variational procedures. The outcome is essentially an additional transport parameter, the so–called slip length (the length at which the macroscopic fluid velocity field extrapolates to zero behind the wall of the flow channel), which scales with the excitation mean free path.

This motivated an investigation of how the slip effect of the (normal) velocity at container surfaces modifies the hydrodynamic description of fluid flow, as observed, in particular in torsional oscillator and sound propagation experiments. The major result was that the slip effect renormalizes both the shear viscosity (which may become dramatically reduced) and the surface impedance.

Assuming the skin depth to be large compared with the size of the flow channel, the case of (quasi–) stationary Poiseuille flow could be treated on a microscopic level for arbitrary Knudsen numbers. It eventually led to a quantitative agreement between experiment and theory, if the Andreev scattering processes were taken into account (quantum slip effect). In the opposite limit of small skin depths also the surface impedance could be calculated on a microscopic level for arbitrary frequencies (however limited to the macroscopic regime) and a couple of elastic wall scattering laws. One of those, namely Andreev scattering, seemed to have an influence also on the velocity shift, that was measured in a first sound experiment.

In the case of spin–dependent transport the discrepancies between theory and experiment could be explained by the fact that the Larmor frequency had been completely eliminated from the hydrodynamic treatment. This can usually be justified by transforming the spin–dependent kinetic equation into a system which rotates at the Larmor frequency. In the case of interacting Fermi liquids, however, the exchange contributions to the Landau quasiparticle interaction lead to (exchange) molecular fields about which the quasiparticle magnetization and magnetization currents may precess even in the Larmor system. In normal Fermi liquids this mechanism is known as the Leggett–Rice effect (or exchange–induced spin current rotation effect) and acts so as to modify the hydrodynamic spin diffusion. In superfluid Fermi liquids this effect survives as a property of the normal component, the Bogoliubov quasiparticles.

An exotic feature emerged from the study of the special case of p–wave superfluids. Here the spin current rotation effect is working even in the absence of exchange Landau molecular fields. This fact could be traced back essentially to the anisotropy introduced by the intrinsic spin structure of the p–wave order parameter. In a system with p–wave pairing, the intrinsic spin relaxation of the normal magnetization component is also affected by exchange precession.

This nonhydrodynamic effect has numerous consequences, of which I have discussed only the modified temperature dependences of the transverse spin diffusion and the transverse intrinsic spin relaxation. Both quantities turned out to be (in most cases) in quantitative agreement with the experimental observation.

Let me conclude by saying that Landau's kinetic theory of Fermi liquids, with no free fitting parameter, appropriately generalized to include Fermi systems with s– and p–wave pairing correlations, can give a *quantitative* account of a large variety of experimental observations, which are related to its (two–fluid) local response and transport properties, and which have particularly been made in normal and superfluid ^{3}He. These observations partly reach down to temperatures where the normal component, the gas of thermal excitations, becomes extremely dilute. The agreement in such a case can be obtained only if the theory accounts, in addition, for certain details of the method of measurement such as finite size effects and large magnetic fields.

In summary the low temperature phases of liquid ^{3}He have proven not only to be some of the very few clean neutral Fermi liquids on earth available for experimental investigations, but also can they be regarded as an ideal model system for studies of the ballistic motion and the quantum–mechanical scattering in the gas of its thermal excitations.

Bibliography

[1] I. M. Khalatnikov in *The Physics of Liquid and Solid Helium*, Part I, K. H. Bennemann and J. B. Ketterson, eds. (Wiley, New York, 1978), p. 1

[2] A. J. Leggett and S. Takagi, Ann. Phys. **106** 79 (1977)

[3] D. Einzel, J. Low Temp. Phys. **54**, 427 (1984)

[4] D. Einzel, Physica **108** B + C, 1143 (1981)

[5] D. Einzel, H. Højgaard Jensen, H. Smith, and P. Wölfle, J. Low Temp. Phys. **53**, 695 1983

[6] C. N. Archie, T. A. Alvesalo, J. D. Reppy, and R. C. Richardson, J. Low Temp. Phys. **42**, 295 (1981)

[7] D. C. Carless, H. E. Hall, and J. R. Hook, J. Low Temp. Phys. **50**, 605 (1983)

[8] D. Einzel and J. M. Parpia, Phys. Rev. Lett. **58**, 1937 (1987)

[9] A. J. Leggett, M. J. Rice, Phys. Rev. Lett. **20**, 586 (1968) and **21**, 506 (1968); A. J. Leggett, J. Phys. C **12**, 448 (1970)

[10] P. M. Platzman and P. A. Wolff, Phys. Rev. Lett. **18**, 280 (1967)

[11] D. Einzel, G. Eska, Y. Hirayoshi, T. Kopp, and P. Wölfle, Phys. Rev. Lett. **53**, 2312 (1984); Phys. Rev. Lett. **54**, 608 (1985)

[12] L. D. Landau, Zh. Eksp. Teor. Fiz. **30**, 1058 (1956); **32**, 59 (1957); **35**, 95 (1958); [Sov. Phys. JETP **3**, 920 (1957); **5**, 101 (1957); **8**, 70 (1959)]

[13] A. A. Abrikosov and I. M. Khalatnikov, Rep. Progr. Phys. **22**, 329 (1959)

[14] G. Baym and C. J. Pethick, in *The Physics of Liquid and Solid Helium*, Part II, K. H. Bennemann and J. B. Ketterson, eds. (Wiley, New York, 1978), p. 1

[15] L. D. Landau, Zh. Eksp. Teor. Fiz. **32**, 59 (1957) [Sov. Phys. JETP **5**, 101 (1957)]

[16] H. Højgaard Jensen, H. Smith, and J. W. Wilkins, Phys. Rev. **185**, 323 (1969); J. Sykes and G. A. Brooker, Ann. Phys. **56**, 1 (1970)

[17] V. J. Emery and D. Cheng, Phys. Rev. Lett. **21**, 533 (1968)

[18] A. Layzer and D. Fay, Int. J. Magn. **1**, 35 (1971)

[19] R. Balian and N. R. Werthamer, Phys. Rev. **131**, 1553 (1963)

[20] A. J. Leggett, Reviews of Modern Physics **47**, 331 (1975)

[21] D. Vollhardt and P. Wölfle, *The Superfluid Phases of He 3* (Taylor and Francis, London, 1990)

[22] Y. Nambu, Phys. Rev. **117**, 648 (1960)

[23] B. Mühlschlegel, Z. Phys. **155**, 313 (1959)

[24] O. Betbeder–Matibet and P. Nozieres, Ann. Phys. (N. Y.) **51**, 392 (1969) P. Wölfle, J. Low Temp. Phys. **22**, 157 (1976)

[25] D. Einzel and P. Wölfle, J. Low Temp. Phys. **32**, 19 (1978)

[26] K. Yosida, Phys. Rev. **110**, 769 (1958)

[27] P. Wölfle, Phys. Rev. **B 14**, 89 (1976)

[28] D. Einzel, P. Wölfle, and P. J. Hirschfeld, J. Low Temp. Phys. **80**, 31 (1990)

[29] P. Bhattacharyya, C. J. Pethick, and H. Smith, Phys. Rev. **15**, 3367 (1977)

[30] C. J. Pethick, H. Smith, and P. Bhattacharyya, Phys. Rev. **15**, 3384 (1977)

[31] Y. A. Ono, J. Hara, and K. Nagai, J. Low Temp. Phys. **48**, 167 (1982)

[32] R. Combescot and H. Ebisawa, Phys. Rev. Lett. **33**, 810 (1974)

[33] D. Greywall, Phys. Rev. **B33**, 7520 (1986)

[34] M. Pfitzner, PhD thesis, Technische Universität München, (1985) (unpublished)

[35] J. C. Wheatley in *Progress in Low Temperature Physics* (North-Holland, Amsterdam, 1978), Vol. **7**, p.1;
J. R. Hook (private communication)

[36] D. Bloyet, E. Varoquaux, C. Vibet, O. Avenel, P. M. Berglund, and R. Combescot, Phys. Rev. Lett. **42**, 1158 (1979)

[37] J. R. Hook, unpublished work and private communication

[38] R. A. Webb, R. E. Sager, and J. C. Wheatley, Phys. Rev. Lett. **35**, 1164 (1975)

[39] G. Eska, K. Neumaier, W. Schoepe, K. Uhlig, and W. Wiedemann, Phys. Lett. **87 A**, 311 (1982)

[40] O. Ishikawa, Y. Sasaki, T. Mizusaki, A. Hirai, and M. Tsubota, J. Low. Temp. Phys. **75**, 35 (1988)

[41] Yu. M. Bunkov, V. V. Dimitriev, A. V. Markelov, Yu. M. Mukharskii, and D. Einzel, Phys. Rev. Lett. **65**, 867 (1990)

[42] J. C. Maxwell, *Scientific papers* (Dover, New York, 1953), Vol. 2, p. 704.

[43] J. Cercignani, *Theory and Application of the Boltzmann Equation*, (Scottish Academic Press, Edinburgh, 1975)

[44] A. F. Andreev, Zh. Eksp. Teor. Fiz. **46**, 1823 (1964) [Sov. Phys. JETP **19**, 1228 (1964)]

[45] J. Kurkijärvi and D. Rainer in *Modern Problems in Condensed Matter Sciences*, W. P. Halperin and L. P. Pitaevskii, eds. (Elsevier, Amsterdam)

[46] H. Højgaard Jensen, H. Smith, P. Wölfle, K. Nagai, and Maak Bisgaard, J. Low Temp. Phys. **41**, 473 1980

[47] H. Højgaard Jensen, H. Smith, and P. Wölfle, J. Low Temp. Phys. **51**, 81 (1983)

[48] D. Einzel, P. Panzer, and M. Liu, Phys. Rev. Lett. **64**, 2269 (1990)

[49] D. Einzel, P. Wölfle, H. Højgaard Jensen, and H. Smith Phys. Rev. Lett. **52**, 1705 (1984)

[50] D. Einzel, P. Wölfle, and P. J. Hirschfeld, J. Low Temp. Phys. **80**, 31 (1990)

[51] G. Eska, K. Neumaier, W. Schoepe, K. Uhlig, and W. Wiedemann, Phys. Rev. **B27**, 5534 (1983)

[52] K. Uhlig, private communication

[53] V. P. Silin, Sov. Phys. JETP **6**, 945 (1958)

[54] N. Masuhara, D. Candela, D. O. Edwards, R. F. Hoyt, H. N. Scholz, D. S. Sherrill, and R. Combescot, Phys. Rev. Lett. **53**, 1168 (1984)

[55] E. P. Bashkin, Zh. Eksp. Teor. Fiz. **33**, 11 (1981) [JETP Lett. **33**, 8 (1981)]
C. Lhuillier and F. Laloe, J. de Physique **43**, 197 (1982); J. de Physique **43**, 225 (1982); J. de Physique **44**, 1 (1983)

[56] P. J. Nacher, G. Tastevin, M. Leduc, S. B. Crampton, and F. Laloe, J. Physique Lett. **45**, L441 (1984)

[57] J. R. Owers–Bradley, H. Chocholacs, R. M. Mueller, Ch. Buchal, M. Kubota, and F. Pobell, Phys. Rev. Lett. **51**, 2120 (1983)

[58] B. R. Johnson, J. S. Denker, N. Bigelov, L. P. Levy, J. H. Freed, and D. M. Lee, Phys. Rev. Lett. **52**, 1508 (1984)

[59] K. Miyake, W. J. Mullin and P. C. Stamp, J. Physique **46**, 663 (1985)

[60] R. Combescot, Phys. Rev. Lett. **33**, 946 (1974)

[61] I. A. Fomin, Zh. Eksp. Teor. Fiz. **88**, 2039 (1985) [Sov. Phys. JETP **61**, 1207 (1985)]

[62] A. V. Markelov, Zh. Eksp. Teor. Fiz. **91**, 2053 (1986) [Sov. Phys. JETP **64**, 1218 (1986)]

GPSR Compliance

The European Union's (EU) General Product Safety Regulation (GPSR) is a set of rules that requires consumer products to be safe and our obligations to ensure this.

If you have any concerns about our products, you can contact us on ProductSafety@springernature.com

In case Publisher is established outside the EU, the EU authorized representative is:

Springer Nature Customer Service Center GmbH
Europaplatz 3
69115 Heidelberg, Germany

Batch number: 10838062

Printed by Printforce, the Netherlands